AF522255

FUNDAMENTALS OF GENE EVOLUTION

ENCYCLOPAEDIA OF GENE EVOLUTION

Vol. I

FUNDAMENTALS OF GENE EVOLUTION

By

Dr. M. Prakash

DISCOVERY PUBLISHING HOUSE
NEW DELHI-110002

First Published-2006

ISBN 81-8356-169-1 (Set)

© Author

Published by

DISCOVERY PUBLISHING HOUSE
4831/24, Ansari Road, Prahlad Street,
Darya Ganj, New Delhi-110002 (India)
Phone: 23279245 • Fax: 91-11-23253475
E-mail:dphtemp@indiatimes.com

Printed at:
Amit Enterprises
Delhi-53

PREFACE

The present title **Encyclopaedia of Gene Evolution** encompasses diverse areas of study from the origin of life and is subsequent record to the molecular and population concepts that explain these events. All biological phenomena derive from evolutionary relationships and past interactions. As the great evolutionary geneticist Theodosius Dobzhansky remarked, "Nothing in biology makes sense except in the light of evolution" yet unifying all biology under an evolutionary theme is still difficult, although the explosive increase in molecular, organismic, and populations information makes the realization of this goal more possible now than ever before.

The aim in writing this text was to bring together the highly dispersed literature on gene structure, function and expression, to integrate this with our emerging knowledge of chromosome and genome structure, and to draw comparison both within and between paralogous and orthologous gene sequences in order to establish the nature of the mutational mechanisms responsible for their evolutionary divergence. Such an approach, encompassing a large number of human genes and their extended families, is essential for any attempt to discern underlying evolutionary principles.

The purpose of this book is to bring together some prevailing knowledge and ideas about evolution, to provide an informed framework of together for under graduates, post-graduate and those of related fields.

Since evolution is the broadest of biological fields, covering the greatest range of disciplines, even the brief survey of evolution offered here has errors and ambiguities. To the extent that this book has been spared many such failings the author owes thanks to many reviewers we commented on one or more sections of this text.

There can be no claim to originality except in the manner of treatment and much of the information has been obtained from the books and scientific journals available in the different libraries.

The author expresses his thanks to his friends and colleagues whose continue inspirations have initiated him to bring out this book.

The author expresses his gratitude to Mr. Wasan and staff of M/s Discovery Publishing House for their whole hearted co-operation in the publication of this book.

Author

Contents

1

ORIGIN OF MOLECULES

How many of the 90 naturally occurring elements are essential to life? After more than a century of increasingly refined investigation, the question still cannot be answered with certainty. Only a year or so ago the best answer would have been 20. Since then four more elements have been shown to be essential for the growth of young animals: fluorine, silicon, tin and vanadium. Nickel may soon be added to the list. In many cases the exact role played by these and other trace elements remains unknown or unclear. These gaps in knowledge could be critical during a period when the biosphere is being increasingly contaminated by synthetic chemicals and subjected to a potentially harmful redistribution of salts and metal ions. In addition, new and exotic chemical form of metals (such as methyl mercury) are being discovered, and a complex series of competitive and synergistic relations among mineral salts has been encountered. We are led to the realization that we are ignorant of many basic facts about how our chemical milieu affects our biological fate. Biologists and chemists have long been fascinated by the way evolution has selected certain elements as the building blocks of living organisms and has ignored others.

The composition of the earth and its atmosphere obviously sets a limit on what elements are available. The earth itself is hardly a chip off the universe. The solar system, like the universe, seems to be 99 per cent hydrogen and helium. In the earth's crust helium is essentially nonexistent (except in a few rare deposits) and hydrogen atoms constitute only about .22 per cent of the total. Eight elements provide more than 98 per cent of the atoms in the earth's crust: oxygen (47 per cent), silicon (28 per cent), aluminium (7.9 per cent), iron (4.5 per cent), calcium (3.5 per cent), sodium (2.5 per cent), potassium (2.5 per cent) and magnesium (2.2 per cent). Of these eight elements only five are among the 11 that account for more than 99.9 per cent of the atoms in the human body. Not surprisingly nine of the 11 are also the nine most abundant elements in sea-water. Two elements, hydrogen and oxygen, account for 88.5 per cent of the atoms in the human body; hydrogen supplies 63 per cent of the total and oxygen 25.5 per cent. Carbon accounts for another 9.5 per cent and nitrogen 1.4 per cent. The remaining 20 elements now thought to be essential for mammalian life account for less than .7 per cent of the body's atoms.

Background of Selection

Three characteristics of the biosphere or of the elements themselves appear to have played a major part in directing the chemistry of living forms. *First and foremost* there is the ubiquity of water, the solvent base of all life on the earth. Water is a unique compound; its stability and boiling point are both unusually high for a molecule of its simple composition. Many of the other compounds essential for life derive their usefulness from their response to water fullness from their response to water: whether they are soluble or insoluble, whether or not (if they are soluble) they carry an electric charge in solution and, not least, what effect they have on the viscosity of water. The *second directing force* involves the chemical properties of carbon, which evolution selected over silicon as the central building block for constructing giant molecules. Silicon is 146 times more plentiful than carbon in the earth's crust and exhibits many of the same properties. Silicon is directly below carbon in the periodic table of the elements; like carbon, it has the capacity to gain four electrons and form four covalent bonds.

The crucial difference that led to the preference for carbon compounds over silicon compounds seems traceable to two chemical features: the unusual stability of carbon dioxide, which is readily soluble in water and always monometric (it remains a single molecule), and the almost unique ability of carbon to form long chains and stable rings with five or six members. This versatility of the carbon atom is responsible for the millions of organic compounds found on the earth. *Silicon* in contrast, is insoluble in water and forms only relatively short chains with itself. It can enter into longer chains, however, by forming alternating bonds with oxygen, creating the compounds known as silicones (-Si-O-Si-O-Si). Carbon-to-carbon bonds are more stable than silicon-to-silicon bonds, but not so stable as to be virtually immutable, as the silicon-oxygen polymers are. Nevertheless, silicon has recently been shown to be essential in a way as yet unknown for normal bone development and full growth in chicks.

The third force influencing the evolutionary selection of the elements essential for life is related to an atom's size and charge density. Obviously the heavy synthetic elements from neptunium (atomic number 93) to lawrencium (No. 103), along with two lighter synthetic elements, technetium (No. 43) and promethium (No. 61), were never available in nature. (The atomic number expresses the number of protons in the nucleus of an atom or the number of electrons around the nucleus). The eight heavy elements in another group (No. 84 and 85 and Nos. 87 through 92) are too radioactive to be useful in living structures. Six more elements are inert gases with virtually no useful chemical reactivities: helium, neon, argon, krypton, xenon and radon.

On various plausible grounds one can exclude another 24 elements, or a total of 38 natural elements, as being clearly unsatisfactory for incorporation in living organisms because of their relative unavailability (particularly the elements in the lanthanide and actinide series) or their high toxicity (for example mercury and lead). This leaves 52 of the 90 natural elements as being potentially useful. Only three of the 24 elements known to be essential for animal life have an atomic number above 34. All three are needed only in trace amounts: molybdenum (No. 42), in (No. 50) and iodine (No. 53). The four most abundant atoms in living organisms—hydrogen, carbon, oxygen and nitrogen—have atomic numbers of 1, 6, 7 and 8. Their preponderance seems attributable to their being the smallest and lightest elements that can achieve stable electronic configurations by adding one to four electrons.

The ability to add electrons by sharing them with other atoms is the first step in forming chemical bonds leading to stable molecules. The seven next most abundant elements in living organisms all have atomic numbers below 21. In the order of their abundance in mammals they are calcium (No. 20), phosphorus (No. 15), potassium (No. 19), sulphur (No. 16), sodium (No. 11), magnesium (No. 12) and chlorine (No. 17). The remaining 10 elements known to be present in either plants or animals are needed only in traces. With the exception of fluorine (No. 9) and silicon (No. 14), the remaining eight occupy positions between No. 23 and No. 34 in the periodic table.

Table 1.1. Comparison between the composition of the human body with the approximate composition of seawater, the earth's crust and the universe at large.

Composition of Universe		*Composition of Earth's Crust*		*Composition of Seawater*		*Composition of Human Body*	
Percent of Total Number of Atoms							
H	91	O	47	H	66	H	63
He	9.1	Si	28	O	33	O	25.5
O	.057	Al	7.9	Cl	.33	C	9.5
N	.042	Fe	4.5	Na	.28	N	1.4
C	.021	Ca	3.5	Mg	.033	Ca	.31
Si	.003	Na	2.5	S	.017	P	.22
Ne	.003	K	2.5	Ca	.006	Cl	.03
Mg	.002	Mg	2.2	k	.006	K	.06
Fe	.002	Ti	.46	C	.0014	S	.05
S	.001	H	.22	Br	.0005	Na	.03
		C	.09			Mg	.01
All others	<.01	All others	<.1	All others	<.1	All others	<.01

It is interesting that this interval embraces three elements for which evolution has evidently found no role: gallium, germanium and arsenic. None of the metals with properties similar to those of gallium (such as aluminum and indium) has proved to be useful to living organisms. On the other hand, since silicon and tin, two elements with chemical activities similar to those of germanium, have just joined the list of essential elements, it seems possible that germanium too, in spite of its rarity, will turn out to have an essential role. Arsenic, of course, is a well-known poison.

Functions of Essential Elements

Some useful generalizations can be made about the role of the various elements. Six elements—carbon, nitrogen, hydrogen, oxygen, phosphorus and sulphur—make up the molecular building blocks of living matter: amino acids, sugars, fatty acids, purines, pyrimidines and nucleotides. These molecules not only have independent biochemical roles but also are the respective constituents of the following large molecules: proteins, glycogen, starch, lipids and nucleic acids. Several of the 20 amino acids contain sulphur in addition to carbon, hydrogen and oxygen.

Phosphorous plays an important role in the nucleotides such as adenosine triphosphat (ATP), which is central to the energetics of the

Table 1.2. 21 out of the first 34 elements in the periodic table which are essential for animal life.

Element	*Symbol*	*Atomic Number*	*Comments*
Hydrogen	H	1	Required for water and organic compounds.
Helium	He	2	Inert and unused.
Lithium	Li	3	Probably unused.
Beryllium	Be	4	Probably unused; toxic.
Boron	B	5	Essential in some plants, function unknown.
Carbon	C	6	Required for organic compounds.
Nitrogen	N	7	Required for many organic compounds.
Oxygen	O	8	Required for water and organic compounds.
Fluorine	F	9	Growth factor in rats; possible constituent of teeth and bone.
Neon	Ne	10	Inert and unused
Sodium	Na	11	Principal extracellular cation.
Magnesium	Mg	12	Required for activity of many enzymes in chlorophyll.
Aluminium	Al	13	Essentiality under study.
Silicon	Si	14	Possible structural unit of diatoms; recently shown to be essential in chicks.
Phosphorous	P	15	Essential for biochemical synthesis and energy transfer.
Sulphur	S	16	Required for proteins and other biological compounds.
Chlorine	Cl	17	Principal cellular and extracellular anion.
Argon	A	18	Inert and unused.
Potassium	K	19	Principal cellular cation.
Calcium	Ca	20	Major component of bone; required for some enzymes.
Scandium	Sc	21	Probably unused.
Titanium	Ti	22	Probably unused.
Vanadium	V	23	Essential in lower plants, certain marine animals and rat.
Chromium	Cr	24	Essential in higher animals; related to action of insulin.
Manganese	Mn	25	Required for activity of several enzymes.

Iron	Fe	26	Most important transition metal ion essential for hemoglobin and many enzymes.
Cobalt	Co	27	Required for activity of several enzymes in vitamin B_{12}.
Nickel	Ni	28	"Essentiality under study.
Copper	Cu	29	Essential in oxidative and other enzymes and hemocyanin.
Zinc	Zn	30	Required for activity of many enzymes.
Gallium	Ca	31	Probably unused.
Germanium	Ge	32	Probably unused.
Arsenic	As	33	Probably unused; toxic
Selenium	Se	34	Essential for liver function
Molybdenum	Mo	42	Required for activity of several enzymes.
Tin	Sn	50	Essential in rats; function unknown.
Iodine	I	53	Essential constituent of the thyroid hormones.

cell. ATP includes components that are also one of the four nucleotides needed to form the double helix of deoxy-ribonucleic acid (DNA), which incorporates the genetic blueprint of all plants and animals. Both sulphur and phosphorous are present in many of the small accessory molecules called coenzymes. In bony animals phosphorous and calcium help to create strong supporting structures. The electrochemical properties of living matter depend critically on elements or combinations of elements that either gain or lose electrons when they are dissolved in water, thus forming ions.

The principal cations (electron-deficient, or positively charged, ions are provided by four metals: sodium, potassium, calcium and magnesium. The principal anions (ions with a negative charge because they have surplus electrons) are provided by the chloride ion and by sulphur and phosphorous in the form of sulfate ions and phosphate ions. These seven ions maintain the electrical neutrality of body fluids and cells and also play a part in maintaining the proper liquid volume of the blood and other fluid systems. Whereas the cell membrane serves as a physical barrier to the exchange of large molecules, it allows small molecules to pass freely.

The electrochemical functions of the anions and cations serve to maintain the appropriate relation of osmotic pressure and charge distribution on the two sides of the cell membrane. One of the striking features of the ion distribution is the specificity of these different ions. Cells are rich in potassium and magnesium, and the surrounding plasma is rich in sodium and calcium. It seems likely that the distribution of ions in the plasma of higher animals reflects the oceanic origin of their evolutionary antecedents. One would like to know how primitive cells learned to exclude the sodium and calcium ions in which they were bathed and to deve!ɔp an internal milieu enriched in potassium and magnesium. The third and last group of essential elements consists of the trace elements. The fact that they are required in extremely minute quantities in no way diminishes their great importance.

In this sense they are comparable to the vitamins. We now know that the great majority of the trace elements, represented by metallic ions, serve chiefly as key components of essential enzyme systems or of proteins with vital functions (such as hemoglobin and myoglobin, which respectively transports oxygen in the blood and stores oxygen in muscle). The heaviest essential element, iodine, is an essential constituent of the thyroid hormones thyroxine and triiodothyronine, although its precise role in hormonal activity is still not understood.

Trace Elements

To demonstrate that a particular element is essential to life becomes increasingly difficult as one lowers the threshold of the amount of a substance recognizable as a "trace." It has been known for more than 100 years, for example, that iron and iodine are essential to man. In a rapidly developing period of biochemistry between, 1928 and 1935 four more elements, all metals, were shown to be essential: copper, manganese, zinc and cobalt. The demonstration can be credited chiefly to a group of investigators at the University of Wisconsin led by C.A. Elvehjem, E.B. Hart and W.R. Todd. At that time it seemed that these four metals might be the last of the essential trace elements.

In the next 30 years, however, three more elements were shown to be essential chromium, selenium and molybdenum. Fluorine, silicon, tin and vanadium have been added since 1970. The essentiality of five of these last seven elements was discovered through the careful, paintstaking efforts of *Klaus Schwarz* and his associates, initially located at the National Institute of Health and now based at the Veterans Administration Hospital in Long Beach, Calif. For the past 15 years Schwarz's group has made a systematic study of the trace-element requirements of rates and other small animals. The animals are maintained from birth in a completely isolated sterile environment.

The apparatus is constructed entirely of plastics to eliminate the stray contaminants contained in metal, glass and rubber. Although even plastics may contain some trace elements, they are so tightly bound in the structural lattice of the material that they cannot be leached out or be picked up by an animal even through contact. A typical isolator system houses 32 animals in individual acrylic cages. Highly efficient air filters remove all trace substances that might be present in the dust in the air. Thus the animals only access to essential nutrients is through their diet. They receive chemically pure amino acids instead of natural proteins, and all other dietary ingredients are screened for metal contaminants.

Since the standards of purity employed in these experiments far exceed those for reagents normally regarded as analytically pure, Schwarz and his co-workers have had to develop many new alalytical chemical methods. The most difficult problem turned out to be the purification of salt mixtures. Even the purest commercial reagents were contaminated with traces of metal ions. It was also found that trace elements could be passed from mothers to their offspring. To minimize this source of contamination animals are weaned as quickly as possible, usually from 18 to 20 days after birth. With these precautions Schwarz and his colleagues have within the past several years been able to produce a new deficiency disease in rates. The animals grow poorly, lose hair and muscle tone, develop shaggy fur and exhibit other detrimental changes. When standard laboratory food is given these animals, they regain their normal appearance. At first it was thought that all the symptoms were caused by the lack of one particular trace element. Eventually four different elements had to be supplied to complete the highly purified diets the animals had been receiving. The four elements proved to be fluorine, silicon, tin and vanadium. A convenient source of these elements is yeast ash or liver preparations from a healthy animal. The animals on the deficiency diet grew less than half as fast as those on a normal or supplemented diet. Growth alone, however, may not tell the entire story. There is some evidence that even the addition of the four elements may not reverse the loss of hair and skin changes resulting from the deficiency diet.

Functions of Trace Elements

The addition of tin and vanadium to the list of essential trace metals brings to 10 the total number of trace metals needed by animals and plants. What role do these metals play? For six of the eight trade metals recognized from earlier studies (that is, for iron, zinc, copper, cobalt, manganese and molybdenum) we are reasonably sure of the answer. The six are constituents of a wide range of enzymes that participate in a variety of metabolic processes. In addition to its role in hemoglobin and myoglobin, iron appears in succinate dehydrogenase, one of the enzymes needed for the utilization of energy from sugars and starches.

Enzymes incorporating zinc help to control the formation of carbon dioxide and the digestion of proteins. Copper is present in more than a dozen enzymes, whose roles range from the utilization of iron to the pigmentation of the skin. Cobalt appears in enzymes involved in the synthesis of DNA and the metabolism of amino acids. Enzymes incorporating managnese are involved in the formation of urea and the metabolism of pyruvate. Enzymes incorporating molybdenum participate in purine metabolism and the utilization of nitrogen. These six meals belong to a group known as transition elements. They owe their uniqueness to their ability to form strong complexes with ligands, or molecular groups, of the type present in the side chains of proteins. Enzymes in which transition metals are tightly incorporated are called metalloenzymes, since the metal is usually embedded deep inside the structure of the protein. If the metal atom is removed, the protein usually loses its capacity to function as an enzyme.

There is also a group of enzymes in which the metal ion is more loosely associated with the protein but is nonetheless essential for the enzyme's activity. Enzymes in this group are known as metal-ion-activiated enzymes. In either group the role of the metal ion may be to maintain the proper conformation of the protein, to bind the substrate (the molecule acted on) to the protein or to donate or accept electrons in reactions where the substate is reduced or oxydized. In 1968 the complete three dimensional structure of the first metalloenzyme, cytochrome *c*, was published. Cytochrome *c*, a red enzyme containing iron, is universally present in plants and animals. It is one of a series of enzymes, all called cytochromes, that extract energy from food molcecules by the stepwise addition of oxygen. The complete amino acid sequence of cytochrome *c* obtained from the human heart was determined some 10 years ago by a group led by *Emil L. Smith* of the University of Califormia at Los Angeles and by *Emanuel Margoliash* of North-western University.

The iron atom is partially complexed with an intricate organic molecule, protoporphyrin, to form a heme group similar to that in

Table 1.3. Wide variety of metalloenzymes required for the successful functioning of living organisms.

Metal	*Enzyme*	*Biological Function*
Iron	Ferredoxin	Photosynthesis
	Succinate Dehydrogenase	Aerobic oxidation of carbohydrates
Iron in Heme	Aldehyde Oxidase	Aldehyde oxidation
	Cytochromes	Electron transfer
	Catalase	Protection against hydrogen peroxide
	(Hemoglobin)	Oxygen transport
Copper	Ceruloplasmin	Iron utilization
	Cytochrome Oxidase	Principal terminal oxidase
	Lysine Oxidase	Elasticity of aortic walls
	Tyrosinase	Skin pigmentation
	Plastocyanin	Photosynthesis
	(Hemocyanin)	Oxygen transport in invertebrates
Zinc	Carbonic Anhydrase	CO_2 formation; regulation of acidity
	Carboxypeptidase	Protein digestion
	Alcohol Dehydrogenase	Alcohol metabolism
Manganese	Arginase	Urea formation
	Pyruvate Carboxylase	Pyruvate metabolism
Cobalt	Ribonucleotide Reductase	DNA biosynthesis
	Glutamate Mutase	Amino acid metabolism
Molybdenum	Xanthine Oxidase	Purine metabolism
	Nitrate Reductase	Nitrate utilization
Calcium	Lipase	Lipid digestion
Magnesium	Hexokinase	Phosphate transfer

hemoglobin. Of the iron atom's six coordination sites, four are attached to the heme group through nitrogen atoms. The other two sites form bonds with the protein chain; one bond is through a nitrogen atom in the side chain of a histidine unit at site No. 18 in the protein sequence and the other bond is through a sulphur atom in the side chain of a methionine unit at site No. 80. Although the cytochrome *c* molecule is complicated, it is one of the simplest of the metalloenzymes.

Cytochrome oxidase, probably the single most important enzyme in most cells, since it is responsible for transferring electrons to oxygen to form water, is far more complicated. Each molecule contains about 12 times as many atoms as cytochrome *c*, including two copper atoms and two heme groups, both of which participate in transferring the elections. More complicated yet is cysteamine oxygenase, which catalyzes the addition of oxygen to a molecule of cysteamine; it contains one atom each of three different metals: iron, copper and zinc.

There are many other combinations of metal ions and unique molecular assemblies. An extreme example is xanthine oxidase, which contains eight iron atoms, two molybdenum atoms and two molecules incorporating riboflavin (one of the B vitamins) in a giant molecule more than 25 times the size of cytochrome *c*. The metal-containing proteins of another group, the metalloproteins, closely resemble the metalloenzymes except that they lack an obvious catalytic function. Hemoglobin itself is an example. Others are hemocyanin, the copper-containing blue protein that carries oxygen in many invertebrates, metallothionein, a protein involved in the absorption and storage of zinc, and transferrin, a protein that transports iron in the bloodstream. There may be many more such compounds still unrecognized because their function has escaped detection.

Newest Essential Elements

Much remains to be learned about the specific biochemical role of the most recently discovered essential elements. In 1957 Schwarz and Calvin M. Foltz. working at the National Institutes of Health, showed that selenium helped to prevent several serious deficiency diseases in different animals, including liver necrosis and muscular dystrophy. Rats were protected against death from liver necrosis by a diet containing one-tenth of a part per million of selenium. Comparably low doses reversed the white muscle disease observed in cattle and sheep that happen to graze in areas where selenium is scarce.

In April a group at the University of Wisconsin under *J.T. Rotruck* reported a direct biochemical role for selenium. Oxidative damage to red blood cells was detected in rats kept on a selenium-deficient diet. This damage was related to reduced activity of an enzyme, glutathione peroxidase, that helps to protect hemoglobin against the injurious oxidative effects of hydrogen peroxide. The enzyme uses hydrogen peroxide to catalyze the oxidation of glutathione, thus keeping hydrogen peroxide from oxidizing the reduced state of iron in hemoglobin. Oxidized glutathione can readily be converted to reduced glutathione by a variety of intracellular mechanisms. There is some reason to believe glutathione peroxidase may even contain some form of selenium acting as an integral part of the functional enzyme molecule.

The physiological importance of chromium was established in 1959 by *Schwarz* and *Walter Mertz*. They found that chromium deficiency is characterized by impaired growth and reduced life-span, corneal lesions and a defect in sugar metabolism. When the diet is deficient in chromium, glucose is removed from the bloodstream only half as fast as it is normally. In rats the deficiency is relieved by a single administration of 20 micrograms of certain trivalent chromic salts. It now appears that the chromium ion works in conjunction with insulin, and that in at least some cases diabetes may reflect faulty chromium metabolism.

After developing the all-plastic trace-element isolator described above, *Schwarz, David B. Milne* and *Elizabeth Vineyard* discovered that tin, not previously suspected as being essential, was necessary for normal growth. Without one or two parts per million of tin in their diet, rats grow at only about two-thirds the normal rate. The next element shown to be essential in mammals by the Schwarz group was vanadium, an element that had been detected earlier in certain marine invertebrates but whose essentiality had not been demonstrated. On a diet in which vanadium is totally excluded rats suffer a retardation of about 30 per cent in growth rate. *Schwarz* and *Milne* found that normal growth is restored by adding one-tenth of a part per million of vanadium to the diet.

At higher concentrations vanadium is known to have several biological effects, but its essential role in trace amounts remains to be established. A high dose of vanadium blocks the synthesis of cholesterol and reduces the amount of phospholipid and cholesterol in the blood. Vanadium also promotes the mineralization of teeth and is effective as a catalyst in the oxidation of many biological substances. The third element most recently identified as being essential is fluorine. Even with tin and vanadium added to highly purified diets containing all other elements known to be essential, the animals in Schwarz's plastic cages still failed to grow at a normal rate. When up to half a part per million of potassium fluoride was added to the diet, the animals showed a 20 to 30 per cent weight gain in four weeks.

Although it had appeared that a trace amount of fluorine was essential for building sound teeth, Schwarz's study showed that fluorine's biochemical role was more fundamental than that. In any case fluoridated water provides more than enough fluorine to maintain a normal growth rate. Although there were earlier clues that silicon might be an essential life element, firm proof of its essentiality, at least in young chicks, was reported only three months ago. *Edith M. Carlisle* of the School of Public Health at the University of California at Los Angeles finds that chicks kept on a silicon-free diet for only one or two weeks exhibit poor development of feathers and skeleton, including markedly thin leg bones. The addition of 30 parts per million of silicon to the diet increases the chicks' growth more than 35 per cent and makes possible normal feathering and skeletal development.

Considering that silicon is not only the second most abundant element in the earth's crust but is also similar to carbon in many of its chemical properties, it is hard to see how evolution could have totally excluded it from an essential biochemical role. Nickel, nearly always associated with iron in natural substances, is another element receiving close attention. Also a transition element, it is particularly difficult to remove from the food used in special diets. Nickel seems to influence the growth of wing and tail feathers in chicks but more consistent data are needed to establish its essentiality. One incidental result of Schwarz's work has been the discovery of a previously unrecognized organic compound, which will undoubtedly prove to be a new vitamin.

Synergism and Antagonism

The interaction of the various essential metals can be extremely complicated. The absence of one metal in the diet can profoundly influence, either positively or negatively, the utilization of another metal that may be present. For example, it has been known for nearly 50 years that copper is essential for the proper metabolism of iron. An animal deprived of copper but not iron develops anaemia because the biosynthetic machinery fails to incorporate iron in hemoglobin molecules. It has only recently been found in our laboratories at Florida State University that ceruloplasmin, the copper-containing protein of the blood, is a direct molecular link between the two metals. Ceruloplasmin promotes the release of iron from animal liver so that the iron-binding protein of the serum, transferring, can complex with iron and transfer it to the developing red blood cells for direct utilization in the biosynthesis of hemoglobin. This represents a synergistic relation between copper and iron.

As an example of antagonism between elements one can cite the instance of copper and zinc. The ability of sheep or cattle to absorb copper is greatly reduced if too much zinc or molybdenum is present in their diet. Evidently either of the two metals can displace copper in an absorption process that probably involves competition for sites on a metal-binding protein in the intestines and liver.

The recent discoveries present many fresh challenges to biochemists. One can expect the discovery of previously unsuspected metalloenzymes containing vanadium, tin, chromium and selenium. New compounds or enzyme systems requiring fluorine and silicon may also be uncovered. The multiple and complex interdependencies of the elements suggest many hitherto unrecognized and important facts about the role and inter-relations of metal ions in nutrition and in health and disease.

2

ORIGIN OF LIFE

The problem of how life could have originated by ordinary chemical means long seemed insuperable. This is hardly surprising if we consider the present complexity of even the most elementary unit of life, the cell. At its simplest, a modern cell is surrounded by a highly selective permeable membrane, composed of lipids and proteins, that regulates the kinds of substances that pass through. Within the cell, the cytoplasm consists of a multitude of structures and substructures involved in the synthesis, storage, and breakdown of a large variety of chemical compounds.

AMINO ACIDS

Foremost among the metabolic agents that enable the cell to function are the many different proteins that catalyze and regulate practically all living chemical reactions. In basic structure, proteins consist of subunits, called *amino acids*, that have the following features:

$$\begin{array}{ccccc} & & H & & O \\ & & | & & | \\ H_2 & - & C^* & - & C-OH \\ & & | & & \\ & & R & & \end{array}$$

- An alpha carbon atom (C*) to which all other parts are attached
- An amino NH_2 group with a potential positive charge (NH_3^+)
- A carboxyl COOH group with a potential negative charge (COO^-)
- An H atom
- An R side chain that varies in structure among the different amino acids

These amino acids link together by chemical bonds called *peptide linkages* into linear *polypeptide chains* that are the constituents of proteins. The highly specific structure of any protein molecule, whether it functions as an *enzyme* (catalyst) or for some other purpose, derives from the exact linear placement of its various component amino acids. These specific amino acid sequences enable polypeptide chains to fold into specific three-dimensional forms that confer specific properties on proteins. It is, in fact, reasonable to claim that most present living phenomena, whether absorption, sensation, motion, structure, or whatever, derive from the enzymatic and regulatory activities of these long sequences of amino acids. Complexity, however, is not limited to proteins, since the amino acid sequences of proteins are actually determined by the nucleotide sequences in another group of basic molecules, nucleic acids.

Nucleic Acids

Nucleic acids are long-chained molecules composed of *nucleotide* subunits, each containing a pentose (5-carbon) sugar, a monophosphate group, and a nitrogenous base. The two kinds of sugar used in nucleic acids, ribose (hydroxylated at the

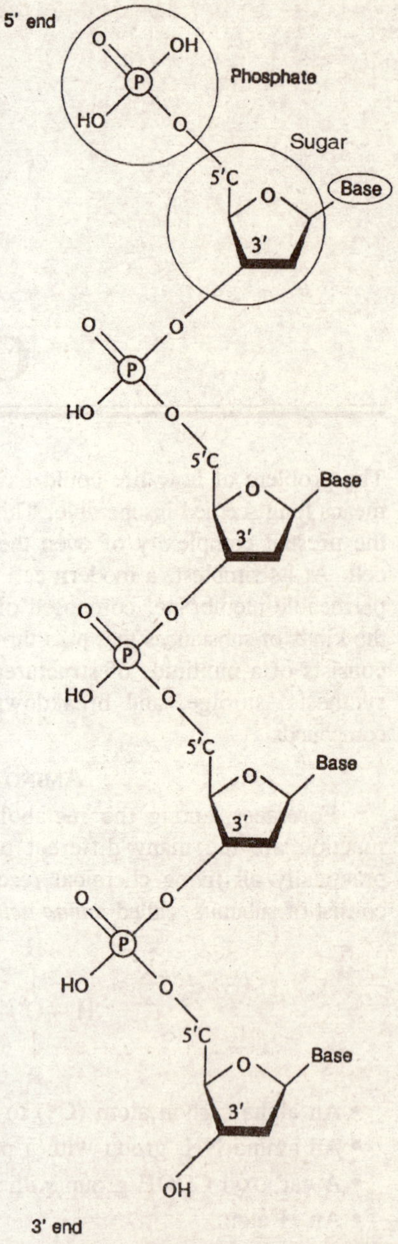

5' end
O
OH
P
HO
O
Phosphate
Sugar
5'C
O
Base
3'
O
O
P
HO
O
5'C
O
Base
3'
O
O
P
HO
O
5'C
O
Base
3'
O
O
P
HO
O
5'C
O
Base
3'
OH
3' end

Fig. 2.1. General structure for DNA and RNA chain

2' carbon position) and deoxyribose (lacks 2' hydroxyl) provide the names for the two kinds of nucleic acids, ribonucleic acid (RNA) and deoxyribonucleic acid (DNA). In both these nucleic acids, the phosphate groups occupy the same position, tying the 3' carbon of one sugar to the 5' carbon of its neighbour via a phosphodiester bond. Connected to the 1' carbon of each sugar is one of four kinds of nitrogenous heterocyclic bases, of which two are purines (A, adenine, and G, guanine, in both DNA and RNA) and two are pyrimidines (C, cytosine, and T, thymine, in DNA and C, cytosine, and U, uracil, in RNA).

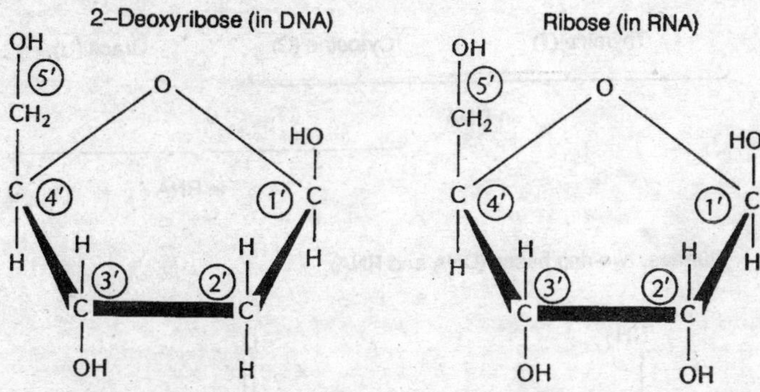

Fig. 2.2. Difference between the sugars found in DNA (deoxyribose) and RNA (ribose).

Since the complexity of proteins derives from the complexity of nucleic acids, you might think that the restriction of nucleic acid composition to only four different kinds of bases would limit the message-bearing capacity of these molecules to only four kinds of messages, but it doesn't. The facts that nucleic acid molecules may be many thousands or millions of nucleotides long, and that each message can be encoded by a unique linear sequence of nucleotides, endow these molecules with the capacity to bear an immense variety of highly complex messages. For any one nucleotide position, 4 different messages are possible (A, G, C, or T); for two nucleotides in tandem, 4^2 or 16 different messages are possible (AA, AG, AC, AT, GG, GC,...); and so on: the rule being simply that for a linear sequence of *n* nucleotides, 4^n different possible messages can be encoded. Thus a linear sequence of only 10 nucleotides can discriminate among more than 1 million (4^{10}) potentially different messages.

All this helps explain the information-carrying role of nucleic acids but does not explain how they replicate and transmit this information. The model presently accepted for nucleic acid replication derives from

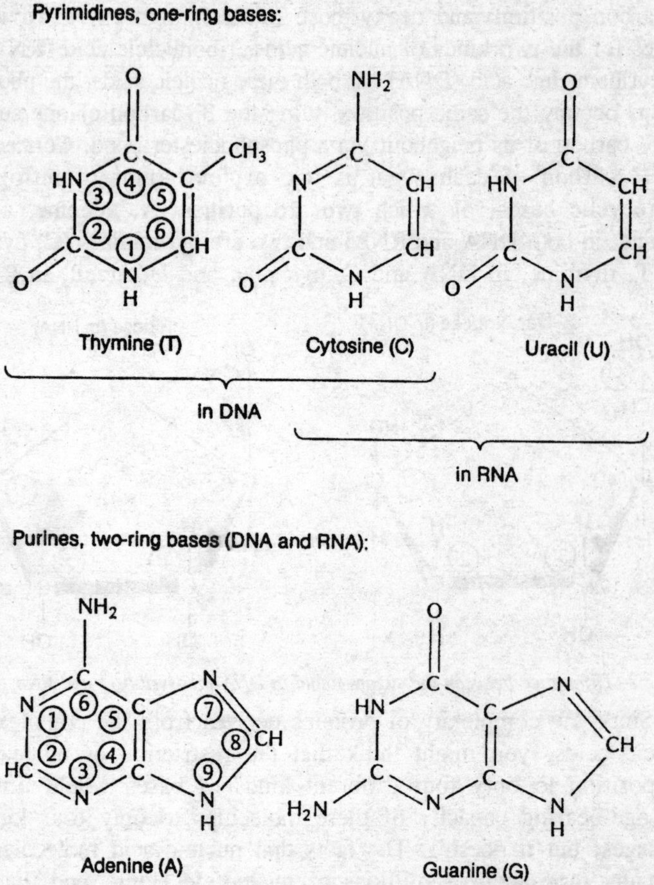

Pyrimidines, one-ring bases:

Thymine (T) Cytosine (C) Uracil (U)

in DNA

in RNA

Purines, two-ring bases (DNA and RNA):

Adenine (A) Guanine (G)

Fig. 2.3. The basic kinds of nitrogenous bases found in DNA (T, C, A, G) and RNA (C, U, A, G).

the now-familiar *double helix* structure that Watson and Crick first offered. The DNA double helix consists of two antiparallel strands coiled around each other in the form of a right-hand screw, with complementary pairing between purine bases on one strand and pyrimidine bases on the other (A-T, G-C). In the familiar B form of DNA, there are approximately 10 base pairs for each complete turn of the helix, and the bases are stacked almost perpendicularly to the helical axis.

The replicatory power of the DNA double helix obviously derives from the ability of each of the two strands to serve as a template for a newly complementary strand, so that two new double helices can

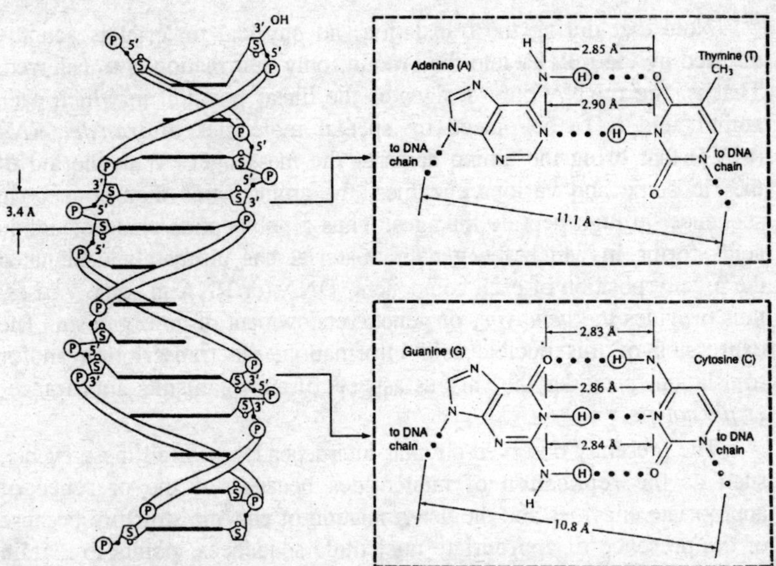

Fig. 2.4. The Watson-Crick model of the standard DNA double helix on the left, with examples of hydrogen-bond pairing between bases on the right (A–T, G–C).

form bearing nucleotide sequences that are identical to each other as well as to that of the parental molecule. This unique quality of exact molecular replication, enabling similar messages to be transmitted from generation to generation, confers on nucleic acids their function as "genetic material."

Fundamental to our understanding of the relationship between genetic material and protein, therefore, is an important concept: the three-dimensional structure of a protein—its form, shape, and subsequent function—is primarily determined by the linear sequence of amino acids of which it consists. This linear sequence of amino acids in turn derives from the linear sequence of bases in nucleic acids by means of a protein-synthesizing apparatus involving three different kinds of RNA.

In brief, the genetic material, through the process of transcription, produces a molecule of *messenger RNA* (*mRNA*) that is, base for base, a complement to the bases on one of the DNA strands. Through the mediation of ribosomes, which themselves consist of *ribosomal RNA* (*rRNA*) and protein, a sequence of bases in mRNA then translates into a sequence of amino acids. This translation follows the triplet rule that a sequence of three mRNA bases designates 1 of the 20 different kinds of amino acids used in protein synthesis.

Note that during the translation, no physical material is actually inserted by the mRNA into the protein; only information is transferred. That is, the mRNA only designates the linear position in which each amino acid is to be placed by special molecules of *transfer RNA* (*tRNA*) that bring the amino acids to the messenger. With the aid of the ribosome and various enzymes, the amino acids then connect in sequence through peptide linkages. Thus a polypeptide chain of amino acids forms in which the genetic material has ultimately designated the precise position of each component. DNA (or RNA in some viruses) thus provides the *genotype*, or genetic endowment of an organism. The expression of this nucleic acid information, via transcription and/or translation, provides the various aspects of an organism's appearance, or *phenotype*.

The presently observed circular interdependency of all these events, such as the replication of nucleotides because of the presence of appropriate enzymes and the determination of enzyme structure because of the presence of appropriate nucleotide sequences, points to certain difficulties in finding a reasonable explanation for the origin of life. Which of the many biochemical agents came first? How did they arise? How could they have functioned before an entire cellular structure formed? Many proposals for the origin of life exist, and we can generally divide such proposals into two broad categories: life on Earth developed from previous life, or life on Earth originated by chemical means.

Life only from Prior Life

The concept that life did not originally arise on Earth is embodied in almost all the creation myths of humans. These myths usually presume the special creation of life on Earth by one or more superior, intelligent, and all-powerful beings who themselves possess attributes of life such as sensation, thought, and purposive movement. This concept has the advantage that it explains the origin of terrestrial life in a simple fashion that is especially attractive to those who believe that natural events are governed by conscious agents. It does not explain the source of the initial creator, and therefore does not explain life's origin.

Another ancient concept is that life can arise spontaneously at any time, such as the presumed origin of insects from sweat and crocodiles from mud. Strangely enough, this view was often held simultaneously with the view that life derives from a conscious creator, and was popular in Europe throughout medieval times. Pasteur and others put spontaneous generation theory to rest in the 1860s, and it has not since been revived in its original form. As we shall see later, the modern concept of the spontaneous origin of life on Earth does not include such simple means as the immediate action of sunlight on liquid or clay but proposes instead the past existence of more complex yet more understandable biochemical processes.

One variation of the theory that life comes only from life is the proposal that life is somehow ingrained in all matter, and the creation of matter by whatever cause is responsible for the creation of life. Although this notion offers the advantage of ascribing life to a natural event, its origin would seem difficult if not impossible to understand. Obviously, many aspects of matter show no evidence of life if we define life to include those attributes possessed by terrestrial organisms, such as metabolism, reproduction, and so on. How did these attributes arise?

Another variation suggests that life on Earth arose elsewhere, perhaps on a distant planetary body circling a distant star, and was then transported to Earth by space-resistant spores or other means. This notion, called *panspermia*, was fostered by the chemist Arrhenius (1859-1927) in the early part of this century and still has some adherents among scientists today. It overcomes the difficulty of seeking a chemical explanation for the origin of life on Earth but does not, of course, explain the distant origin of life.

Proponents of the panspermia hypothesis point to the discovery of a number of different organic compounds in carbon-containing meteorites (*carbonaceous chondrites*), ranging from carbohydrates to amino acids. Looked at closely, however, the structural forms or isomers of amino acids in these carbonaceous chondrites possess the two different kinds of *optical activity* (dextro- and levorotary) in equal amounts, therefore comprising a *racemic mixture* that shows no optical activity. In contrast, the amino acids of living forms generally show optical activity of only one type, levorotary. Furthermore, a number of the amino acids in meteorites do not appear in proteins. Along with other observations, these findings indicate that organic compounds probably formed through random chemical reactions in the meteorite itself or in its parent body, rather than through ordered living processes.

Opposition to the panspermia hypothesis also notes the difficulty of envisaging how spores or "bugs" from outer space can get to Earth without the help of conscious agents in spaceships. If the bug is too large, it cannot be easily ejected from its home planet nor subsequently

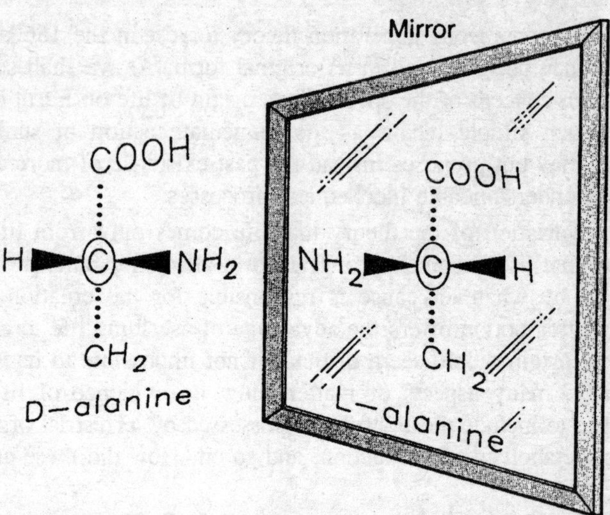

Fig. 2.5. Structures of an L-form and a D-form amino acid. The dark, wedge-shaped bonds indicate that the attached NH_2 and H groups project above the plane of the paper.

pushed out by sun radiation in its particular solar system. If it is too small, it will be kept from entering Earth's field by radiation pressure from our sun. Shklovskii and Sagan suggest that bugs larger than 0.6 μm cannot escape from Earth, while bugs smaller than 0.6 μm would be pushed away from Earth. So the donor planet must have been capable of ejecting bugs that we cannot eject (e.g., about 1 μm); that is, the sun of the donor planet must have been very hot (high radiation pressure). But if such a sun were hot enough for this purpose, it would destroy ejected particles by radiation.

In addition, the hazards of interstellar travel are many. For example, ultraviolet radiation (UV) from our sun will kill ejected particles from Earth within about one day of interplanetary travel. If the bugs were shielded from UV, they would be too heavy to be ejected. Among other space hazards are the hot, ionized gases that surround early-type stars, the presence of cosmic rays, and the absorption of bugs into passing suns by gravitational attraction.

However, even assuming that bugs survived all other obstacles, the vastness of space would disperse these spores so widely that their chance of reaching Earth would be infinitesimal. Shklovskii and Sagan have calculated that 100 million life-bearing planets in our galaxy would each have had to eject about 1,000 tons of spores in order for Earth to have received a single microorganism during its first billion years of history. Such difficulties make questionable whether an event as seemingly rare as panspermia is more probable than the chemical origin of life on this planet: Why should the origin of life on other bodies have had a greater probability than its origin on Earth? Such misgivings, along with a rapid increase in our understanding of molecular biology and biochemistry, have influenced most scientists to concentrate their attention on a terrestrial origin of life.

Terrestrial Origin of Life

The difficulty in visualizing life originating on Earth is essentially one of visualizing the molecular environment and events that occurred in a long-distant past. Unfortunately, we have as yet no certainties about the details of our molecular past, and we may never have such knowledge because molecular fossils are indeed sparse. At best, we can try to deduce the general nature of some of the original molecular events from present living structures and reactions and try to reconstruct such events experimentally under controlled conditions. However, before undertaking such a molecular review, it is important to consider the framework in which people usually pose the question of the origin of life from a terrestrial source. That is, we must attempt to deal with the problem of whether a highly complex, ordered phenomenon such as life could have arisen at all from the molecular chaos assumed to have existed during the Earth's early history.

Obviously, the probability for a modern, self-reproducing cell arising from complete disorder is embarrassingly small. To use an oft-quoted example, can a monkey, given even billions of years, produce the works of Shakespeare by randomly pressing the keys of a typewriter? Even if we restrict Shakespeare's writings to 1 million (10^6) alphabetical letters and limit the typewriter to 26 keys, the chance for such an event is $(1/26)^{10^6}$. This means that even if a monkey could type 1 million words a second, we could expect such an event to occur only once in $7 \times 10^{1,414,965}$ years!

By similar reasoning, the chances for most complex organic structures to arise spontaneously are infinitesimally small. Even a small enzymatic sequence of 100 amino acids would have only one chance in 20^{100} ($= 10^{130}$) to arise randomly, since there are 20 possible kinds of different amino acids for each position in the sequence. Thus, if we randomly generated a new 100-amino-acid-long sequence each second, we could expect such a given enzyme to appear only once in 4×10^{122} years! In terms of the volume necessary to generate all such possibilities, the difficulty appears just as immense: if an entire universe 10 billion light-years in diameter were densely packed with randomly produced polypeptides, each 100 amino acids long, the number of such molecules—10^{103}—would not equal their 10^{130} possibilities.

These arguments long seemed formidable and were further strengthened by the suggestion that nature itself would deteriorate any complex organization of matter even if such complexity were to arise accidentally. Theorists often pointed out that according to the *second law of thermodynamics*, the energy in a system tends toward diffusion rather than concentration; that is, *entropy* (disorder) increases rather than decreases. Thus, there appeared to be only a negative answer to the question that Pasteur had posed in the nineteenth century: "Can matter organize itself?" It seemed either that the living organization of matter must be explained as arising from a mystical nonnatural source, or that this event, if it were of natural origin, was so improbable that any attempt at comprehension or reconstruction would be meaningless.

In answer to these apparent difficulties, many scientists today point to two important considerations:

- The likelihood that primeval "living" organisms did not have many of their present complexities.
- The formation of organic molecules and subsequent organic structures was not the result of completely random events, although such events were nevertheless natural in the sense that they followed chemical and physical laws.

The first consideration, that early organisms were more primitive than those of today, is extremely important. Perhaps the most basic quality of life we would recognize in even a primitive living organism is its ability to perform those reactions necessary for it to grow and replicate, certainly the endless loop of metabolism and information transfer that we now see embodied in the intricate relationship between proteins and nucleic acids need not always have been of the same complexity. As we shall see later, protein-like compounds may arise in reaction mixtures of amino acids without intervention by nucleic acids and, although formed randomly, these compounds may then function in a variety of enzymatic ways.

Although it is true that chemically generated proteins do not have the repeatable and precisely ordered sequences of amino acids in cellular proteins, it is probably also true that the metabolic functions necessary for survival and growth were much simpler in the past. The relative simplicity of early "life" and its precursors, especially in the absence of competition with the more sophisticated later forms, seems a reasonable assumption to make.

The second consideration, that life did not arise from absolute chaos, has been supported in many ways. Many scientists have pointed out that the evolution of our solar system offered a number of essential prerequisites that enabled the development and sustenance of life:

1. Our planet possessed a sun of moderate size that was on the main sequence of stellar evolution. This provided a steady rate of emitted radiation over a long enough period of time for life to develop. The amount of energy (available from solar radiation seems to have always been far greater than any other source provided, although energy from other sources may also have been important in initiating particular chemical reactions.
2. A fairly large sampling of different elements existed, such as H, O, C, N, S, P, Ca, and others. These elements must have provided considerable chemical diversity, enabling reactions to occur that were necessary to form organic molecules involving carbon. The important chemical attributes of carbon, its ability to form four covalent bonds and the tetrahedral arrangement of its outer electrons, provided the opportunity for the formation of a large number of different kinds of stable molecules with considerable three-dimensional variety and complexity. Interestingly, the terrestrial presence of such molecules is not unique: by means of spectroscopy, researchers can now observe a variety of organic molecules in the dense interstellar clouds that give rise to stars and planets. Such observations indicate that a number of compounds necessary for the origin of life were already present both before and during the formation of our solar system, and their synthesis was *abiotic*-independent of living systems.
3. The Earth followed a nearly circular orbit at a fairly uniform distance from the sun. Such an even orbit would eliminate temperature extremes in which organic molecules would be incapable of forming or functioning.
4. There was present on Earth large amounts of an excellent solvent, water, which is stable in liquid form over a relatively wide range of temperatures and enables both acids and bases to ionize and react. Water also has the advantage that it floats in its crystalline frozen form (ice), so that bodies of water containing organic matter may remain liquid under a surface of ice, rather than freezing because of the subsurface accumulation of ice. Geochemists now

believe that water must have been present early in the Earth's history in the cold planetesimal condensations and appeared in liquid form as soon as the lithosphere reached appropriate temperatures. Additional water has been continually emitted into the atmosphere through volcanic activity, which was probably greater in the past than at present. Since water causes crustal erosion that leads to sedimentary rock formation, the presence of significant amounts of water dates back to the beginning of the geological record.

5. Hydrogen-containing gases existed for a long initial period in the Earth's history, derived from the high cosmic abundance of hydrogen and the consequent abundance of its compounds. Even though free hydrogen was probably lost early in the Earth's history, outgassing from the Earth's interior would have led to at least a partially reducing atmosphere in some or many localities in which hydrogen protons could be donated to a variety of proton-accepting elements, especially carbon. Such gases, along with the energy from solar and ultraviolet radiation, therefore enabled the formation of a variety of organic molecules that could, in turn, provide both structure and energy for living processes; for example, amino acids, sugars, fatty acids, purines, and pyrimidines.

In summary, we can say that there was considerable *molecular preadaptation* for the biochemical events leading to the formation of life even before life appeared. That is, there were appropriate energy sources, chemicals, temperature, and solvent—the foundations of a "universal organic chemistry." What kinds of reactions would then have taken place?

Origin of Basic Biological Molecules

In the 1920s, Oparin, a Russian biochemist, and Haldane, an English geneticist, independently suggested that the primitive atmosphere of the Earth was reducing and that organic compounds formed in such an atmosphere might be similar to those presently used by living organisms. However, almost 30 years elapsed before someone undertook an experimental test of this hypothesis. In 1953, Miller placed together in a glass apparatus methane, ammonia, and hydrogen gases. He generated an electric spark in a large 5-liter flask, and boiled water in a smaller flask to provide vapour to the spark as well as to circulate the gases. Compounds formed by sparking were then condensed, or recirculated if they were volatile. After one week of continuous electrical discharge, he chromatographed and analyzed the products

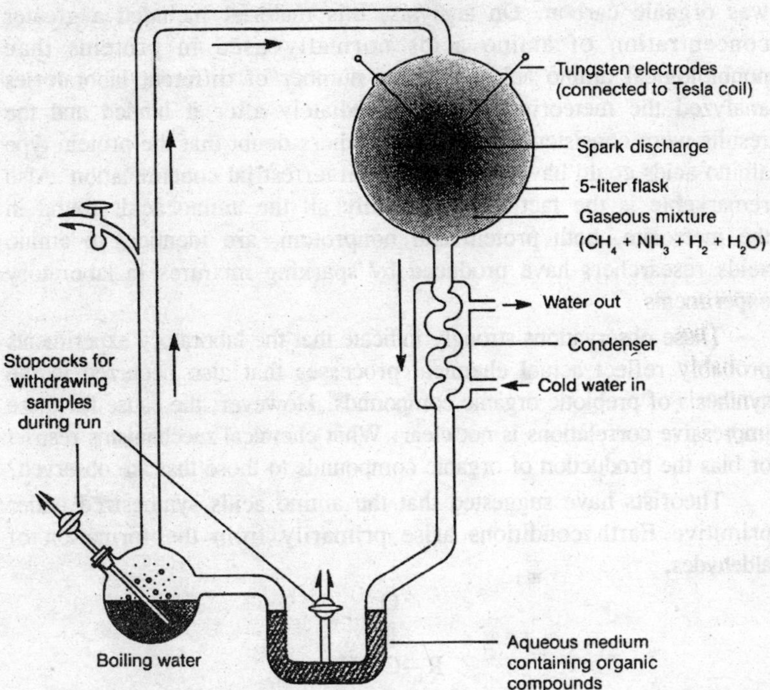

Fig. 2.6. Apparatus Miller (1953) used to demonstrate the synthesis of organic compounds by electrical discharge in a reducing atmosphere.

accumulated in the aqueous phase. Note that a large portion of these compounds are relatively simple and include both amino acids and other substances, such as urea, found in living organisms. In fact, of the wide array of possible complex molecules that such apparently random chemical reactions could have produced, it is remarkable that significant amounts of such relatively simple compounds essential to life actually formed. These experiments and others that followed therefore point strongly to the likelihood that the chemical environment that existed before the origin of life was probably not "chaos." Rather, the Earth had a significant amount of simple organic molecules that could participate in forming living organisms.

Moreover, astronomers can see such compounds in interstellar clouds in our galaxy and also in various carbonaceous meteorites that they believe represent material remaining in space from the original solar condensation 4.6 billion years ago. One such example, the Murchison meteorite that fell in Australia in 1969, contained more than 80 kilograms of carbonaceous material, of which about 1 percent was organic carbon. On analysis, this material included a greater concentration of amino acids normally used in proteins than nonbiological amino acids. Since a number of different laboratories analyzed the meteorite almost immediately after it landed and the results were consistent overall, researchers doubt that the protein-type amino acids could have originated from terrestrial contamination. Also remarkable is the fact that practically all the amino acids found in the meteorite, both protein and nonprotein, are identical to amino acids researchers have produced by sparking mixtures in laboratory experiments.

These observations strongly indicate that the laboratory experiments probably reflect actual chemical processes that also occurred in the synthesis of prebiotic organic compounds. However, the cause for these impressive correlations is not clear: What chemical mechanisms restrict or bias the production of organic compounds to those that are observed?

Theorists have suggested that the amino acids synthesized under primitive Earth conditions arise primarily from the formation of aldehydes,

$$\mathrm{R{-}\overset{\overset{\large O}{\|}}{C}{-}H}$$

(where R may represent any group), which then interact with ammonia and cyanide compounds. These reactive chemicals may have arisen from a variety of simple gases or from their further interactions:

$$N_2 + H_2 \rightarrow NH_3 \text{ (ammonia)}$$
$$N_2 + H_2O \rightarrow NH_3 \text{ (ammonia)}$$

$$CH_4 \longrightarrow C_2H_2 \text{ (HC}\equiv\text{CH, acetylene)}$$

$$CH_4 + N_2,\quad CH_4 + NH_3,\quad CO + NH_3,\quad C_2H_2 + N_2 \rightarrow HCN \text{ (HC}\equiv\text{N, hydrogen cyanide)}$$

$$CH_4 + H_2O,\quad CH_4 + CO_2,\quad CO_2 + H_2,\quad CO_2 + H_2O \rightarrow HCHO\ (\mathrm{H\overset{\overset{O}{\|}}{C}H}\text{, formaldehyde})$$

$$CH_4 + H_2O \longrightarrow CH_3CHO\ (\mathrm{H_3C{-}\overset{\overset{O}{\|}}{C}H}\text{, acetaldehyde})$$

According to one of the possible pathways of amino acid synthesis (the Strecker synthesis), subsequent steps are as follows:

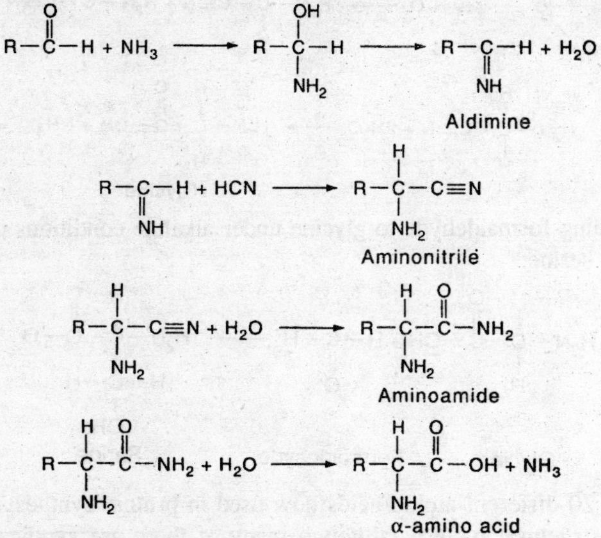

$$R{-}C({=}O){-}H + NH_3 \longrightarrow R{-}C(OH)(NH_2){-}H \longrightarrow R{-}C({=}NH){-}H + H_2O$$

Aldimine

$$R{-}C({=}NH){-}H + HCN \longrightarrow R{-}C(H)(NH_2){-}C{\equiv}N$$

Aminonitrile

$$R{-}C(H)(NH_2){-}C{\equiv}N + H_2O \longrightarrow R{-}C(H)(NH_2){-}C({=}O){-}NH_2$$

Aminoamide

$$R{-}C(H)(NH_2){-}C({=}O){-}NH_2 + H_2O \longrightarrow R{-}C(H)(NH_2){-}C({=}O){-}OH + NH_3$$

α-amino acid

If R in the preceding reactions is a hydrogen atom—that is, if the initial molecule,

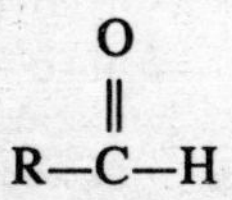

$$R{-}C({=}O){-}H$$

is formaldehyde (HCHO)—then the resulting amino acid is glycine. Glycine can also result from adding water (hydrolysis) to cyanide polymers:

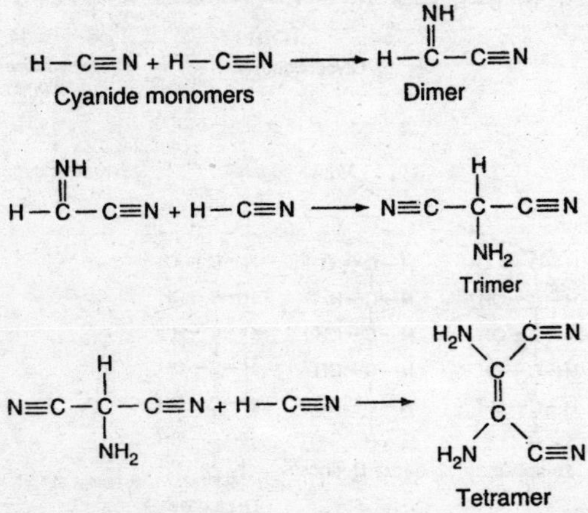

$$H{-}C{\equiv}N + H{-}C{\equiv}N \longrightarrow H{-}C({=}NH){-}C{\equiv}N$$

Cyanide monomers — Dimer

$$H{-}C({=}NH){-}C{\equiv}N + H{-}C{\equiv}N \longrightarrow N{\equiv}C{-}C(H)(NH_2){-}C{\equiv}N$$

Trimer

$$N{\equiv}C{-}C(H)(NH_2){-}C{\equiv}N + H{-}C{\equiv}N \longrightarrow (H_2N)(N{\equiv}C)C{=}C(NH_2)(C{\equiv}N)$$

Tetramer

$$(H_2N)(N{\equiv}C)C{=}C(C{\equiv}N)(NH_2) + H_2O \longrightarrow H_2N{-}CH_2{-}C{\equiv}N + H_2N{-}\overset{O}{\overset{\|}{C}}{-}C{\equiv}N$$

$$H_2N{-}CH_2{-}C{\equiv}N + 2H_2O \longrightarrow \underset{\text{Glycine}}{H_2N{-}CH_2{-}\overset{O}{\overset{\|}{C}}{-}OH} + NH_3$$

Adding formaldehyde to glycine under alkaline conditions can then produce serine:

$$\underset{\text{Glycine}}{H_2N{-}CH_2{-}\overset{O}{\overset{\|}{C}}{-}OH} + \underset{\text{Formaldehyde}}{H{-}\underset{O}{\underset{\|}{C}}{-}H} \longrightarrow \underset{\text{Serine}}{H_2N{-}CH(CH_2OH){-}\overset{O}{\overset{\|}{C}}{-}O}$$

All 20 different amino acids now used in protein synthesis have a similar structural pattern, although many of them are synthesized by different biochemical pathways. Among other basic organic molecules

$$2H{-}\overset{O}{\overset{\|}{C}}{-}H \longrightarrow H{-}\overset{O}{\overset{\|}{C}}{-}\overset{OH}{\overset{|}{C}}H{-}H$$

Formaldehyde Glycoaldehyde

$$H{-}\overset{O}{\overset{\|}{C}}{-}\overset{OH}{\overset{|}{C}}H{-}H + H{-}\overset{O}{\overset{\|}{C}}{-}H \longrightarrow H{-}\overset{O}{\overset{\|}{C}}{-}\overset{OH}{\overset{|}{C}}H{-}\overset{OH}{\overset{|}{C}}H{-}H \rightleftharpoons H{-}\overset{OH}{\overset{|}{C}}H{-}\overset{O}{\overset{\|}{C}}{-}\overset{OH}{\overset{|}{C}}H{-}H$$

Glyceraldehyde → Aldose sugars

Dihydroxy-acetone → Ketose sugars

Aldose sugars → D-ribose, 2-deoxy-D-ribose, D-glucose

D-ribose: H–C=O / H–C–OH / H–C–OH / H–C–OH / H–C–OH / H

2-deoxy-D-ribose: H–C=O / H–C–H / H–C–OH / H–C–OH / H–C–OH / H

D-glucose: H–C=O / H–C–OH / HO–C–H / H–C–OH / H–C–OH / H–C–OH / H that could easily be synthesized under fairly simple conditions are the *sugars*. Significant yields of glucose, ribose, and deoxyribose, for example, occur from condensing formaldehyde.

The *purine* and *pyrimidine* bases that are essential components of nucleic acids can also be synthesized under prebiotic conditions. For example, Oro and co-workers have shown that heating aqueous solutions of ammonium cyanide (prepared by the reaction of HCN with NH_4OH) produces up to 0.5 percent yield of adenine. Similarly, ultraviolet radiation acting on hydrogen cyanide solution produces a number of purines, including adenine and guanine. Condensation reactions in forming adenine have been studied in some detail, and researchers have suggested that one sequence may be as follows:

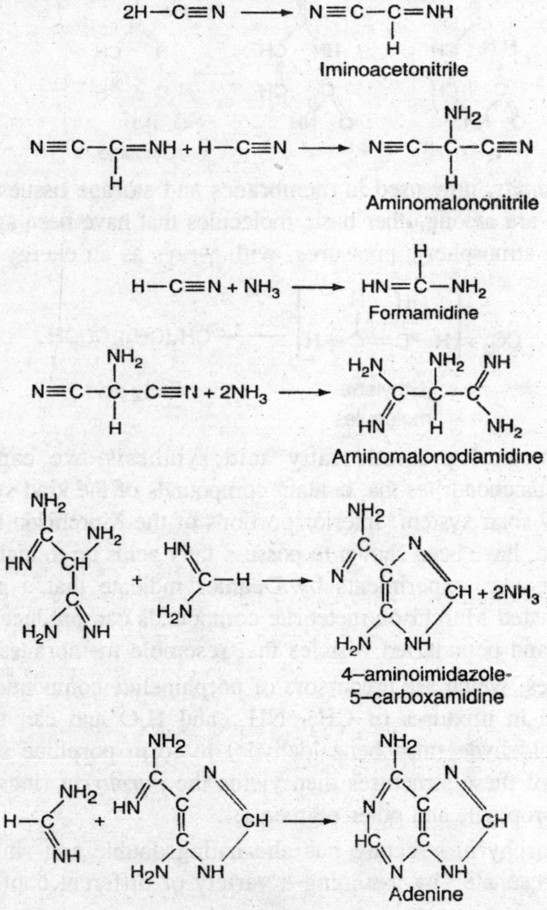

$$2H{-}C{\equiv}N \longrightarrow N{\equiv}C{-}CH{=}NH$$

Iminoacetonitrile

$$N{\equiv}C{-}CH{=}NH + H{-}C{\equiv}N \longrightarrow N{\equiv}C{-}CH(NH_2){-}C{\equiv}N$$

Aminomalononitrile

$$H{-}C{\equiv}N + NH_3 \longrightarrow HN{=}CH{-}NH_2$$

Formamidine

$$N{\equiv}C{-}CH(NH_2){-}C{\equiv}N + 2NH_3 \longrightarrow H_2N(HN{=})C{-}CH(NH_2){-}C({=}NH)NH_2$$

Aminomalonodiamidine

$$\text{Aminomalonodiamidine} + HN{=}CH{-}NH_2 \longrightarrow \text{4-aminoimidazole-5-carboxamidine} + 2NH_3$$

4–aminoimidazole-5–carboxamidine

$$H{-}C(NH_2){=}NH + \text{4-aminoimidazole-5-carboxamidine} \longrightarrow \text{Adenine}$$

Adenine

The reaction of cyanoacetylene with cyanates such as urea has produced the pyrimidine cytosine as shown next, and researchers have proposed similar synthetic procedures for the other pyrimidines uracil and thymine:

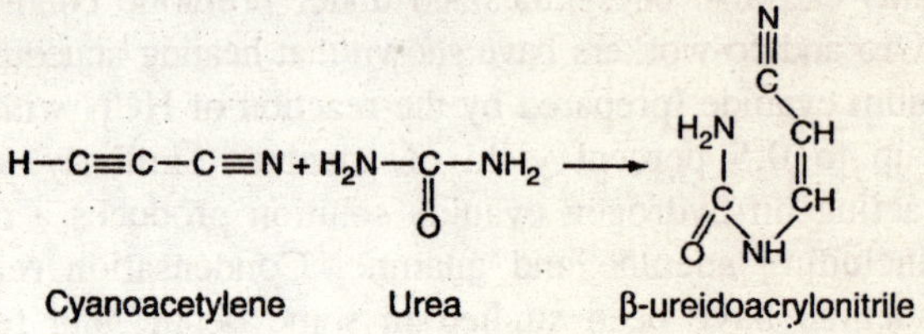

$$H{-}C{\equiv}C{-}C{\equiv}N + H_2N{-}\underset{\underset{O}{\|}}{C}{-}NH_2 \longrightarrow$$

Cyanoacetylene Urea β-ureidoacrylonitrile

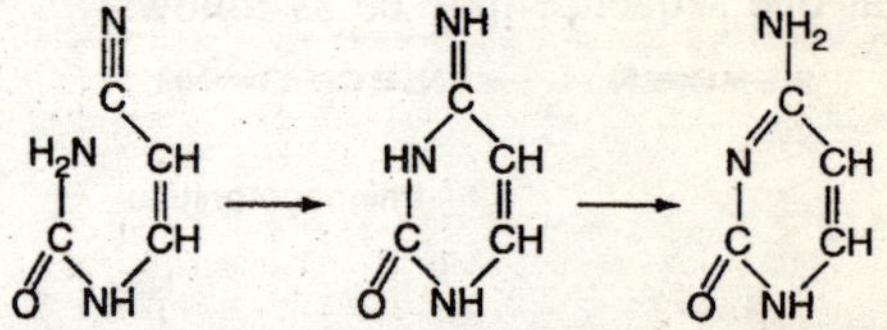

Cytosine

Fatty acids, now used in membranes and storage tissues of living organisms, are among other basic molecules that have been synthesized under high atmospheric pressures, with γ-rays as an energy source:

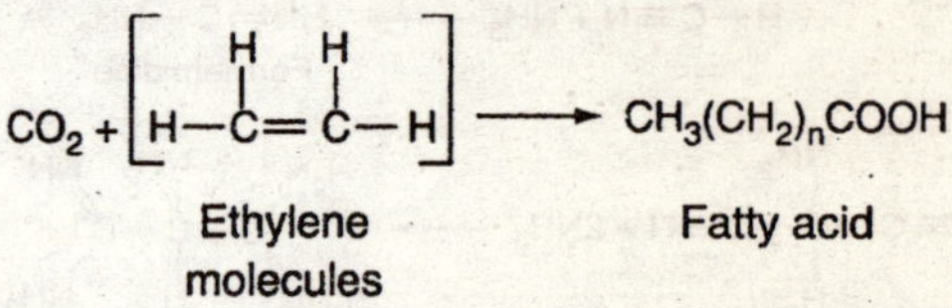

$$CO_2 + \left[H{-}\overset{\overset{H}{|}}{C}{=}\overset{\overset{H}{|}}{C}{-}H \right] \longrightarrow CH_3(CH_2)_nCOOH$$

Ethylene molecules Fatty acid

For evidence of prebiotic fatty acid synthesis, we can look to carbonaceous chondrites that contain compounds of the kind synthesized in the early solar system. Interior portions of the Murchison meteorite, for example, have been shown to possess fatly acids up to eight carbons long. Moreover, experiments by Deamer indicate that a portion of uncontaminated Murchison meteorite compounds can produce fatty-like structures and boundaried vesicles that resemble membranes.

Pyrroles, which are precursors of porphinelike compounds, can be synthesized in mixtures of CH_4, NH_3, and H_2O and can then react with formaldehyde (also benzaldehyde) to form porphine structures. Oxidizing of these structures then yields the *porphyrin* rings found in heme, chlorophyll, and other pigments.

The porphyrin structure has alternating double and single bonds that can "resonate" by assuming a variety of different configurations

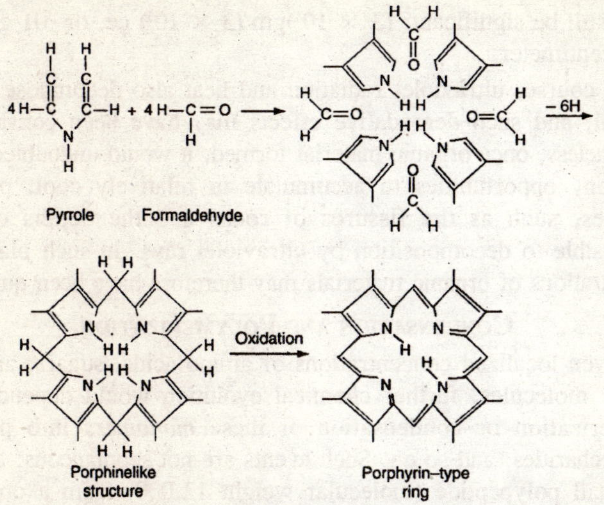

Pyrrole Formaldehyde

Oxidation

Porphinelike structure

Porphyrin–type ring

without changing the position of their constituent atoms. Such resonance confers stability on porphyrins, enabling them to hold extra electrons and thus to function as electron acceptors (oxidation) or electron donors (reduction). Similar oxidative and reductive functions can be performed by nucleotide derivatives such as nicotinamide adenine dinucleotide (NAD), called *coenzymes* because they act in union with protein enzymes to catalyze a wide variety of chemical reactions.

We can thus see that many of the basic organic molecules used in living organisms form relatively easily in many reactions. The amounts per reaction are usually small, but the overall quantities of such substances may have been quite large. Shklovskii and Sagan, for example, point out that 1 photon of ultraviolet radiation produces a quantum yield of about 1/100,000 to 1/1,000,000 of a simple organic molecule. If we take 10^{-22} grams as the average mass of such a molecule, then the quantum yield per photon is about $10^{-5} \times 10^{-22} = 10^{-27}$ grams.

Shklovskii and Sagan estimate the number of photons at the top of the Earth's atmosphere in primitive times at 3×10^{14} photons/cm^2/sec, so the quantum yield per square centimeter per second may have been $10^{-27} \times (3 \times 10^{14}) = 3 \times 10^{-13}$ grams. Thus, if the reducing atmosphere lasted for 300 million years (about 10^{16} seconds), enough energy would have formed to produce $(3 \times 10^{-13}) \times 10^{16} = 3 \times 10^{3}$ grams of matter per square centimeter of the Earth's surface. Furthermore, even if this material were diluted in as deep an ocean as the present (3×10^{5} centimeters), the concentration of the solution would still be significant: (3×10^3)gm/(3×10^5) cc, or .01 gram per cubic centimeter.

Of course, ultraviolet radiation and heat also decompose organic material, and such degradative effects may have been considerable. Nevertheless, once organic material formed, it would undoubtedly have had many opportunities to accumulate in relatively cool, protected localities, such as the fissures of rocks and the depths of pools inaccessible to decomposition by ultraviolet rays. In such places, the concentrations of organic materials may therefore have been quite high.

Condensation and Polymerization

Given localized concentrations of amino acids, sugars, and other organic molecules, further chemical evolution would depend on the polymerization or condensation of these *monomers* into peptides, polysaccharides, and so on. Such events are not spontaneous: to obtain one small polypeptide (molecular weight 12,000) from a one molar aqueous amino acid solution, in the absence of any other chemical forces, would require a volume of amino acids 10^{50} times that of the Earth. How then could such polymerizations occur?

Most polymerizations depend on the removal of water molecules from the monomers to be condensed. Peptide bonds, for example, ordinarily form in the cell on ribosomes through *phosphate bond* energy. Outside the cell, the task is more difficult, but bonds can nevertheless form either in aqueous medium or under anhydrous conditions.

So far, compounds identified as condensing agents that could have existed early in the Earth's history include the following:

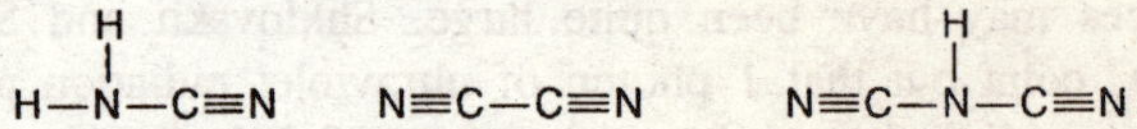

$H{-}N(H){-}C{\equiv}N$ Cyanamide

$N{\equiv}C{-}C{\equiv}N$ Cyanogen

$N{\equiv}C{-}N(H){-}C{\equiv}N$ Dicyanamide

$(H{-}N{=})(H{-}N(H){-})C{-}N(H){-}C{\equiv}N$ Dicyandiamide

$H{-}N{=}C{=}O$ Cyanic acid

$H{-}C{\equiv}C{-}C{\equiv}N$ Cyanoacetylene

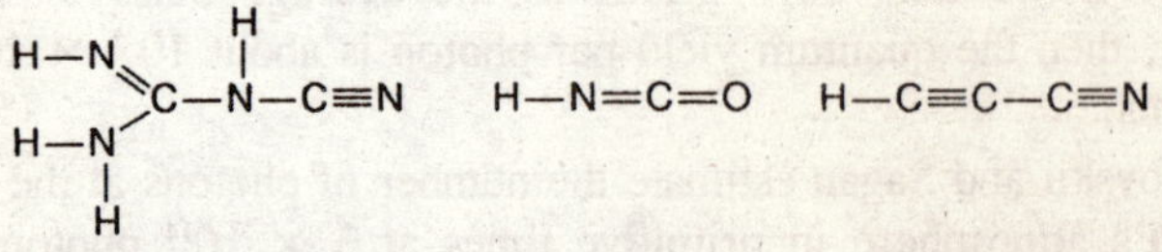

In each case, the unsaturated cyano-carbon-nitrogen bonds enable the condensing agent to combine with water and release energy during this hydration. For example,

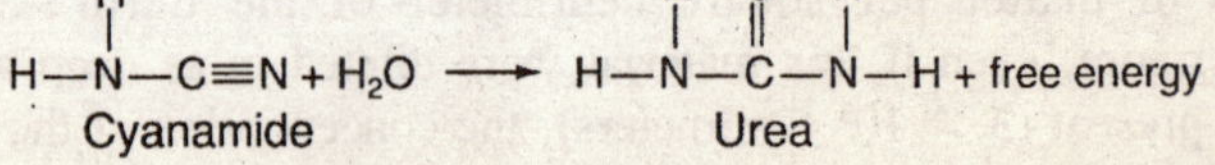

$$\underset{\text{Cyanamide}}{H{-}N(H){-}C{\equiv}N} + H_2O \longrightarrow \underset{\text{Urea}}{H{-}N(H){-}C({=}O){-}N(H){-}H} + \text{free energy}$$

Thus, the condensation of two amino acid into a dipeptide can couple to the *hydrolysis* of cyanamide:

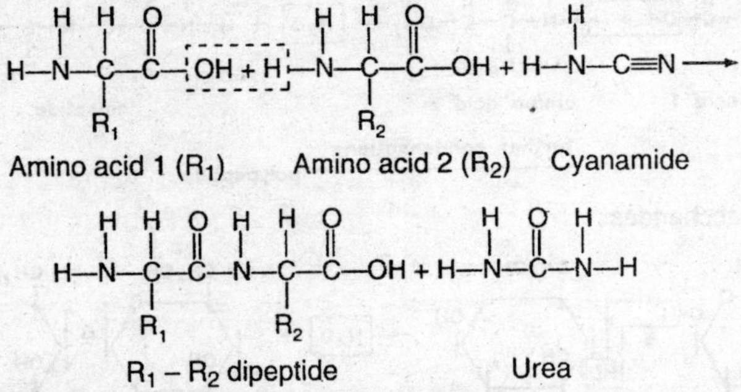

$$\mathrm{H{-}N(H){-}C(H)(R_1){-}C(=O){-}[OH + H]{-}N(H){-}C(H)(R_2){-}C(=O){-}OH + H{-}N(H){-}C{\equiv}N \longrightarrow}$$

Amino acid 1 (R_1) Amino acid 2 (R_2) Cyanamide

$$\mathrm{H{-}N(H){-}C(H)(R_1){-}C(=O){-}N(H){-}C(H)(R_2){-}C(=O){-}OH + H{-}N(H){-}C(=O){-}N(H){-}H}$$

$R_1 - R_2$ dipeptide Urea

Because of their apparent preference for reacting with organic molecules carrying anions (e.g., phosphate HPO_4^-), many of the cyanic condensing agents produce peptide bonds between amino acids even in aqueous solutions. Some, such as cyanogen and cyanamides, also cause nucleotides to form by the phosphorylation of adenosine, uridine, and cytosine; for example,

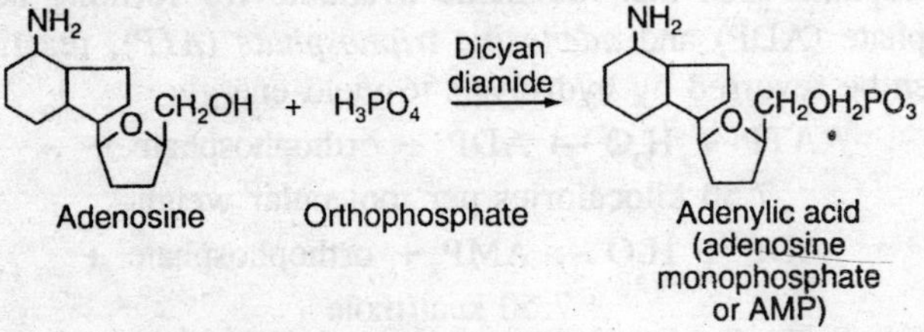

$$\text{Adenosine } (\mathrm{NH_2}, \mathrm{CH_2OH}) + \mathrm{H_3PO_4} \xrightarrow{\text{Dicyan diamide}} \mathrm{CH_2OH_2PO_3}$$

Adenosine Orthophosphate Adenylic acid (adenosine monophosphate or AMP)

Under anhydrous conditions, with no or few water molecules, heat can promote condensation and polymerization by causing the loss and evaporation of water molecules even in the absence of specific condensing agents. One such reaction, accomplished by heat in the laboratory and by enzymes in living organisms, is the formation of high-energy phosphate bonds from orthophosphate:

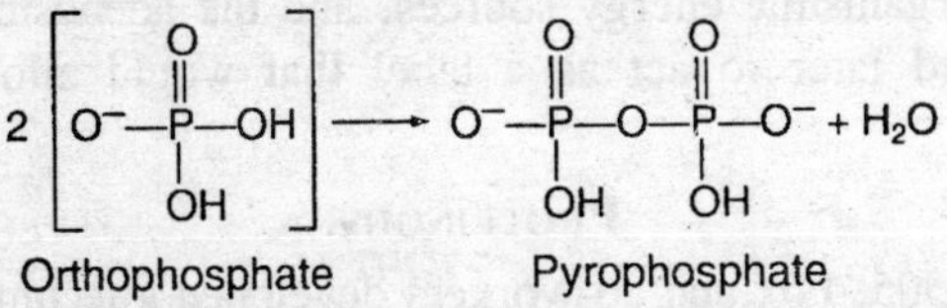

$$2\left[\mathrm{O^-{-}P(=O)(OH){-}OH}\right] \longrightarrow \mathrm{O^-{-}P(=O)(OH){-}O{-}P(=O)(OH){-}O^- + H_2O}$$

Orthophosphate Pyrophosphate

High yields of pyrophosphate can also be synthesized by the condensing agent cyanic acid (cyanate) reacting on precipitated hydroxyapatite [$Ca_{10}(PO_4)_6(OH)_2$], a major phosphate mineral. Such

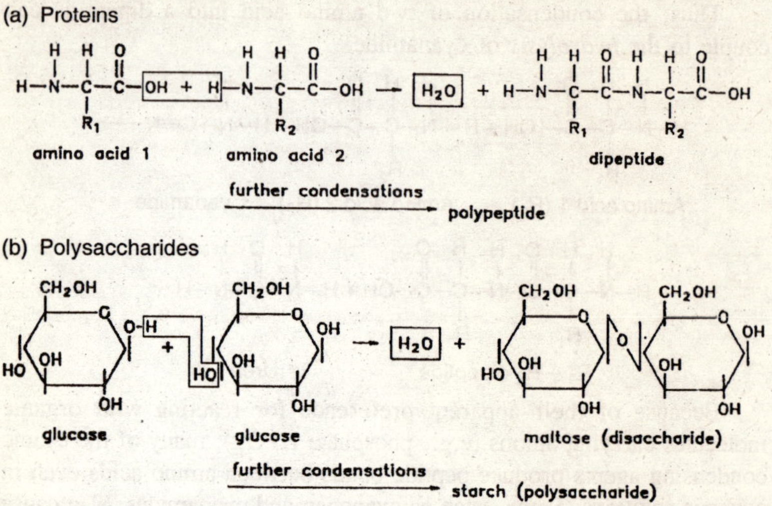

(a) Proteins

amino acid 1 + amino acid 2 → H_2O + dipeptide

further condensations → polypeptide

(b) Polysaccharides

glucose + glucose → H_2O + maltose (disaccharide)

further condensations → starch (polysaccharide)

Fig. 2.7. Examples of condensation reaction leading to the formation of peptides and polysaccharides.

pyrophosphates can then be made available for forming adenosine diphosphate (ADP) and *adenosine triphosphate* (*ATP*), reactions that can then be reversed by hydrolysis to yield energy:

$$\text{ATP} + H_2O \rightarrow \text{ADP} + \text{orthophosphate} + 7.50 \text{ kilocalories per molecular weight}$$

$$\text{ADP} + H_2O \rightarrow \text{AMP} + \text{orthophosphate} + 7.50 \text{ kcal/mole}$$

$$\text{AMP} + H_2O \rightarrow \text{adenosine} + \text{orthophosphate} + 3.40 \text{ kcal/mole}$$

When one or more steps in this hydrolytic sequence occurs, the cell gets its main source of energy and the primary means of removing further water molecules during condensation reactions. Biochemists have, in fact, suggested that polyphosphate chains may have provided some of the first organismic energy sources, and the adenosine component in ATP added later to act as a label that would allow enzymatic recognition.

PROTEINOIDS

In the 19505, Fox and co-workers developed a technique in which heat could also be used to produce peptides from dry mixtures of amino acids. Depending on the kinds of amino acids in the mixture, they found that temperatures of 150° to 180°C could produce as much as 40 percent yield of peptidelike products with molecular weights between 4,000 and 10,000 daltons. Fox called these polymers *proteinoids* (also *thermal proteins*), and he and his group have shown that these compounds bear remarkable proteinlike features. According to their analyses, the proteinoids possess nonrandom proportions of amino acids; that is, their compositions are not simply based on the frequency of the different amino acids in the initial mixture.

They also suggest that the positions of the amino acids in the polymer are not based on their overall frequencies in the chain, since some amino acids preferentially occupy the N- and C-terminals of the proteinoids. The nonrandomness of proteinoid structure also seems supported by the fact that these polymers all show similar properties as tested by sedimentation rates, electrophoretic techniques, column fractionation, and other measurements. Thus, some preferential interaction between amino acids in proteinoid formation seems to dictate their position and frequency and lead to some degree of uniformity in the kinds of molecules produced.

Although not all the amino acid bonds formed in such proteinoids are of the peptide variety, nor do the shapes of these molecules follow the familiar α-helix of protein structure, there still seem to be enough peptide linkages to characterize them as proteins in many tests. Thus, proteinoids give positive colour tests with the same reagents that proteins do; their solubilities resemble proteins; they are precipitable with similar reagents and have other proteinlike traits.

Most importantly, proteinoids engage in a number of enzymelike activities that can increase the rates of various organic reactions. For example, they help split apart certain molecules by addition of water (hydrolysis), they catalyze the condensation of nucleotides such as ATP into di- and trinucleotides, and they help to remove carboxyl groups or amino groups from various structures. Moreover, they can improve catalytic activity of molecules, such as heme, that aid hydrogen peroxide in removing hydrogen from reduced compounds in oxidation reactions.

Fox and Dose have suggested that some of these reactions, combined into a particular sequence, may have served as the beginnings of later metabolic systems. Thus, decarboxylation of oxaloacetic acid can be followed by decarboxylation of its product, pyruvic acid, leading to acetic acid and carbon dioxide; or amination of pyruvic acid can lead to alanine. Furthermore, some proteinoids even show relatively sophisticated hormonal activity and can stimulate the production of melanin-producing cells.

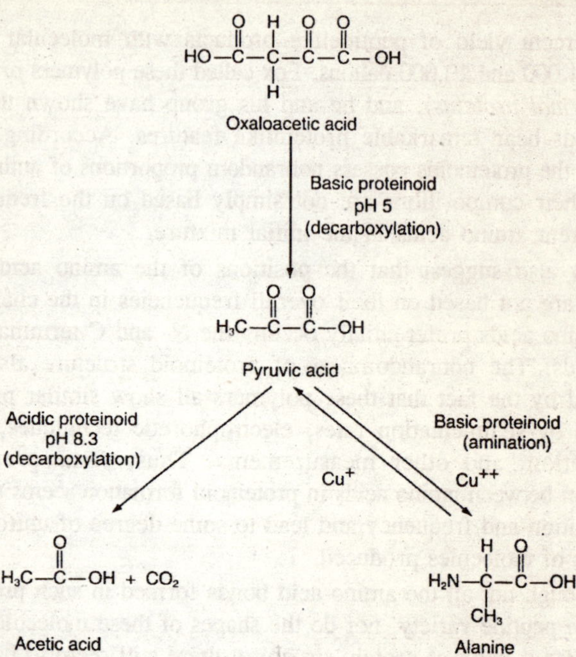

Fig. 2.8. Some sequential reactions catalyzed by different proteinoids or proteinoid complexes.

Although researchers have debated whether the thermal synthesis of proteins could occur extensively in present natural surroundings, the exact conditions encountered on the primitive Earth are certainly not known. Surfaces near some volcanic regions may have maintained appropriate temperatures for the condensation of amino acids, and cooling rains may have dispersed such thermally produced proteinoids to places where further interactions could lake place.

In any case, a wide-enough array of condensation mechanisms have been established, one or more of which most probably occurred in the past. Paecht-Horowitz and co-workers, for example, have shown that phosphate-activated amino acids such as aminoacyl adenylates will condense to form high yields of polypeptide chains on layered clays such as montmorillonite:

$$n\left[\mathrm{H{-}N(H){-}C(H)(R){-}C(=O){-}O{-}P(=O)(OH){-}O{-}adenosine}\right] \xrightarrow{\text{clay}}$$

Aminoacyl adenylate

$$H{-}\underset{|}{\overset{H}{N}}{-}\underset{R_1}{\overset{H}{C}}{-}\overset{O}{\overset{\|}{C}}{-}O{-}\overset{H}{N}{-}\underset{R_2}{\overset{H}{C}}{-}\overset{O}{\overset{\|}{C}}{-}O{-}\cdots + nAMP$$

Polypeptide $(R_1 - R_2 \cdots R_n)$

The amino acid ends of the adenylates apparently penetrate the narrow layers of the clay, and the condensation reactions take place there. Some clays may also polymerize nucleotides, and Burton and co-workers report that certain nucleotide compounds that are strongly absorbed by montmorillonite clays form small amounts of dinucleotides. Thus, the early availability of cyanamides, heat, clays, and other condensing agents make it highly probable that polypeptides, polysaccharides, lipids, and perhaps even polynucleotides were present early in the Earth's history, and could have been used for primitive organismlike reactions and structures.

Origin of Organized Structures

The presence of appropriate organic monomers and polymers is only a first step in the origin of life. Living processes of metabolism and function occur because the materials of which organisms consist are highly organized. How did such organization come about?

At its earliest, interactions between molecules must have led them to assume relative positions based on forces such as hydrogen bonding, ionization, solubility, adhesion, surface tension, and so on. Phospholipids, for example, are organic molecules with a phosphorus-containing polar group at one end and nonpolar fatty acid groups at the other end. In water, a polar solvent, the polar ends of these molecules are oriented toward water (hydrophilic), while their nonpolar ends are oriented

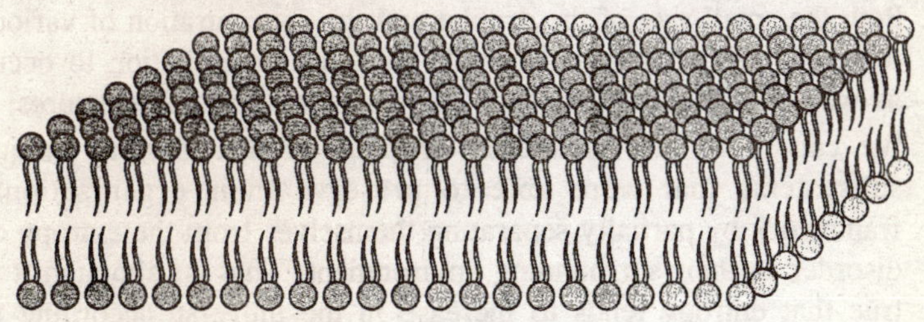

Bimolecular layer

Fig. 2.9. A diagrammatic view of a bimolecular sheetlike double layer of phospholipid molecules that have self-assembled with their hydrophilic phosphate heads facing the water solvent, and their hydrophobic hydrocarbon tails facing each other.

toward each other, away from water (hydrophobic). As a result, phospholipid membranous structures can form quickly yielding *vesicles* composed of bimolecular layers in which the nonpolar surfaces of each of the two layers "dissolve" in each other. Carried further, such vesicles can encapsulate inclusions in tide pools that undergo drying and wetting cycles.

Membranous droplets or vesicles composed of lipids, polypeptides, or other molecules undoubtedly formed in great quantities, produced by the mechanical agitation of molecular films on liquid surfaces or even spontaneously. The attainment of such droplet levels of organization was an important step in the origin of life for a number of reasons:

1. Depending on its structure and permeability, the membrane surrounding the droplet can selectively choose which compounds can enter from the environment and exit from the droplet.
2. Such *selective permeability* allows concentrations of particular compounds to differ across the membrane, enabling reactions to occur within the droplet that would not have occurred outside the droplet.
3. The presence of a basic protein causes a 100-fold increase in the entrapment of nucleic acids into such droplets. As Jay and Gilbert point out, "Protein-mediated encapsulation creates high local concentrations of protein and nucleic acids within the vesicular volume. . . . This would enhance the interaction of molecules with low affinities, potentiating the formation of aggregates with biological function"
4. The small size of the droplet can permit a chain or network of reactions to occur, the products of one reaction being available to serve as the substrates for another reaction.
5. Both the small size of the droplet and the concentration of various materials within it would permit localized precipitation to occur as well as the organization of compartments and substructures.
6. We can think of "advanced" droplets of this kind as unique subsystems that were able to preserve their organizational framework by partially separating themselves from the entropy or disorder in their surrounding environment. That is, although it is true that entropy tends to increase in the universe according to the second law of thermodynamics, it can nevertheless decrease in such subsystems during their life spans. Because of their semipermeable membranes they can use entering energy and matter to retain, and even enhance, their organizational and informational

structures as long as they can continue to perform biological processes.

It seems presumptuous and unrealistic to assume that the only kinds of organization capable of growth, metabolism, and reproduction are the structures found in present-day organisms. Although present forms are highly efficient, early forms could have functioned at a much lower level of efficiency, because they were not then competing with the more advanced forms. For a primitive form to show some (but certainly not all) "living" attributes, it would have been sufficient if it could merely grow (e.g., increase in size), maintain its individuality, and divide. It is therefore interesting to note that some authors ascribe such properties to bimolecular vesicles, and there are, in addition, at least two types of fairly simple laboratory-produced structures that seem to possess some aspects of these basic prerequisites: Oparin's coacervates and Fox's microspheres.

Coacervates

Coacervates have long been known to occur when dispersed colloidal particles separate spontaneously out of solution into droplets because of special conditions of acidity, temperature, and so on. If there is more than one type of macromolecular particle in the colloid, complex coacervates can form that show a number of interesting properties:

- They possess a simple but persistent organization.
- Although they are mostly unstable, some coacervates can maintain themselves in solution for extended periods.
- They can increase in size.

Oparin, the first to draw serious attention to these droplets, developed artificial coacervate systems that could incorporate enzymes that performed functions such as the synthesis and hydrolysis of starch

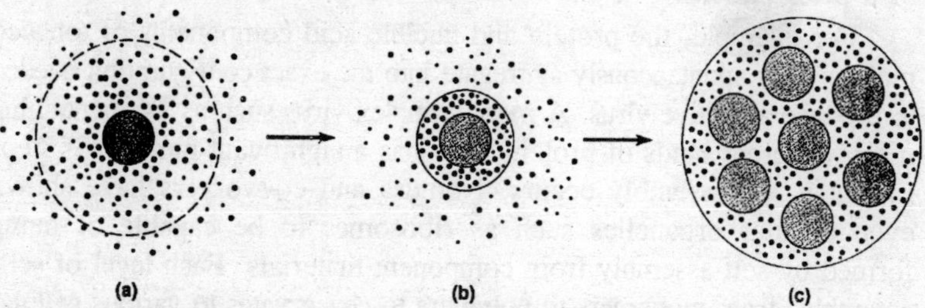

Fig. 2.10. Formation of coacervates by the exclusion of water molecules from associated colloidal particles.

as well as synthesizing polynucleotides. In a coacervate system containing chlorophyll irradiated with visible light, Oparin and co-workers showed that there can be a constant inflow of reduced ascorbic acid and oxidized methylene red, which then converts into a constant outflow of oxidized ascorbic acid and reduced methylene red. The chlorophyll picks up electrons from the ascorbic acid and then supplies those for the methylene red reduction—a process similar to common noncyclic photosynthesis in which water molecules supply electrons for reducing the coenzyme $NADP^+$ to NADPH.

Microspheres

Fox showed that these small spheres formed when the thermally produced proteinoids were boiled in water and allowed to cool. The microspheres are uniform in size, stable, bounded by double membranes that appear somewhat cell-like, and can undergo fission and budding. They appear in large numbers: 1 gram of proteinoid material can produce 10^8 or 10^9 microspheres. Among microsphere qualities indicating active internal processes are their selective absorption and diffusion of certain chemicals but not others, their growth in size and mass, and observations demonstrating osmosis, movement, and rotation. Moreover, microspheres show the potential for transferring information, in that proteinoid particles pass through junctions between them.

The spontaneous *self-assembly* of macromolecules into vesicles, coacervates, and microspheres indicates that the occurrence of similar entities under primitive conditions would probably not have been an unusual event. Such entities are not cells, of course, and considerable time may well have elapsed before more elegant structures with more complex metabolic capabilities could develop. Nevertheless, considerable evidence shows that the component materials of even more complex structures can self-assemble without the immediate presence of a prior pattern.

For example, the protein and nucleic acid components of tobacco mosaic virus spontaneously aggregate into the exact configuration needed to produce an active virus. A more complex virus such as T4, containing many different kinds of protein, also has a significant number of steps in which self-assembly occurs. Nomura and co-workers have shown even cellular organelles such as ribosomes to be capable of being formed by self-assembly from component materials. Each level of self-assembly, from monomers to polymers to coacervates to various cellular organelles, may thus have derived from nonrandom events, in that certain combinations form more quickly and easily than others.

Nonrandom self-assembly, however, is hardly sufficient to account for more than a few complexities of life. In their most essential aspects, such as the precisely ordered monomer sequences found in proteins and nucleic acids, present forms of life did certainly not result from mere chemical attractions between component amino acids or nucleotides. At the same time, we also have good reason to believe that, because of their specificity, the positioning of monomers in these precisely ordered sequences may nevertheless have had a nonrandom basis. To recapitulate a theme raised previously: How can the nonrandom biological order of amino acid and nucleotide sequences arise from disorder? The answer lies in selection.

Origin of Selection

Primitive structures, whether coacervates, microspheres, or other localized organizations, would have had one important evolutionary feature: they would serve as the first distinctive, multichemical *individuals*, or *protocells*, that could interact as units with their environment. Together with their various neighbours and progenies, such individuals would form a group or *population* on which selection could act. That is, protocells incorporating those organizations and metabolic activities most successful in growth and division would increase most in relative frequency or in area occupied.

We can thus say selection arises when the following conditions are reached:

- A population of individuals exists.
- The properties of these individuals are governed by reactions in which they absorb and transform environmental material into their own material.
- Individuals differ in the efficiency with which these processes take place.
- Availability of materials and energy is limited so that not all types of individuals can form, nor can all types of individuals formed survive.

The mechanism that enabled the formation of protocells is a crucial issue in understanding selection itself. If protocells could have originated only by self-replication, this would indicate that selection had always operated on the same efficient basis as it does now, with advantageous traits rapidly transmitted to succeeding generations by fairly exact replicative mechanisms. However, if protocells initially formed only through acts of prevailing environmental chemistry, as many biologists now believe, then selection was not very efficient in the past and would itself have undergone evolution from chemical nonreproductive selection to the more modern biological natural selection.

That is, early selection would have been confined to the survival of nonreproductive individuals who could wrest the most material from their environment and transform it for their own benefit with the least expenditure of energy. Although differences among such individuals could not be precisely transmitted, the fact that some such individuals survived and others did not would undoubtedly have affected the composition and further interactions of succeeding groups.

Inheritance would therefore, at first, have been mostly a matter of transmitting molecular "things" that permit survival, rather than transmitting the nucleic acid patterns that produce the "things." The earliest forms of "living" individuals may well have replicated themselves poorly, yet passed on some of their metabolic and enzymatic properties, which continued to be selected and improved.

From a materialistic point of view, unless we postulate an accident of immense proportions and infinitely low probability, selection must have bridged the gap between chemical evolution (changes in the composition of nonreproductive or poorly reproductive molecules, coacervates, microspheres, and so on) and biological evolution (changes in inherited differences among reproductive organisms). So far, selection is the only natural mechanism we know that can account for the creative changes among nonreproductive individuals that could have led them in the direction of living organisms. Although the events may be complex, the device is simple: organisms that react to their environment with improved functional information replace those that lack such information.

Nevertheless, selection is not merely a passive agent that sifts the good from the bad, the adaptive from the nonadaptive, but, because of its historical continuity, enables a succession of adaptations to accumulate that lead to something entirely new. Selection thus acts as a creative force that has made possible biological organizations that would otherwise have been highly improbable. To use a previous example, a polypeptide chain consisting of a specific sequence of 100 amino acids has an extremely low probability (10^{-130}) of occurring spontaneously without selection. However, if each step in the growth of the chain attains a selective advantage when the correct amino acid inserts, then the probability of achieving a functionally advantageous polypeptide is almost immeasurably increased.

Once the game of life has begun, the evolutionary replacement of the players, whether they are genes, organisms, races, species, or other entities, becomes inextricably bound to their ability to play the game further—an ability that selection has previously molded and is now in turn measured anew by selection. Some random replacement of the players certainly occurs by accident rather than by selection, but participation in life is selective by its very nature since the resources of life are always limited in one way or another. Thus, it is not merely their origin by selection that characterizes living systems but their continued ability to subject themselves to selection. This ability has led to a coupling between function and reproduction that provides living forms with their relatively rapid evolutionary rates.

3

ORIGIN OF GENE

Discoveries of the elaborate structure of genes raise evolutionary questions that lie outside the scope of traditional theory. Darwinism rationalizes how structures evolve, but does not address the more significant issue of the roles that specialized gene structures play in the process of evolution itself. I suggest that molecular genetics can be integrated into evolutionary theory only through the paradigm that underlies all biological explanation; the structure-function principle. Organisms carry out evolutionary processes as biological functions. To do so, species evolve special structures with *evolutionary functions* just as they evolve other phenotypic structures to carry out various adaptive functions. I shall discuss the evolutionary functions of a variety of structures, ways that these functional capacities evolve, and some implications of evolution being a process that organisms become specialized to carry out instead of just change imposed on the species from the external environment.

Heredity lies at the heart of all theories of evolution. Indeed, evolution is the time dimension of genetics. Darwin's greatest handicap in explaining evolution was a complete ignorance of genetics. This forced him to accept two interwoven causal processes in evolutionary change. Natural selection drove evolution by survival of the fittest but an internal hereditary process generated the individual variability on which selection acts. Darwin was unable to evaluate the relative importance of these two agencies. As time went on, he admitted an increasing role for heredity in directing the process of evolution through some sort of Lamarckian mechanism. The discovery of Mendelian genetics rescued natural selection. Recombination of particulate genes, and ultimately blind mutation, produced continuing variation. Since both processes could be completely random and spontaneous, natural selection no longer had to be diluted with organismal hereditary activities. The stark objectivity of the new Darwinism allowed it to overwhelm its vitalistically tainted competitors. Its causal simplicity also bedazzled the 'Neo-Darwinist' into believing that the simplest possible way that evolution could take place must also be the way that evolution actually did proceed.

This misconception might have been exposed by the growing doubts from the fossil record as to whether Neo-Darwinism addresses the key features of the actual history of life on the planet. In fact, it was insight into heredity that again proved pivotal. The revolution of molecular biology has made it possible to examine directly the structure and physiology of genes. Genetic determinants turn out to be very different from the ineffectual abstractions upon which Neo-Darwinism was based. We still cannot foresee their full role in evolution, but it is probably substantial. I suggest that evolution is primarily an internal genetic process. Natural selection, as we know it, occurs and is essential, but it is the junior partner to heredity. Not only the mechanisms but also the goals of evolution are defined by the genetic message. To describe this interpretation I will first sketch the main characteristics of genes as functional structures, then describe the observed evolutionary dynamics of genes, and finally suggest that the phenotype evolves through these same categories of changes.

Nature of Gene

Genes are Units of Molecular Structure

The modern geneticist extracts genes as molecules from cells, looks at them with the electron microscope, measures their physical properties, such as length, and describes their chemical composition. This is entirely different from the approach that prevailed when Neo-Darwinism was being formulated. T. H. Morgan observed in his 1933 Nobel lecture that: 'There is no consensus of opinion amongst geneticists as to what the genes are—whether they are real or purely fictitious—because at the level at which the genetic experiments lie it does not make the slightest difference whether the gene is a hypothetical unit, of whether the gene is a material particle'. Since then, the main concern of genetics has shifted from inheritance of genes to their structure. This approach is fruitful because genes are surprisingly elaborate in structure and this structure is the basis for their function.

Genes are Highly Individual

Until this decade, the quest of genetics was for universal truths about inheritance—Mendel's laws, Fisher's fundamental law of evolution, Crick's central dogma. This made the simplest genes the ones of choice for study. The resulting lowest common denominator orientation reduced genes to no more than runs of codons in DNA molecules, distinctive only in their sequences. Such forms of genes do exist, and probably are abundant, but they seem to be used only for elementary coding functions. Typically, genes with complex information and active evolutionary dynamics are elaborate, and their significant capacities stem directly from their individualistic structure. This realization has a profound evolutionary implication. To understand how important phenotypic traits evolve we must describe the actual dynamics of the complex genes involved, and not merely the stereotyped behaviour of simple genes.

For example, it once seemed all but impossible that the mammalian genome could code for hundreds of millions of different antibody molecules that encompass every possible foreign antigen. Yet vertebrates are able to do so and even make antibodies against artificial molecules to which their species had never been exposed in its entire evolutionary history. Both the capacity to code for the immune response and the ability to evolve this information are made possible only by the very special and unique structure of the immune genes.

Even lowly bacteria rely on highly distinctive gene structures. Although the run of the mill genes in bacteria have a stable configuration on the chromosome, those genes that are actively evolving tend to develop into configurations such as transposons. These structures can move around in the genome and can even hitch-hike from one cell to another by associating with other specialized DNA structures called transmissible plasmids. By exchanging transmissible genes among other species in the ecosystem instead of relying on mutations, bacteria are able to evolve far more rapidly and effectively than would be possible through simple mutation and selection.

It took decades for bacteriologists finally to accept the empirical data that bacteria evolve resistances to antibiotics far too fast for a process of mutation and selection and it has taken evolutionists even longer to appreciate how micro-organisms actually do evolve. As late as 1970, a treatise on evolution still insisted that 'Microorganisms… have the capacity to develop strains resistant to antibiotics and other drugs. This resistance results from the selection of a few resistant mutations of gene combinations exactly as in higher organisms'. With the growth of modern genetics we now know that this is a secondary mechanism for bacterial adaptation. Adaptive evolution in bacteria does not centre on mutation and selection. It is foremost a process of restructuring genes into transposable and transmissible forms and then moving them about the cell and the ecosystem. Thus, for example, the important information for evaluating the potential of a bacterial pathogen to evolve resistance to an antibiotic is not the mutation rate or the frequency of resistant individuals, but the *structure* of the genes that provide that characteristic.

Genes are Working Molecules of the Cell

The third cornerstone of modern genetics is what I call the 'profane' side of the DNA molecule. Classical genetics was predicated on the genotype having a sacred status. Genes were inviolable messages from ancient ancestors passed on faithfully from generation to generation, except for rare mutational corruptions. They were transcendental to the activities and metabolism of the organism carrying them. We now realize that this simply is not true. Instead cells actively manipulate the structure of their DNA molecules for both physiological and evolutionary reasons. They engineer changes in their genes with an impressive array of 'gene-processing enzymes'. Cells seem to have enzymes to catalyze almost every imaginable sort of change in the structure of the DNA molecule. Topoisomerase unties knots in DNA strands, recA protein inserts single strands of DNA into homologous double-stranded helices, acquisitionase grabs bits of DNA and intercalates them into chromosomes and so on. These enzymes for manipulating DNA molecules are not only diverse but also numerous. For example, there are thought to be at least 20 different sorts of DNA glycosylases.

Complex genes, such as multigene families, owe much of their distinctiveness to special enzymic pathways that manipulate their DNA sequences. For example, the antibody and T-cell receptor multigene families have highly specialized enzymic pathways to rearrange their DNA in particular somatic cells and the duplicated genes in the haemoglobin families are enzymically 'corrected' one against another in germ line cells during evolution.

The ultimate authority over the information content of its DNA would be for the cell actually to write gene segments from scratch, and to edit the result in order to achieve a useful product. Remarkably, even this capacity is realized! Mammals are able to polymerize *de novo* sequences of nucleotides at crucial interior sites in the genes for their antibodies. Those physiologically written segments code for particular portions of antibody molecules that recognize their antigens. The creations that are useful are selected from others by an impressively complex multicellular system. A decade ago, geneticists would have considered this capacity unimportant for evolution because it is confined to a small number of highly specialized genes. It is not a general property of genes. However, we now realize that every important gene is probably exceptional in one way or another, and that it is these unique genetic mechanisms that have allowed the important accomplishments of life to evolve.

The full contribution of gene-processing enzymes to evolution remains uncertain. One obvious role is to produce mutants. Probably the majority of mutations important in evolution result from the activities of enzymes. This is significant because enzymes are deliberate and precise, the antithesis of the supposed 'randomness' of mutation. Indeed, enzymically induced changes in the structure of multigene families often are referred to as gene 'switching', 'correction', 'rectification', 'excision', and so on, rather than 'mutation'. I also find it suggestive that geneticists have chosen names for gene-processing enzymes that could be used almost as well for the keys on a word processor.

Genes are Aware of their Surroundings

Genes require information about the organism and its environment in order to express their message appropriately. To meet this demand, complex bacterial genes (operons) have evolved special sensory proteins called 'repressors', which inform them of particular relevant conditions. A typical repressor has two binding sites, one specific for some meaningful metabolite in the cell and a second for a recognition site in the operon. The two binding capacities are mutually exclusive, so that the repressor tells the operon about the concentration of the metabolite by binding to the gene's DNA. Sensor proteins of bacteria are versatile. Some activate rather than repress genes, and some inform genes about hormonal signals generated by the bacterium, or about the status of other genes in the genome. Also, some operons receive information from several sensory systems simultaneously.

Eukaryotic genes are more complex and less well understood than those of bacteria. They are tucked away in a membrane-bound nucleus, and germ line cells are isolated in gonads from the soma. Despite these specializations, receptor proteins still carry hormone molecules directly to sites on the chromosome within the nucleus. Even germinal cells of vertebrates produce receptors on their cell surfaces for a variety of hormones, growth factors, neuromodulators and so forth. Higher organisms presumably have a far greater potential than bacteria for informing their genes but we still know almost nothing of its evolutionary significance.

Types of Evolutionary Change in Genes

In addition to revising our ideas about the fundamental nature of the gene, molecular geneticists also have discovered the types of change that occur in genes during their evolution. Each of these has been chosen as a definition for 'evolution' by one school or another.

Fluctuations in Gene Frequencies

Fifty years ago it was known that genes were located on chromosomes and could mutate to alternative states. Their structure was otherwise mysterious. The only information about genes that could be interpreted (other than map positions) concerned the mathematical *frequencies* of their detectable allelic forms in populations. The obvious basis for a genetic theory of evolution was change in this measurable quantity. Neo-Darwinists therefore defined evolution as simply a change in frequencies of gene alleles in a population from one generation to the next. Such quantitative change is fully reversible for a gene, and the frequencies of individual alleles can shift back and forth to adapt organisms in the population to their changing environment. New alleles occasionally arise by mutation or become 'fixed' by completely replacing their alternative alleles but these processes are very slow and irrelevant to immediate adaptation. In fact, species evolve numerous and diverse mechanisms especially to minimize both the loss of variation by gene fixation and the formation of new mutations, most of which are deleterious. The Neo-Darwinian process concerns fluctuations in the frequencies, but not the forms, of genes.

Change in Gene Structure

When it became technically possible to read the amino acid sequences of primary gene products (and later the nucleotide sequences of the genes themselves), a new paradigm for molecular evolution became the comparison of homologous sequences taken from related species. Species were found to diverge from one another, and from their ancestors, by the accumulation of small substitutions in the molecular structure of their genes. Irreversible permanent change in DNA structure replaced population dynamics as the definition of gene evolution.

'Evolutionary distance' is now measured as the number of substitutions accumulated in the structure of a gene or protein with time. Curiously, this yardstick gives anomalous readings of relatedness among species. It measures the length of time since two species diverged from a common ancestor rather than how much their biology has changed during that time. This is because most base substitutions that accumulate in genes have little effect on phenotype. They are selectively neutral, or nearly so, and driven inexorably into the gene by mutation pressure. Comparative biochemists recognize the distinctiveness of this predominating process of relentless gene change by calling it 'non-Darwinian evolution'. This implies that Darwinian adaptation through natural selection is a separate, and quantitatively lesser, process of change. The main debate in molecular evolution today is about the relative contributions of neutral and adaptive processes to change in gene structure.

Adaptation

There have been two main approaches to the study of adaptive changes in gene structure. One is to expose laboratory populations of micro-organisms to artificial selection pressures and then to analyze the changes by which they adapt. The other is to identify the functionally important parts of genes or proteins and to describe how these vary among related species.

In pioneering the former method, John Langridge spread large numbers of *Escherichia coli* on agar that contained lactobionic acid as the only energy source. This sugar is a derivative of lactose, a disaccharide that *E. coli* metabolizes using the enzyme β-galactosidase. However, the enzyme cannot hydrolyse the analogue sugar effectively. Mutant colonies were readily obtained and genetic analysis showed that many of them had genetic alterations in their β-galactosidase gene.

Langridge's major discovery was the striking variety of *alternative* ways in which the genes were able to acquire this new function. The adapted state is not a unique one, and when two populations or species adapt to a common environment they can be expected to do so in different ways. This raises a key unresolved question. What determines how a gene will evolve when multiple alternative pathways for adaptation are available? The factors that decide the choice between alternatives probably are as important in determining what will evolve as are the environmental stresses to which the species is adapting. So far, few inquiries about gene evolution have reached the degree of sophistication necessary to address this question.

The other approach, of examining the particular parts of macromolecules responsible for functional properties, takes advantage of molecular biology's strong emphasis on structure-function relationships in macromolecules. It is now possible to picture protein molecules in three dimensions and to identify the local regions responsible for substrate binding, catalysis, subunit interaction and so on. Comparing these particular parts of a protein between species shows that adaptation of genes is distinct from just change in structure. Perhaps the most significant revelation from these studies is that the more functionally important a segment of gene is, the slower it changes. In fact, the principal way to locate the functional elements in a gene or protein is to search for regions that have not changed during evolution. This means that selection acts mainly *against change*—a defensive role quite distinct from the Neo-Darwinian view of selection as the driving force *for change* in gene frequencies. Natural selection has been likened to an editor whose main activity is to proofread for errors rather than to promote change.

Advancement

Genes do slowly acquire adaptive changes, but it is clear that not all such changes are of equal importance. Although the majority of selected mutations optimize structural details for the exact functional demands of the environment, a smaller number *advance* the organization of the gene. They create functional capacity instead of adjusting it, and the gene or protein is developed, not just adapted.

Haemoglobin illuminates this process. Most adaptive alterations in this protein probably adjust its oxygen-binding curve to compensate for changes in the organism's size, habitat and so on. In addition, a smaller number of entirely different sorts of changes have *advanced* the haemoglobin molecule from a primitive monomer to its derived tetrameric form. This oxygen-binding protein originally consisted of a single polypeptide chain with a haeme group, a form that still persists in myoglobin. When the chains evolved the capacity to aggregate, they acquired the possibility for allosteric interactions between subunits. This allowed oxygen-loading curves to become sigmoidal with a threshold and a narrowed range between the oxygen tensions for loading and unloading. Allostery also permitted oxygen loading to be regulated by concentrations of other ligands, especially hydrogen ion and diphosphoglycerate. Allosteric regulation is essential to the function of modern haemoglobin, and to enzymes in general. Indeed, nearly all enzymes with sophisticated functions have evolved into oligomers. In the evolutionary history of haemoglobin, the emergence of the capacity for allostery was a landmark advance distinct from its subsequent adaptive fine-tuning.

Several other developments further advanced haemoglobin. One was the evolution of two different types of subunits, alpha and beta, which join together to form the molecule. Heterodimerism allows subunits individually to evolve specialized functions. The importance of this capacity again is evident from the large number of other proteins which also have dissimilar subunits.

Another profound advancement was the organization of the haemoglobin genes into multigene families. Mammals have tandem arrays of genes for the alpha-type and beta-type subunits. The gene copies in these families are differentially expressed during ontogeny to match the haemoglobin molecule to the individual demands of the embryo, foetus and adult. Some species have developed additional haemoglobin gene copies for various other special purposes. Clearly, the vertebrates have actively utilized the opportunities for adaptation opened up by the organization of haemoglobin genes into multigene families. However, the significant evolutionary step was that which created this special adaptability in the first place.

The distinction between advancement and adaptation stands out in many genes, especially in those with demanding biological functions. Significantly, the two processes depend on different kinds of mutations. Genes adapt mainly through base substitutions and occasional small deletions. In contrast, they advance by a wide variety of rearrangements and additions as well as by point mutations. For example, the genes for antibodies acquired their advanced structure by capturing transposons, shuffling exon gene segments, developing special enzymes for rearranging and mutating gene sequences and repeatedly duplicating segments of genes, whole genes and entire multigene families. Molecular geneticists interested in structure-function relationships repeatedly have found it important to distinguish the process that originates and develops structure from the lesser process of its subsequent adaptation. Adaptation obviously plays a role in the origin of novelty, but the two processes are categorically different.

Principles of Evolutionary Advancement

We have a reasonable understanding of how genes fluctuate in allele frequency, change in structure, and adapt. If progression of form is distinct from these processes it is important to determine just how it occurs. Four extra principles seem to be emerging from studies of gene structure. They are discussed here in order of increasing causal complexity.

Tautology of Advancement

Structures that are the most advanced are those that were able to advance the most. This really says no more for evolution than it does for a handful of peas spilled on the ground. Some peas will roll further than others and there must be a reason for the variation. Three can be imagined: (1) some peas might be intrinsically better able to roll, perhaps due to a rounder shape or smoother skin; (2) the surface of the ground may vary from place to place, blocking some peas with bumps and rough spots and facilitating others with depressions or grooves that channel their velocity; (3) chance variation as peas hit the ground and jostle one another will move some peas further than others. The analogous factors in evolution are (1) the genetically determined organization of the species, (2) the selective environment, and (3) whim.

Obviously it is important to evaluate which of these factors have caused some forms of life to advance so much further than others during their 3×10^9 or so years of existence. However, the interactions between phenotype and ecology of higher organisms seem to be vastly too complex for resolving this issue. Fortunately, advancement at the level of gene structure is more tractable. Here, an important observation is that genes that have evolved conspicuously in a particular direction often have identifiable physiological mechanisms that enhance the ability to change in that direction. This association is strong enough for molecular geneticists to rely on it to discover significant physiological mechanisms. For example, a number of multigene families with outstandingly large numbers of gene copies are associated in some way with special mechanisms for reproducing gene copies. These include 5S RNA genes, rRNA genes, pseudogenes, IAP sequences of mice and other retrovirus-like genes, transposons, VAT genes of trypanosomes, possibly satellite DNAs, and unique sequences intercalated into satellite DNA.

As other examples, the extraordinary uniformity of gene copies in some multigene families imply special enzymic mechanisms for removing mutational variation, and one can predict that specialized genes that occur in similar form among unrelated species of bacteria probably are organized as transposons or associated with transmissible plasmids to aid their dissemination. A species also may progress notably in a particular evolutionary direction because it has overcome some factor that constrains such evolution in other species. This represents another way by which a gene or species can specialize its phenotype for a particular kind of evolution. There is a good deal of speculation about structural constraints on the evolution of macromolecules but neither clear examples nor a comprehensive analysis of this issue has yet appeared.

Preadaptation

A particularly important problem for evolutionary advancement concerns the origin of new form. Once a structure advances enough to be functional it can develop further by selection; but how does it initially develop enough to begin to function? This problem worried Darwin and has yet to be resolved. Two principles are probably important. One relates to the well accepted concept of preadaptation. A structure that had evolved originally for one function subsequently acquires a second function. Selection for the original function elaborates the structure until its side effects happened to become useful in other ways. The swim bladder, from which the lung arose, and the reptilian foreleg, which evolved into the avian wing, are familiar preadaptations. The term preadaptation today represents only the opportunism of the evolutionary process without any of the teleological overtones implied by certain earlier evolutionists.

Genes are particularly favourable as preadaptations because they can be duplicated by a single mutation. One copy can preserve the original function, while the other is unhindered to acquire a new function. Moreover, a supernumerary gene can become silent and drift to a new structure unimpeded by selective constraints. Gene duplication and divergence have been dominating processes for evolving new proteins. Most proteins of eukaryotes are members of families that arose from single ancestral genes. These genes of common descent are homologous in the same sense that the wing of the bird is homologous to the leg of the reptile. Of course, not all duplicated genes acquire a new function. The genome of higher organisms is rife with redundant gene copies that have failed to acquire a new function and are degenerating back to random sequence as 'pseudogenes'.

It is significant, but not surprising, that when a duplicated gene acquires a new function that new usage is usually related to the original one. Some groups of homologous enzymes span a wide range of catalytic activities, but most related genes have related functions. This suggests that when genes for a function proliferate, the preadaptive potential for further advances of that general function may increase as well. Some genes have spawned many more new functional derivatives and pseudogenes than others, but it is not clear whether this reflects structural features of the genes or just the functions that they serve.

Autopoiesis

A second mechanism for the origin of novelty involves the physical principle of autopoiesis. This is the spontaneous emergence of organization out of nothing—or, more precisely, out of noise. Closed physical systems and open ones near equilibrium gravitate towards definite simple states of maximal entropy and minimal rate of entropy formation, respectively. Dissipative open systems far from equilibrium have entirely different behaviour. They can spontaneously became more organized with time. A system that is highly dissipative gives fleeting existence to an astronomical number of random or chaotic local microstructures. Any of these twinkling organizational states has the opportunity to stabilize itself from decay and to extend its organization catalytically past its borders if it is able. Just a microscopic seed of such 'autopoietic' organization in an energy dissipating system may grow to macroscopic dimensions, typically as a periodic structure. For example, a thin layer of liquid traversed by a large heat flux characteristically develops a regular periodic pattern of convection that is stable whenever the temperature gradient is applied. Analogous self-forming organization also develops in appropriate dissipative chemical systems and is conjectured to emerge spontaneously even in economic and social systems.

Autopoiesis seems particularly applicable to evolution, which generates new self-propagating organization in a highly dissipative system. A major difference between biological and physical processes is that biological systems are extremely organized to begin with in contrast to the typical system for a physics experiment, such as a container of liquid, on a heat source. The physicist's demonstrations are important to show that dissipating energy does have the capacity to organize matter spontaneously but the appropriate examples to clarify the role of autopoiesis in evolution must come from biology and not thermodynamic paradigms.

Autopoiesis plays an essential generative role in various forms of genetic organization. For example, multigene families easily evolve once they get started because after a gene duplicates it has a propensity to vary further in copy number due to non-reciprocal crossing over. Moreover, the larger a family grows, the more intrinsically variable it becomes. If every additional copy in a multigene family were as difficult to achieve as the first or second, one never would expect to see huge multigene families. However, immense families are common and evolve extraordinarily rapidly. These satellite DNAs give every indication of being products of a runaway autopoietic process.

Autopoiesis has an even more obvious role in the evolution of transposons. A large number of genes scattered along chromosomes and bathed in nucleoplasm rich in enzymes represents the substantial level of organization that characterizes a genome. For example, most cells code for transposases that can synthesize extra copies of a gene and plant them elsewhere in the genome. Brewer's yeast uses a transposase to change sex by copying a silent gene for one or the other sex into a special location where it is expressed. The two sex-determining genes can be copied because they are flanked by special short DNA sequences that the transposase enzyme recognizes as its substrate. Other organisms use transposases for other specialized functions. As long as a transposase, or other gene-processing enzyme, is unrelated to the genes that it acts upon, its activities are causally simple. However, it is possible for the continual shuffling of genes along the DNA molecule to sandwich a transposase gene between recognition sequences for its own enzyme. The enzyme now acts *self-referently* on its own genetic determinant and, in the case of a transposase, a transposon is born. The transposon propagates autonomously by copying itself. Moreover, since its components are genetically specified, the unit can mutate and evolve. Variant copies accumulated in the cell or gene pool can compete with each other for resources, differential fecundity, or *lebensraum* with the 'fitter' form supplanting the others. This evolution is quite distinct from that of other genes because it is *selfish*. Self-reference can allow a gene to evolve for its own sake without dependence on the environment of the organism. Transposons evolve both by modifying the genes they already have and by acquiring new genes.

'Selfish evolution' has been likened to a Darwinian process, with the transposon enjoying the status of an individual 'organism'. Actually the process is considerably more powerful because the 'environment' for the gene is not the inanimate physical world but a milieu determined by a genetic programme. We generally believe that organisms adapt to their environment. However, since transposons can (and actively do) accumulate relevant genes of the host, they are able to adapt their environment to suit themselves. For example, if the amount or specificity of an enzyme in the cell is suboptimal for transposition, a transposon need not adapt itself to that restrictive condition. It can pick up the gene for that characteristic of the 'environment' and selfishly evolve it. Once the gene becomes part of a selfish element it will automatically evolve to optimize its contribution to the fitness of that element.

The transposons that have been characterized (mainly from bacteria) include elaborate forms, attesting to the creative power of their autopoietic evolution. Accumulating such genes does not merely complicate the transposons. It also gives them new emergent properties. Most notably, transposable elements in bacteria have an inherent tendency to become increasingly autonomous of the host. It is believed that genes shuffle actively from one level of autonomy to another as an important process in evolution of bacteria and possibly of eukaryotic genes as well. Thus autopoiesis can create new levels of functional activity and can transfer the determinants of evolution from the external environment to the internal evolving structure.

Evolutionary Function

Some changes in genes are so deliberate that they can only be understood through the biological concept of function. We are well acquainted with adaptive functions, those deliberate biological activities that are carried out to aid survival. In addition, various enzymes that operate on genes have *evolutionary functions*. They do not enhance the fitness or survival of the organism, but instead they facilitate its evolution.

An example of an evolutionary function is the rectification of gene copies in a multigene family. Some families contain dozens or hundreds of gene copies with the same function. It is impossible for the individual copies to evolve directly by natural selection because the redundancy makes the contribution of any one copy totally dispensable. For example, only the most dominantly deleterious mutation in a 5S RNA gene of *Xenopus* could possibly be sensitive to selection with 10,000 sister gene copies in the family. In order to expose sequences in a multigene family to natural selection, organisms must change large groups of the genes to a common structure. If they rectify a block with variant genes to wild type, mutations are purged. If they expand a mutant configuration, natural selection can then favour or eliminate that variant allele of the family within the species.

Some multigene families maintain extraordinary levels of homogeneity. Hundreds or thousands of gene copies are kept almost identical even though they readily evolve in concert. Obviously such uniformity can result only from a very effective rectification mechanism, of which eukaryotes have several. A gene copy can be directly corrected to the sequence of another in some multigene families. Saltatory expansion and contraction, sister chromatid exchange, and extrachromosomal replication and integration probably occur in others. These mechanisms are prerequisites for using multigene families as genetic determinants. Without them not only would a multigene family be unable to evolve its functions, but it would inexorably erode into individual genes. The important point is that rectifying a multigene family promotes its evolution but does not increase fitness. (In fact, the process generates genetic load.) The function of the process is an evolutionary one and not adaptive.

The transposases of bacterial transposons are another set of enzymes with important evolutionary functions. Moving transposons around in a cell does not enhance the cell's fitness at all. In fact, it is probably deleterious on the whole because occasionally a transposon gets inserted into an essential gene. However, transposition is essential to bacterial evolution.

Enzymes are not the only structures with evolutionary functions. Sensory machinery also can regulate cellular processes for evolutionary ends. Repressor systems direct most transposase genes to be active only when appropriate for evolution. Some bacteria and higher organisms have evolved very elaborate sensory systems to induce generalized heritable alterations in their genomes in response to stresses.

Since structures with evolutionary functions have entirely different roles in evolution from those with adaptive functions it is useful to have a specific name for them. Elsewhere I have suggested the term 'evolutionary driver' as the counterpart of the noun 'adaptation'. An evolutionary driver is a genetically determined structure with a function of propelling or steering the evolution of a gene, phenotypic trait or species. Enzyme pathways that rectify multigene families, transposases and their repressor systems, the recA system for inducible mutagenesis in bacteria and so forth, are examples of evolutionary directors which participate in the evolution of genes. If trends continue, we will probably discover a diversity of other evolutionary functions during the next few years. Most of the special organizational features of eukaryotic genes that have surprised geneticists during the past several decades are suspected of having evolutionary functions.

Although the concept of evolutionary function extends some of our ideas about evolution, it retains the conventional notion of 'function'. If the function of the enzyme maltase is to hydrolyse maltose, then surely the function of transposase must be to transpose segments of

DNA, and that of acquisitionase is to intercalate bits of foreign bits of DNA into the chromosome. We would call an effect of a structure its 'function' if it provided part or all of the selective advantage for the development or maintenance of that structure. Organisms can evolve structures with evolutionary functions through the same mechanism of *selection by consequence* that underlies the evolution of adaptive structures. All selection is based on the consequences of a structure rather than the structure itself. Species with greater capacity to adapt to their changing environment will benefit from the consequences of that capacity and can eventually prevail.

As discussed above, preadaptation and autopoiesis greatly increase the ability to evolve new adaptive functional capacities. Most evolutionary drivers probably evolved from adaptive precursors that acted as preadaptations. The repressor on transposon Tn917 is a case in point. Its adaptive function is to repress the erythromycin resistance gene of the transposon when erythromycin is absent from the environment. The repressor also has acquired, probably secondarily, control over the expression of the transposase gene. As a result, the presence of erythromycin induces the erythromycin-resistance transposon to move in the cell in addition to expressing a drug-resistant phenotype. Also, many enzymes used for DNA repair have additional functions in recombination and mutagenesis.

Perhaps the foremost discovery about evolution to come from modern genetics is that organisms carry out evolutionary activities as biological functions. Just how many genes have evolutionary functions is unclear, but the number may be large. Eukaryotes may actually have a greater amount of DNA concerned with evolution than with phenotypic fitness! A wide variety of structures with evolutionary functions should come as no surprise considering the wealth of adaptations that even the simplest species have evolved. Any adapted structure must be considered a possible potential preadaptation for an evolutionary driver. Thus the more a species or gene advances, the more opportunities it has for recruiting adaptive structures into evolutionary roles.

EVOLUTION OF PHENOTYPE

As described above, genes change in four distinct ways with time. The frequencies of variant forms fluctuate reversibly in populations, structural alterations become fixed, genes acquire adaptive modifications and they advance in their structure. It is significant that each of these dynamics has its counterpart in the evolution of phenotype.

For example, the distribution of body sizes of European moles fluctuates from one year to another depending on the weather, although there is no evidence for a lasting change. Harsh winters selectively eliminate larger individuals, but milder years allow them to reaccumulate in the population. Presumably body size, like most quantitative traits in animals, is maintained as a long-term balance between intermittent pressures for increase and for decrease. The overwhelming preponderance of phenotypic changes rapid enough to be seen from generation to generation almost certainly are of this nature and leave no mark on evolution over geological time. They are just 'noise' in the long-term process. Most selection on gene frequencies is probably of the stabilizing variety that leads to long-term constancy instead of change in phenotype.

Permanent change in phenotype also does occur, of course. It can be inferred from comparative biology and in some cases can be followed directly in the fossil record. The difficulty is not in detecting change, but in distinguishing changes with no selective importance from adaptations. Obviously changes may be adaptive even though we may not be able to figure out their usefulness. Conversely, and more disconcerting, the selectionist's exercise of devising possible adaptive explanation for a change does not prove that that or any other selective factor caused its evolution. A variety of mechanisms including allometry, mutation pressure, founder effects, fabricational noise, pleiotrophy, drift of various sorts, and gene introgression can cause a phenotypic character to change neutrally or even deleteriously. In recent years, important evolutionists have cautioned against the mistake of assuming that any existing character or phenotypic change automatically must be adaptive. Also, some palaeontologists believe that most morphological change occurs as part of the process of speciation. The role of selection in the formation of new species is poorly understood.

Probably the best guide for picking out the adaptations among evolutionary changes is common sense. The change in the colour of the pepper moth (*Biston betularia*) in England to a darker hue where industrial soot has blackened the tree trunks on which they hide is reasonably inferred to be adaptive. The red colour of blood probably is not. The colour patterns of mollusc shells are more problematical and may include both selective and non-selective factors.

It is easier to document selection for changes over short periods of time than long ones. However, the gulf between the 'ecological time' of the Neo-Darwinist and 'geological time' is so great that there is little reason to believe that the proportion of changes that are adaptive will be the same for the two. It is not even clear which time scale of change is the more constrained by adaptation. One reason for the current interest in the evolution of gene and protein structure is chat it should be so much easier to distinguish adaptive from neutral change where relationships between structure and function are simpler. Yet, even here, selectionists and neutralists bitterly disagree.

Progressive advancement to higher forms is also apparent in phenotypes of organisms. A dominant pattern in the long-term history of life is the appearance of successively higher and higher forms. This does not mean that all lineages inexorably advance. They decidedly do not. The pattern is for most species to get trapped in stasis and for only a minority to continue to advance. Nevertheless, by almost any criterion for 'advancement' that one might choose, life has repeatedly progressed from one record height to another during evolutionary history with the highest forms existing today. Such advancement of phenotype probably involves all four of the principles discussed above for genes.

Tautology of Advancement

The truism that the most advanced forms of life come from the lineages that were able to advance the fastest and farthest suggests that higher organisms evolve by different, and more effective, sorts of changes than do more primitive forms. This is indeed the case. Higher forms evolve in ways that were difficult or impossible for earlier forms, A striking progression has been for life to adapt through modifications first in enzyme catalysis, then in morphology, and ultimately in behaviour. Primitive organisms still evolve largely by changing their biochemical pathways. The main adaptive strategy of bacteria, for example, is to meet changes in their environment with new enzymic activities. To this end they have developed the complex system of transposons and plasmids discussed above. Eukaryotes, in contrast, evolve most notably in their morphology, generating new relationship between cells instead of new types of enzymes. Higher animals continue to evolve the occasional new enzyme pathway, for example, a mammalian lactose synthetase for making milk sugar, but a striking discovery from comparative biochemistry is that *Escherchia coli* has essentially the entire repertoire of basic enzyme activities of human cells. By and large, vertebrates, mammals, man and insects are reflections of special developmental pathways. Morphology offers vastly greater scope for change than do enzymes and, from what we know, is easier and quicker to evolve.

Behaviour allows even more varied and rapid adaptation. The most advanced animals evolve largely through their behaviour. For example, ants are counted among the higher arthropods. Their morphology is not conspicuously advanced but their behaviour makes them outstanding. Ants build large architectural structures, conduct agriculture with both plants and animals, cooperate in societies based on altruism, and take slaves. These functional capabilities emerged from advances in behaviour, although with support from specialized morphology and biochemistry.

Another progression is in the source of variation on which selection can operate. Early forms of life depended on spontaneous mutation. Their advanced descendents developed special enzymic pathways for editing alterations in the DNA. Bacteria evolved the more effective capacity to steal new genes from neighbouring species. Eukaryotes advanced in a different direction, evolving a diploid sexual cycle that generates continual quantitative variation by recombination.

Thus, in various ways, the forms of life that have advanced the farthest in phenotype are commensurately endowed with the greatest capacities for evolving.

Preadaptation

This concept is so familiar in classical evolution that it requires little comment here. Most new functional capacities of higher organisms probably originated opportunistically from pre-existing phenotypic structures.

Autopoiesis

Autopoiesis fits less comfortably with current evolutionary thought because it ascribes the cause of change to endogenous structure instead of interferences from the external world. This is a mode of explanation that Neo-Darwinists have explicitly rejected (although it has gained some recent advocacy from followers of so-called non-equilibrium evolutionary theory). Neo-Darwinism arose in large measure as a repudiation of competing interpretations that evolution was directed from within, such as aristogenesis, orthogenesis, evolution by law, mutationism and so forth. Nevertheless, autopoiesis and self-reference are integral both to the general process of evolution and to specific cases. In fact, the very origin of life was an autopoietic event, with its chicken-and-egg conundrum that faithful replication of the nucleic acid molecule is a prerequisite for the evolution of genetic information, but information for polymerase activity could evolve only after nucleic acids could be faithfully replicated.

Sexual selection is a more specialized type of autopoietic evolution. If female peacocks tend to choose males with large tails as mates, then the genes promoting this attractive masculine character will enjoy a selective advantage. As a consequence, genes that enhance the attraction of females to large-tailed males also will be at an advantage. They will prompt females to mate with the favoured males and hence produce male offspring that are more desirable. Circuits of genes that favour themselves through sexual, social or adaptive processes can erupt into runaway evolution unrelated to economy and ecology. Nevertheless, the directions of such evclution obviously are not random. They must depend in complex ways upon which organizational features of the species are potentially most sensitive to snowballing autopoiesis, and upon the availability of triggers to set the self-reinforcing cycles off. It is suggestive that certain aspects of morphology, such as wing bars, tails and head crests, seem especially prone to exaggeration as sexual adornments of some birds and it is interesting to consider the cause of this pattern.

Far more important self-reference emerges from the action of the organism on its own environment. Neo-Darwinism depends on a Newtonian-like distinction between species on the one hand, and its surroundings as an independent agency on the other. Yet, as Lewontin (1982, 1983) has stressed, this is not realistic. The characteristics of an environment are not independent of the species that inhabit it, but are intimately shaped by the occupants. Probably the most important selective feature of the environment for any higher species is the species itself. Thus, selection pressures are not just dictates of the external world but emerge in large part as the manifestations of the genetic programme of the species itself. This means that the genotype plays two roles in evolution. It responds with adaptive change and it also generates the cause of that change. What a genotype adapts to is largely its own expression. This self-referent cause and effect increasingly dominates evolution as organisms become more competent to manipulate their environments.

Since even the most doctrinaire evolutionist cannot ignore this self-reference completely, Neo-Darwinism has developed special branches of 'frequency-dependent' and 'density-dependent' selection. These have given rise to some really exciting insights into evolution, such as the 'Red Queen paradox' and 'evolutionary stable strategies'. Yet, the most conspicuous feature of deterministic mathematical sojourns into self-reference is their overwhelming complexity. Predictability, optimum and determinism become subordinate to instabilities, alternatives, critical points, feedback and oscillations. If there is one overriding lesson to be learned from the evolution of transposable genes it is that the process of evolution has access to dynamic relationships far beyond the trivia of Newtonian cause and effect. The tautology of advancement implies that significant evolutionary progress is a flag for the more powerful forms of evolutionary cause. I suggest that the important issue is not how the phenotype changes due to the effect of evolution, but how it has become specialized to cause evolution.

Evolutionary Function

An organism's most powerful mode of action is to carry out important activities as biological functions. Recent discoveries about the physiology of genes prove that evolution is able to produce structures to enhance the evolutionary process as well as to aid immediate survival. Much of the phenotype of higher organisms probably functions as evolutionary drivers. For example, the half of eukaryotic organization concerned with sexuality, recombination and mutation did not evolve to enhance the survivability or fecundity of the individual, but for its evolutionary function.

The evolution of sexuality poses a problem for the strict adaptationist because it is a burden on the fitness of the individual. We know that it can be dispensed with because occasional species ranging from plants to insects and vertebrates have become asexual. Scrapping sexuality for asexual reproduction probably aided short-term fitness. Nevertheless, higher organisms have preserved the burden of sexuality except for these sporadic cases because asexuality probably leads to *evolutionary* blind alleys.

Recognizing that phenotypic structures can have evolutionary functions is a separate issue from identifying the ones that actually are evolutionary drivers. I have already discussed the difficulty of sorting out adaptations from selectively neutral changes. Compounding this difficulty is the third possibility of structures evolving to perform evolutionary functions. Even at the level of genotype it is difficult to distinguish among these three possible reasons for a structure's existence. This difficulty is reflected in the fact that geneticists have proposed all three sorts of explanations for each of the organizational features of eukaryotic genes. The problem of assigning functions to structures becomes further complicated when adaptive structures secondarily acquire roles in evolution. They can then serve both sorts of functions simultaneously. An example cited above of a protein with such double function is the repressor of transposon Tn917 which simultaneously regulates both the erythromycin-resistance gene and the transposase gene of the transposon.

A critical task before us now is to assess the extent to which phenotypic traits of higher organisms have acquired evolutionary functions. For example, how widely have canalization systems for reducing noise in development been recruited to guide evolution? A second essential question is how many features of evolution that we take for granted as 'just the way evolution is' actually are results of selection for that mode of evolution? An immediate case is punctuated equilibrium. Is this a specially evolved mode of evolution? Have species developed the capacity or propensity occasionally to reorganize their genomes on a major scale in order to break out of the stasis that otherwise 'straightjackets' established species? Do species evolve special populational, morphological, behavioural or genetic organization to throw off new species? If so, can we explain how and why those features evolved? Also, what about stasis: is the pronounced tendency of many established species to persist with virtually no change an attribute that evolved from selection?

Relationship of Advancement to Adaptation and Change

Both gene structure and phenotype show the four dynamics listed in order of their dependence on each other. Each type depends on change at some underlying level, but may occur with or without producing any higher level dynamics. For example, adaptation presupposes change either in the structure of genes or in the frequencies of gene alleles in the gene pool. However, neither a fluctuation in the composition of the gene pool nor a base substitution in a gene implies that adaptation has taken place.

Advancement is also predicated upon some sort of change occurring but clearly does not inevitably accompany it. Superficially, advancement might seem to require adaptive change but this seems not to be so. For example, coupling a multigene family to an enzyme system that actively corrects one gene copy against another could be a critical advance in the organization of the family even though it had no adaptive importance. Probably complex interactions between the ecology and the structure of organisms determine how fast and in which ways species advance as they change. It is possible that the environment is more important for phenotypic advancement than for advancement of gene structure, but we have yet to understand what situations lead the species to advance instead of simply shifting 'sideways' from one niche to another.

Rates of Evolution

Descriptions of the molecular evolution of genes, and their extrapolation to phenotype, have two implications for rates of evolutionary change. Most notably, *biological structures* play active roles in evolution and must be major determinants of how fast or slow genes and species evolve. For example, their elaborate structures for rearranging and transmitting genes allow bacteria to evolve antibiotic resistance much faster than otherwise would be possible. Genetic and phenotypic organization probably also contribute to slowing down evolution in other cases, although we currently know much less about the mechanisms for stasis. The significance of these genetically determined factors is that they are evolved traits instead of just uncontrolled dictates of the outside environment. To the extent that biological mechanisms participate in evolutionary change they allow its rate to be programmed internally as an attribute of the species.

Secondly, molecular studies show that genes, and presumably phenotype, change in at least four entirely different ways during evolution. These types of change are substantially disconnected from one another and can proceed at independent rates. Analyses of evolutionary rates must begin with a clear description of the sort of change being considered and then proceed to the mechanisms that produce that form of change.

Endogenous versus Exogenous Cause for Evolution

The discovery that genes are inherently dynamic instead of passive structures undermines the basic Darwinian proposition that evolution is change forced on the species by the outside environment. The null hypothesis for Neo-Darwinism is that if a species were boxed into an environment so favourable that it exerted no selection pressure, evolutionary change would cease. This is unrealistic. No matter how much a species was environmentally pampered, its genes would evolve. Species are powerfully influenced by their environment, but environmental forces are superimposed on endogenous engines for change and advancement.

The relative importance of endogenous organization and external environment is perhaps the central question in evolution today. It does not have a simple quantitative answer. One reason is that the species influences its surroundings as discussed above. A more important reason is that the evolutionary role of the genetic message varies from one species to another. In particular, genetic constitution becomes increasingly important as the species advances to a higher form.

Consider the most primitive forms at the dawn of life. Their genes had virtually no capacity to 'do' anything and were maximally exposed to environmental conditions. Before enzyme systems arose to detect and repair lesions and to catalyze change in DNA structure, the mutation spectrum was dictated by physical principles and agents. Evolution probably approached the Darwinian model more closely then than at any time since. Subsequent species evolved increasingly elaborate ways to control mutation, suppressing deleterious types and promoting more productive types. They also developed phenotypic wraps to direct the way that environmental insults forced change on their genes. Higher organisms modified their environment and clustered into aggregates in order to be surrounded by a genetically regulated micro-environment instead of raw nature. Ever more sophisticated adaptations became potential preadaptations for developing ever more sophisticated evolutionary functions. Life asserted greater and greater control over its own evolution as it advanced; the forms progressing the fastest and farthest being those that optimized the factors important to evolutionary advancement. The increasing influence of the species in evolution not only results from advancement but also provides the mechanism for the process of advancement. Evolutionary progress means to gain command over one's evolutionary destiny.

This perspective, that the essence of evolution concerns the capacity to evolve rather than survival of the individual, resolves the enigma of 'higher' forms of life. Theories of evolution restricted to survival of the fittest have to dismiss the whole concept of advancement as illusionary. We might call ourselves and our relatives 'higher' than the protista or prokaryotes but for the survivalist this is just an arbitrary value judgement. By the goal of survival, bacteria are just as fit as we are—perhaps even fitter since they maintain far higher populations and have a phenotype that has survived without significant correction far longer than ours. Certainly we are more *complex*, but is it not just an anthropocentric bias to equate complexity with advancement? Complexity seems to be as irrelevant to fitness as any other general characteristic that varies among extant organisms such as size, life-span, metabolic rate, or degree of specialization.

In contrast, complexity is intimately related to the capacity to evolve. Organization and complexity are measures of both the cause and the effect of evolutionary advancement. An increased capacity to evolve effectively should automatically manifest itself as a more sophisticated phenotype. Reciprocally, phenotypic organization which expands the potential of a species to evolve must advance that process.

This is not just a truism because some phenotypic elaborations can adapt a species exquisitely to a particular environmental factor but impede, instead of facilitate, its subsequent evolution. They decrease evolutionary status even though they make the species more specialized. There is nothing arbitrary or anthropocentric about viewing advancement as a relationship of the organism to its evolution instead of to its external environment. Moreover, it turns the problem of evaluating how advanced a particular species is into an evolutionary question demanding an understanding of evolutionary behaviour, instead of just an ecological issue.

Defining evolution as the process of creating capacities for further evolution has a degree of circularity that is entirely appropriate. The ultimate significance of evolving life is not that it can change and adapt, but that it continually achieves new orders of function. Progressive advancement corresponds more closely to the literal meaning of the word evolution, 'to unfold', than does simple change. What is unfolding, however, is not the organism's fitness for Darwinian survival, but its command over its own evolutionary destiny.

4

Structure of Gene

Recent studies of gene expression in higher organisms highlight the possible significance of change in control sequences for the evolution of new phenotypes in which co-ordinated activity of a number of gene-action systems is required. Three examples are briefly reviewed: (l)a set of genes involved in the control of transcriptional activity in embryogenesis and under stress in adult life; (2) a gene complex involved in a co-ordinated behavioural activity in a mollusc; and (3) the genes involved in segmental organization and compartmentalized gene function in determination and differentiation during insect development. These three examples illustrate the significance of regulatory sequences in allowing for the evolution of gene function to cover more than one role, and also show how the time of activation of a set of genes and their relative productivity may be co-ordinated.

In seeking to explain the basis of major evolutionary divergence of form and function, it is necessary to analyze the mode of action of genes which have an especially significant role in mediating morphological and/or functional (including behavioural) organization. Such genes include those affecting transcriptional patterns at developmental stages crucial for morphogenesis (as in early cleavage and during metamorphosis in holometabolous insects), and those influencing alternative states of cells compartmentalized in cellular fields. The basic processes involved in body organization of the segmented animal phyla may have been highly conserved through evolution. Homoeotic mutations take on new significance as showing how compartmentalization of gene action allows major change in phenotype without concomitant disruption of body function. It has been suggested that rapid evolutionary change may have been favoured in primitive Metazoa by a higher degree of focusing of gene action at the compartmental level than is observed in more advanced groups. The contention in this chapter is that gene action is focused at the compartmental level in advanced as well as in primitive groups. What has changed in the course of evolution of at least the segmented Metazoa is the imposition of tagmatal and other higher order functional groupings of body structures on a primitive, compartmentalized body plan, resulting in an apparent slowing down of the rate of appearance of large-scale' morphological variations in later geological time.

One approach to analysis of evolutionary processes is to start by identifying the time of origin and mechanismal basis of innovations in the phenotype, especially where the appearance of a new higher taxon is involved. The first step, then, is to define the new gene functions. The second stage involves tracing back to establish the changes in gene structure, the alterations of nucleotide sequence and organization that brought about the new genotype, relating the new to the old. The third step in the process is the elucidation of the mechanism whereby the genetic reorganization took place, while a fourth concerns how and why the new genotype came to be perpetuated in the population, or participated in speciation. It is with the first two of these stages, i.e. with the delimitation of kinds of gene function and then with the relation between function and structure of genes that this chapter is concerned. The third and fourth stages suggested are covered by other contributions to this book.

Where there is a simple and direct relationship between gene, protein and phenotype, and especially where the protein product is the phenotype studied, it is often reasonably clear from knowledge of gene structure at the nucleotide level how a new or altered function was derived and what structural changes in the genome were needed to produce it. For instance, the source in the progenitors of the Triassic mammalian radiation of the α-lactalbumin subunit of lactose synthetase needed in milk production was presumably an ancestral lysozyme gene.

It is, however, the appearance of new phenotypes involving complex, integrated patterns of interplay between a number of nucleotide sequences and their products, between a number of cell types and between discrete organs, that constitutes a more interesting and challenging problem. This is nearer the idea of 'macroevolution', of the origin of distinctive new life forms, of new species and, by accumulation, new higher taxa. We have only just begun the task of analyzing the ways in which such integrative control of, and evolutionary change in, interacting gene-action systems can be brought about.

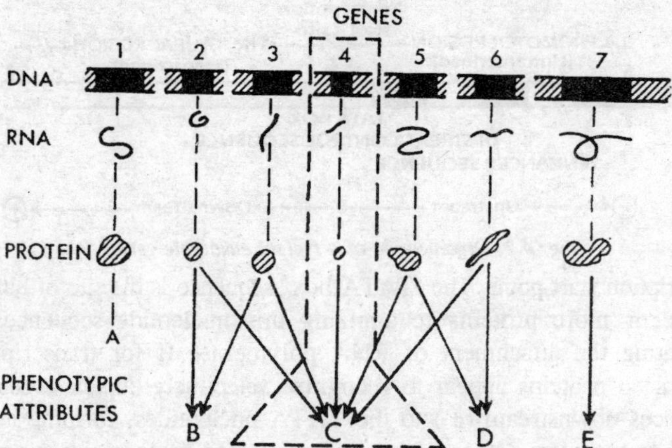

Fig. 4.1. Co-ordinated gene interaction in the production of complex phenotypic attributes.

Recognition of the importance of cellular compartments as units of genetic control in development has led to the conclusion that they are also significant units with respect to evolutionary change. The need is now to establish in detail how the early cleavage products of the zygote come to receive differing signals which operate control sequences that bring about co-ordinated, localized and differential patterns of gene activation. When this mechanism is fully analyzed, many of the major difficulties in the way of deeper understanding of evolutionary change in body architecture, as well as in development, will have been solved.

The idea that '...evolutionary change in regulatory systems accounts for evolution at and beyond the anatomical level' has been increasingly emphasized in the last decade. Regulatory controls must determine the timing of developmental events, the on-off binary choice of developmental path according to whether a gene product is made or not, and the integrated switching of banks of genes in the co-ordinated production of a spectrum of proteins. For an understanding of these processes it is necessary to review our present knowledge of what regulatory sequences are, what they can do, and how such sequences are put together in the evolution of gene control systems.

Patterns of Gene Control in Eukaryotes

Gene Structure in Eukaryotes

Control of the time and amount of transcription from the structural protein-coding nucleotide sequence of the gene is mediated by at least three kinds of untranscribed promoter sequence upstream from the

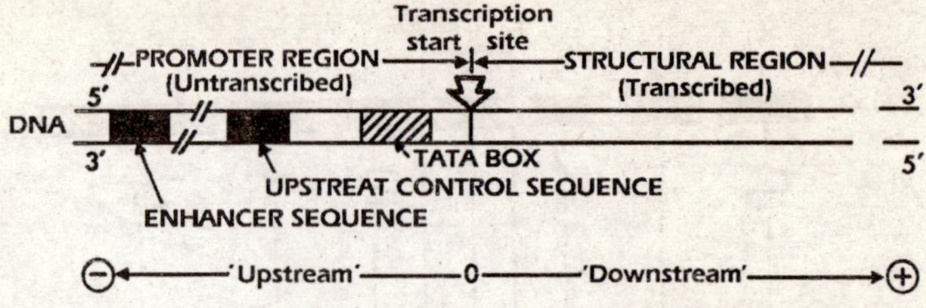

Fig. 4.2. Organization of a typical eukaryote gene.

transcription start-point. The 'TATA box' sequence is the site of binding of one or more proteins recognizing this nucleotide sequence and facilitating the attachment of RNA polymerase II for transcription. Some such proteins appear to recognize selectively duplex consensus sequences downstream (3') to the TATA nucleotides, forming a pre-initiation complex. Other general transcription factors also associate with TATA sequences along with RNA polymerase II. Upstream from the TATA box, promoter regions typically contain a control sequence recognized by DNA binding proteins specific to that gene or to a set of genes and capable of controlling the level of transcription. Even further upstream, enhancer sequences capable of modulating transcription provide yet another level of control. The first two of the regulatory elements are well exemplified by the heat shock genes discussed below.

Further levels of transcriptional control may be mediated by downstream sequences within the transcribed structural region of the gene; such control may be one function of the intron sequences that are transcribed but not represented in the mRNA produced from the transcript. The introns of certain co-ordinately controlled, neurone-specific genes active in the rat brain, for example, contain sequences in common that may reflect this function.

Many lines of evidence contribute to the present understanding of the role of the promoter sequences is gene control. One of the most striking examples is provided by the now well known experiments on enhanced growth in mice which incorporated in their genomes an artificially constructed gene consisting of the strong promoter from the mouse metallothionein gene with the structural region of the rat growth hormone gene. Enhanced activity of this artificial gene compared with that of the normal mouse growth hormone gene led to growth hormone levels 100 to 800 times that of control mice, and to the increased size of the affected mice. The detailed consideration of three genetic systems will illustrate the role of regulatory sequences in the control of gene action affecting metabolism, development and morphogenesis as well as behaviour.

Three Case Histories

Heat shock system

The eukaryotic heat-shock protein (*hsp*) genes were first identified in studies of the effect of elevated temperature on puffing in salivary polytene chromosomes of *Drosophila*. The *hsp* genes fall into three distinct groups within each of which the protein products, the heat-shock proteins (HSPs) are homologous; the genes are named from the molecular mass of their products in kilodaltons (kd) The *hsp* loci puff rapidly when the insect, or isolated polytene tissues *in vitro*, are exposed to a range of stresses including elevated temperature, ethanol or a variety of other chemicals, certain viral infections, radiation, anoxia and so on. Except for the product of *hsp*-82, which is detected in the cytoplasm, especially in association with the Golgi apparatus, the HSPs are largely concentrated in the nucleus. Two major functions of these genes are to modify the pattern of transcription in cells under stress, thus presumably suppressing the formation of abnormal mRNAs, and to modify translational activity during stress to an extent and by mechanisms that vary at least quantitatively from species to species. In some cases, polypeptide chain elongation is blocked, in others most of the mRNAs on ribosomes at the time of stress are discharged, but may be re-engaged when the normal pattern of protein synthesis resumes after recovery from the stress, at which time the nuclear HSPs return to the cytoplasm.

Cloning of the *D. melanogaster hsp*-70 gene allowed the homology of the *hsp*-70 and *hsp*-68 genes to be established, and revealed the general existence of homologous stress-inducible loci in other eukaryotes from yeast to man. Very importantly, however, use of the cloned *hsp*-70 gene as a probe established the existence in many organisms, including *Drosophila*, of genes related to the *hsp* loci but not inducible by heat—the *heat shock cognates*. Presumably it is these cognate genes that are differentially activated in the early *Drosophila* zygote and in imaginal disc cells to produce the 'heat shock' peptides found at these stages and which seem likely to be related to regulation of transcription in normal development.

Correlation of the activation of the *hsp* gene set with induced protection from subsequent stress is evident in two Australian drosophilids feeding on the bracken fern, *Pteridium*. *Drosophila megagenys* and *D. notha* have quite different thermotolerance patterns, but both species are readily protected from otherwise lethal short-term temperature stress by prior exposure to a sub-lethal temperature shock.

Using *D. melanogaster*, Lindquist (1980) showed that production of HSP-82 is induced at a lower temperature than that of HSPs 68 and 70. Relative levels of induction of the individual *hsp* loci are highly dependent on conditions of treatment. *Drosophila megagenys* and *D. notha*, when compared under standardized heat stress, consistently show species differences in the temperature pattern of induction of the HSPs, most markedly of HSP-82, which is readily induced as in *D. melanogaster*, and in HSPs 68 and 70 considered together.

The regulation of the *Drosophila hsp* genes and their cognates illustrates many features of selective and integrated gene-control systems in eukaryotes and of their evolution. First, all the *hsp* genes are induced in response to a common stimulus derived from certain kinds of stress,

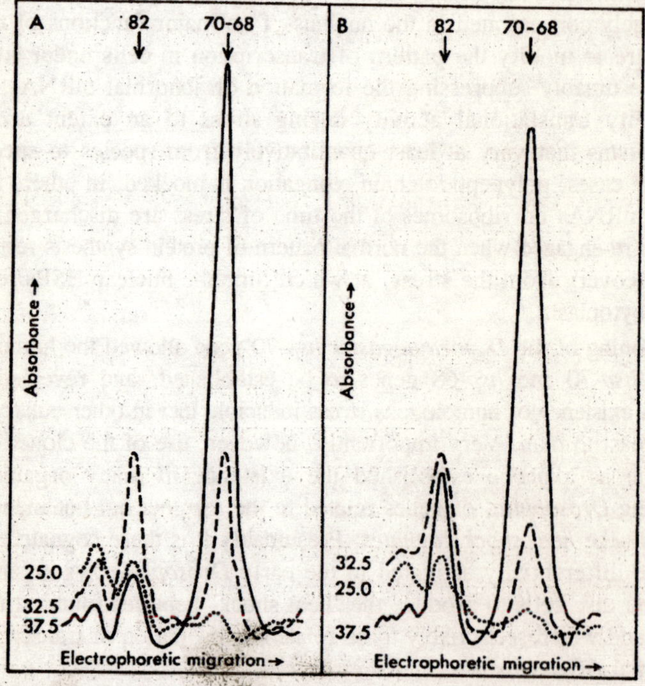

Fig. 4.3. Densitometer tracings of fluorographs showing the pattern of incorporation of ^{35}S-methionine into two major heat-shock proteins of larval ganglia of (a) Drosophila megagenys and (b) D. notha following heat shock for 30 minutes in vivo at 32.5°C and 37.5°C.

so these loci must share a common control element. But, as we have seen, each heat-shock gene has a characteristic and individual pattern of switch-on in any species so that regulatory controls show discrimination between the genes of the *hsp* set. Further, the heat-shock cognates are activated only at specific life stages, for specific times. Production of the heat proteins outside that developmental range, as when the *hsp* loci are induced under conditions of stress, leads to changes of normal transcription and translational patterns, so breaking the continuity of normal metabolic functions. Once the heat-shock cognate genes have played their early developmental role, whether in embryogenesis or in imaginal disc differentiation, there is a cost to the organism in producing HSPs. A trade-off is necessary in evolving gene controls maximizing survival chances. Triggering of *hsp* genes at too low a stress threshold will lead to unnecessarily frequent and disadvantageous disruption of normal function. Triggering at too high a threshold will leave many individuals unprotected against a potentially lethal stress by delaying the reprogramming of transcription and translation in favour of synthesis of gene products needed to counteract the effects of the stress. Detoxifying enzymes in the case of chemical stress, and repair enzymes in the case of radiation exposure, for instance, may be produced too late.

The work of Wu (1984a, b) and of Parker and Topol (1984a, b) has recently provided evidence of how these various controls are achieved. These studies employed the ability of DNA binding proteins to protect a double-stranded DNA sequence from exonuclease digestion, permitting control sequences covered by regulatory proteins to be separated from the rest of the genome. Wu's model of the function of control sequences identified in *hsp* genes upstream (5') from their transcription start points.

Wu's model suggests that a TATA box binding protein (TAB) must complex with the TATA sequence before the upstream controlling sequence common to the *hsp* genes can become inducible by heat shock. According to Wu's proposal heat shock or other stress involves binding of a second protein, heat-shock activator protein (HAP) to the upstream control element. HAP could not on this model bind to the promoters of heat shock cognate loci, for which a separate inducing signal would be required.

Within the *hsp* gene set, Wu ascribes the variation of response to stress of differing kind and intensity to the differing affinity of HAP for the specific control sequences of each gene. Wu (1984b) has extended

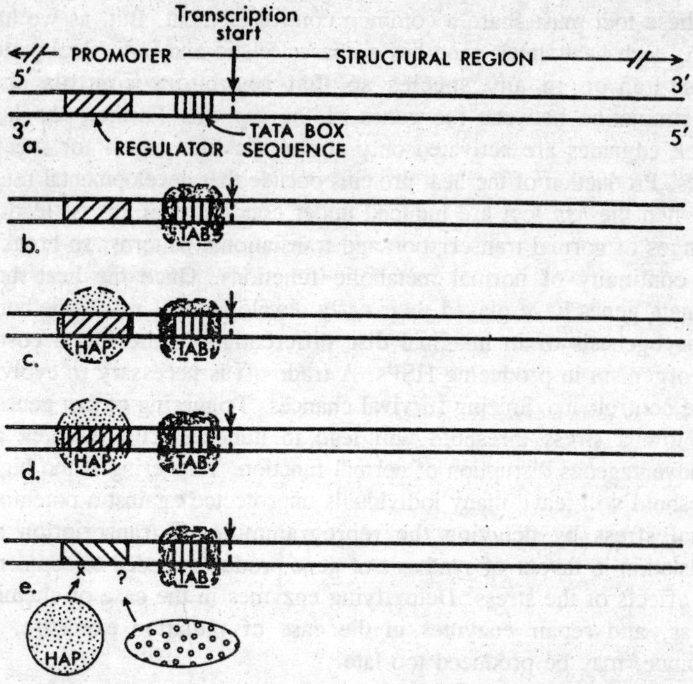

Fig. 4.4. Model of proposed control of heat-shock and cognate genes in Drosophila melanogaster.

analysis of the control sequence to 28 bases. The control sequence of *hsp*-82, induced at the lowest temperature and showing the highest affinity for HAP, has almost perfect internal dyad symmetry:

GAAGCCTCTAGAAG.TTTCTAGACACTTC

The less easily induced *hsp*-70 gene has nucleotide substitutions destroying the near perfect internal symmetry characterizing *hsp*-82, and shows a lower affinity for HAP. Thus Wu's model proposes that each gene in the co-ordinately controlled set is either on or off, but each has its own threshold of sensitivity to the inducing stimulus mediated by its binding affinity for a regulatory protein.

In the heat-shock gene system of *Drosophila* we see the re-use in evolution of a set of genes, the original function of which seems to have been set embryonic transcription patterns, to control transcription under stress conditions in later life. The latter function evolved in such a way to maintain compatibility with the former. The protein products of the two systems remain highly conserved, but separate control systems permit realization of the two functional roles.

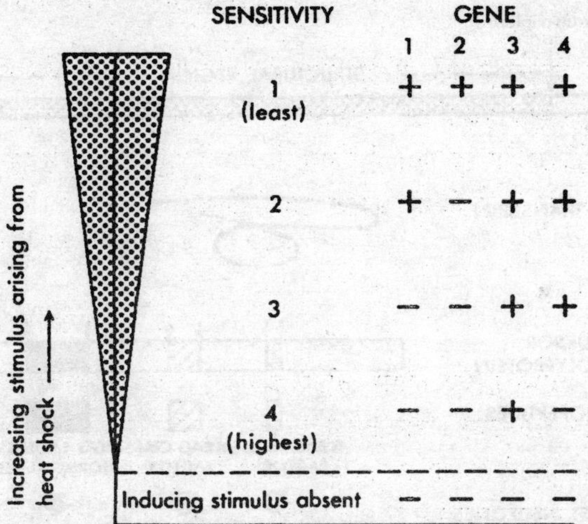

Fig. 4.5. General scheme for co-ordinate regulation of gene set as the hsp and cognate loci.

Other examples of shared 5' control sequences in the co-ordinate regulation of gene sets are well documented, for instance, in setting transcription patterns required for reproduction. Conalbumin, lysozyme and ovalbumin synthesis in the chicken appear to be co-ordinated in this way, as are the late and middle functioning chorion genes in the silkmoth, *Antheraea*.

Genetic control of a molluscan behaviour sequence

Elucidation of the nature and structure of a gene complex mediating innate behaviours in the tectibranch mollusc *Aplysia*, results largely from work of Scheller, Kandel, Axel and colleagues at the Centre for Neurobiology and Behaviour at Columbia University. This study provides an exciting insight into one mechanism of physiological co-ordination in the production of an integrated, in this case behavioural, phenotype.

Egg laying in *Aplysia* involves a set sequence of behaviours. First, movement and feeding of the animal is inhibited, then characteristic head waving takes place as the mollusc grasps the emerging egg string and assists its expulsion from the genital aperture. Finally, the egg string is attached to a solid substratum with the aid of a mucoid buccal secretion.

One neuroactive peptide, the egg laying hormone (ELH), is involved in eliciting some, but not all, of this behavioural sequence. Strurnwasser et al. (1980) showed that ELH and three companion peptides, the alpha and beta bag-cell factors and an acidic peptide, are released

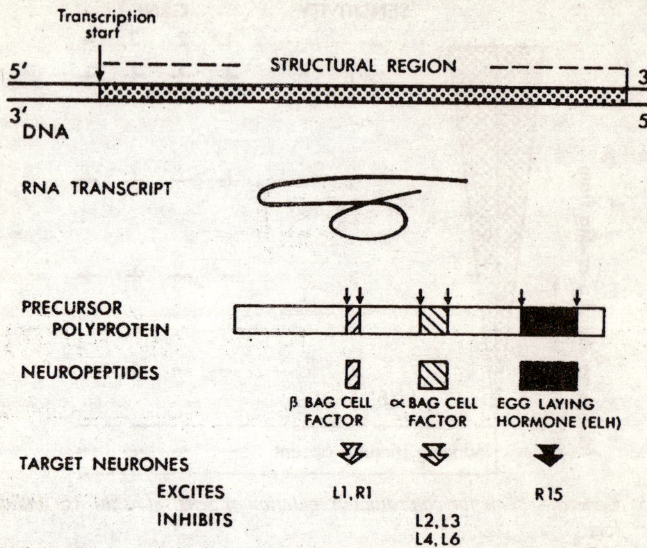

Fig. 4.6. The Aplysia ELH gene, its initial polyprotein product and its functional neurotransmitter polypeptides.

from specialized neurons known as bag cells attached to the abdominal ganglion into which these peptides move. ELH is an excitatory transmitter selectively augmenting the firing of a specific neuron in the abdominal ganglion, while the beta factor excites other neurons and the alpha factor inhibits a third group of neurons. In addition to the effect of ELH as a neurotransmitter, this peptide diffuses into the ovotestis to act hormonally on the smooth muscle of follicles in the initiation of egg laying.

A number of ELH-producing gene clones were identified by screening a genomic library of recombinant clones, using cDNA from mRNAs of bag cell and non-neuronal tissue to locate clones which hybridized with bag-cell cDNA but not that from non-neuronal tissue. Sequencing suggested that the mRNA from the ELH-producing gene would code for a precursor polypeptide of 271 amino acids of which 36 amino acids are those of ELH. Detailed examination of the ELH sequence enabled Scheller et al. (1983) to deduce that the precursor is better regarded as a polyprotein. It contains 10 presumptive sites for endopeptidase cleavage, four of the fragments produced being ELH itself, alpha and beta bag-cell factors and the acidic bag-cell peptide. Further, it became evident that the ELH and the bag-cell factors are related and presumably derived by duplication and subsequent divergence of a common ancestral sequence. These sequences are also related to those of the A and B neuropeptides produced by genes active in atrial neurons: the A and B neuropeptides control the initiation in the bag cells of ELH production and thus the egg-laying sequence.

These genetic studies reveal the key to the evolution of an extraordinarily elegant system of co-ordinate control for a physiological and behavioural phenotype. Related genes coding for similar but selective neurotransmitters have been linked into a linear series within one transcription unit, so that the synthesis and subsequent processing of a single polyprotein co-ordinates the timing of related behaviours in the array. Further, no single behaviour will be released without the related elements preceding and subsequent to it.

From the point of view of evolution of a co-ordinated system of gene activities, the ELH case may be considered a highly sophisticated system in which the production of a set of gene products from separate structural sequences is integrated by splicing them together under the control of a single regulatory sequence, rather than by separate *cis*-acting control sequences in front of each. Control under a single promoter, with a single transcript, processed to give a message subsequently translated into a polyprotein, may be evolutionarily preferable when the emphasis on the activities of a gene set is on distinctive quality of their products, on their relative amounts, and on the time of release. Certainly similar mechanisms have evolved repeatedly as in the cases of the amphibian and avian yolk proteins, phosvitin and lipovitellin, processed from the precursor vitellogenins; the production of insulin A and B chains: and the corticotropin (β-lipoprotein-related peptides, α-melanotropin, endorphin and methionine-enkephalin.

Control of the segmental body plan in animals

In contrast to integration through linear control, co-ordinated regulation of gene activities mediated by separate *cis*-acting control sequences may most reliably produce qualitatively and quantitatively distinct spectra of proteins. The genetic analysis of embryogenesis and morphogenesis in *Drosophila* illustrates the significance of such controls in eukaryotic development and differentiation, as well as the probable evolutionary significance of genetic change in the regulatory mechanisms involved.

The body plan of *Drosophila* is laid down by genes referable to several classes according to their mode of action. The first of four broad groups of genes considered here controls the axis of the embryo, determining in the oocyte before fertilization the future orientation of the zygote. Presumably such genes control the distribution in the egg of morphogens specifying subsequent developmental processes. Females homozygous for *bicaudal* produce embryos with posterior structures at each end; homozygotes for *dorsal* give rise to embryos with two dorsal surfaces.

Subsequent to specification of polarity and orientation, segment number and pattern is defined in the zygote by a second class of genes, of which some 20 have so far been identified, and which fall into two groups: 'gap' loci and 'pair-rule' loci. Mutants of the gap group produce changes in the number of segments in the body tagmata. Homozygotes for *knirps*, for instance, have a normally segmented head and thorax, but only one large 'segment' in place of abdominal segments 1 to 7. Mutants of the pair-rule group show phenotypes with half the normal number of apparent segments, reflecting organization of the germ band into units two segments wide. These double units are defined by boundaries in the normal segmental positions, although in different mutants the pairing of segments into these double units may be out of phase.

A third group of genes appears to be involved in specifying the polarity of segments as anterior or posterior half-segment units. Evidence from the fate of cells marked genetically by induced mutation shows that, at the blastoderm stage, particular cells are already determined as belonging to anterior or posterior compartments in specific segments. The segment polarity loci confirm that the anterior and posterior half-segment units participate in delimitation of the ultimate compartmental building blocks of *Drosophila*. Homozygotes for *gooseberry*, for instance, showed structural alterations limited to the posterior half of every segment.

Genes affecting the kind of tissue organization in the subsegmental compartments comprise a fourth group. Included here are the homoeotic loci, mutations in which lead to the appearance in one segment of structures characteristic of another segment. Genes of this group control which of several possible pathways a given lineage of cells will take, so producing the special structures typical of a particular segment rather than of segments in general. Examples include the *Antennapedia* (*Antp*) complex that affects primarily cephalic and thoracic segments and the *bithorax* (*bx*) complex, a group of some eight mutationally distinct sites that determines features of thoracic and abdominal segments.

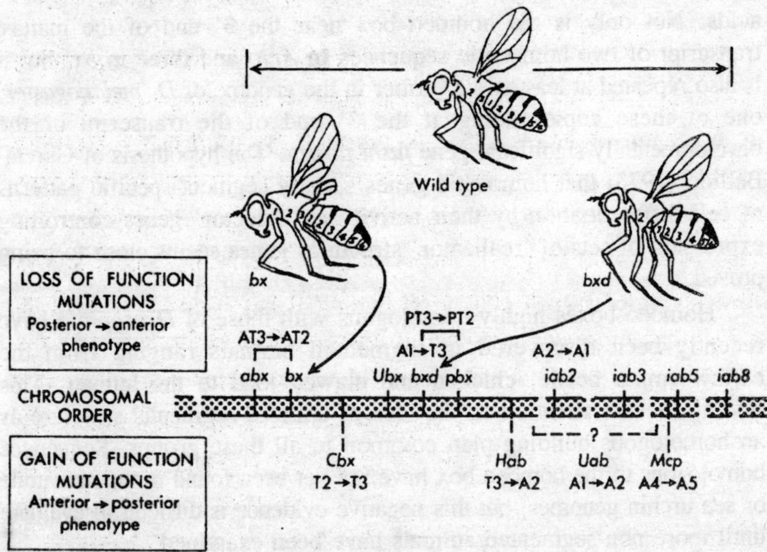

Fig. 4.7. Organization of the bithorax (bx) complex in Drosphila, showing colinearity of chromosome position to phenotypic effect.

Evidence from mutation studies, and especially from deletion mapping, shows that a series of related and closely linked sites in the cluster specify segmental characteristics in similar but distinctive ways, so that these genetic units evolved by duplication and later divergence of a common ancestral sequence. The nucleotide sequence in the *bx* complex is generally colinear with the segmented pattern of expression: the left to right order of the genetic material largely corresponds with the anterior to posterior order of the body structures affected. Loss of function mutations (typically recessive) shift the phenotype of an adult segment towards a more anterior type. Gain of function (typically dominant) mutations in this series shift the phenotype of an adult segment towards a more posterior type.

Struhl (1984) briefly summarizes the many similarities between the *Antp* and *bx* homoeotic complexes. The transcripts of both complexes appear at the same early stage of embryogenesis (cellular blastoderm) in a pattern showing tight localization to particular presumptive segments. The initial transcripts of each are large and then processed into smaller mRNAs. Most importantly both homoeotic complexes contain highly homologous copies of a short nucleotide sequence of 180 bases now known as the homoeo 'box' or domain, representing an open reading frame corresponding to a basic polypeptide of 60 amino acids. Not only is the homoeo box near the 3' end of the mature transcript of two homoeotic sequences in *Antp* and three in *bx*, but it is also repeated at least twice further in the genome of *D. melanogaster*, one of these copies being at the 3' end of the transcript of the developmentally significant gene *fushi tarazu*. The hypothesis of Garcia-Bellido (1975) that homoeotic genes specify segment-specific patterns of cell differentiation by their activity as 'selector' genes controlling expression of sets of 'realizator' structural genes seems close to being proved.

Homoeo boxes highly homologous with those of *Drosophila* have recently been discovered in segmented animals ranging from the earthworm, a beetle, chicken and clawed toad to the human. This finding focuses attention on the idea of unitized segmental structure as an homologous building plan common to all these groups. Sequences homologous to the homoeo box have not yet been found in the nematode or sea urchin genomes, but this negative evidence is difficult to evaluate until more non-segmented animals have been examined.

The genetic control of development, as now understood, involves sequential binary decisions specified by the on or off condition of the relevant nucleotide sequence in terms of its transcriptional and translational products. In succession these alternative gene states must determine anterior *versus* posterior, dorsal *versus* ventral, then a series of successive transverse partitionings of the body down to subsegmental compartments, followed by the control of tissue type and organization involving the homoeotic series of genes.

Three additional points are relevant to considerations of gene switching in bringing about the developmental sequence just outlined. The sharply delimited and non-overlapping boundaries between compartments of differently determined cells arise from two features. First, the initial event setting the choice along one or other of the alternative pathways for a given gene is deterministic rather than stochastic, reflecting the ability of each progenitor cell to distinguish sharply between levels of the controlling morphogen, i.e. in a gradient of morphogen concentration, each gene seems likely to respond according to a clearly defined threshold for its induction or repression. A series of duplicate genes with different thresholds, mediated for instance by different affinities for an inducer as in the model, could then control a compartment-by-compartment pattern of determination and differentiation. Secondly, cells of different compartments do not normally mix in later development. Thirdly, gene settings are fixed

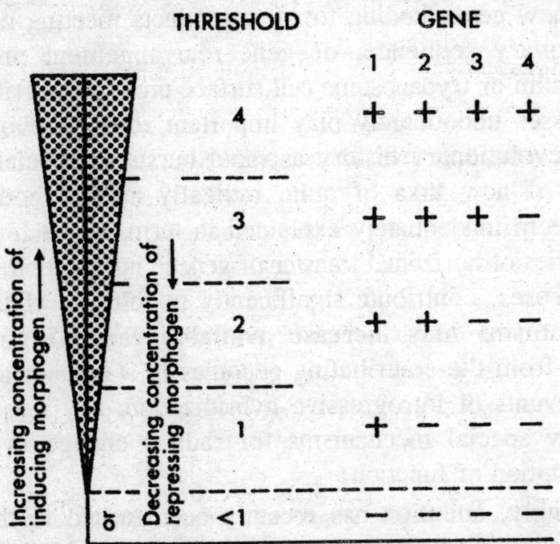

Fig. 4.8. General scheme for co-ordinate regulation of a gene set by hierarchical switching of a series of loci recognizing distinct threshold levels in a gradient of an inducing or repressing morphogen.

early in development when distances are small, and the contrast in the level of signal received by two adjacent cells along a gradient may be relatively great. In *Drosophila*, at the cellular blastoderm stage when the crucial events of determination are already complete, each presumptive segment is only three to four cells wide.

Genetic studies of development, such at those centering on *Drosophila*, now enable formulation of models of gene control that seem to offer new insight into the evolution of change in developmental pattern, and hence into change of phenotype with respect to morphological attributes of considerable scale.

Evolutionary Aspects of Gene Structure in Relation to Function

Rapid growth in our understanding of the structure and function of genes coding for the proteins of eukaryotic cells has highlighted the many processes of evolution that are still far from adequately explained. Nucleotide substitutions leading to change in amino acid sequence resulting in altered kinetic properties of enzymes, for instance in substrate affinity, thermostability or relation to inhibitors, are clearly relevant to adaptation and to change in allele frequencies in populations. So are the intron-exon structure of the eukaryotic transcription unit, the conservation and re-use of domain sequences in families of proteins, duplication and subsequent mutational divergence of nucleotide sequences to provide new genes coding for new products meeting new needs in the evolutionary sequence, or gene rearrangement providing for immunoglobulin or trypanosome cell-surface protein diversity. Although these processes undoubtedly play important roles in evolution, such features of evolutionary history as rapid bursts of speciation and the appearance of new taxa of quite radically altered body plan and behaviour seem inadequately explained in terms of such phenomena. Nor do theories of horizontal transfer of genetic information, for instance via RNA viruses, contribute significantly to solution of the paradox. Such mechanisms may increase available variability by pooling components from the contributing genomes in a manner analogous to occasional events of introgressive hybridization, but it is difficult to see here any special mechanisms for radical change in body form without disruption of function.

Increasingly, attention has recently been turned to the idea that change in regulatory elements controlling genes, and integrating groups of genes, may have a particular significance in evolutionary diversification of form and function. In ascribing evolutionary significance to regulatory components in the genome two problems need consideration. The first concerns the danger of merely pushing back the key events by one step to invoke yet another (hypothetical) gene. Many contemporary discussions of genome function in relation to evolution seem to employ the terms regulatory and structural in an exclusive dichotomous classification of genes. But the regulator of one locus is likely to be itself the product of a structural gene or to require complexing with one or more such products to become functional in the induction or repression of a gene activity, so that there is here a semantic difficulty. Thus genes producing the components specifying the anterior-posterior axis and dorsal-ventral surfaces of the *Drosophila* embryo are structural genes, coding for products that subsequently interact with the regulatory sequences of compartment-specific genes to control the programming of later development. It is exciting that, for *Drosophila*, we seem now to have identified most of the genetic loci involved, so that postulation of additional hierarchies of genes is increasingly unnecessary.

The second problem to be considered in evaluating the role of regulatory elements in evolutionary change is the perception that life processes will necessarily be adversely disrupted by change in co-ordinated gene activities. If a gene set is integrated by way of a common regulatory element, does it allow that mutational change in that regulator sequence bringing about a substantial shift in phenotype will generally cause chaotic disorganization? Erwin and Valentine (1984) write that:

> 'Viable mutations with major morphological or physiological effects are exceedingly rare and usually infertile: the chance of two identical rare mutant individuals arising in sufficient propinquity to produce offspring seems too small to consider as a significant evolutionary event'.

Homoeotic mutants of *Drosophila* and other species have often been regarded as genetic peculiarities of limited developmental interest. It is now emerging that homoeotic mutations may provide us with two of the keys to analysis of the evolution of at least the segmented animal phyla. Firstly, these mutations provide evidence of how change in nucleotide sequences regulating developmental patterns can produce gross phenotypic change. Secondly, such mutations commonly bring about quite extreme changes in body structure which are not inconsistent with viability and even fertility.

Under the subtitle 'Homoeology recapitulates phylogeny', Raff and Kaufman (1983) outline a speculative scheme of arthropod evolution based on the kinds of genetic step compatible with homoeotic gene functions identified in the analysis of *Drosophila* development. One example used by Raff and Kaufman to illustrate these ideas is the onychophoran to myriapod to apterygote to pterygote series. A deletion of bx^+ and pbx^+ in *Drosophila* transforms the fly to a condition mimicking the more primitive four-winged condition. The loss of the $Antp^+$ function leads to a *Drosophila* phenotype resembling the apterygote condition; the three thoracic segments are similar and non-wing bearing. If the entire *bx* complex is deleted, the posterior segments in the *Drosophila* embryo develop as a series of similar thoracic or trunk segments to give a body organization mimicking that of the trignathous myriapods. Deletion of both the *Antp* and *bx* complexes produces an embryo with three head segments and a series of similar trunk segments on the onychophoran pattern. This genetic series in *Drosophila* appears interpretable as providing a model in reverse of the major features resulting from successive steps of gene duplication and acquisition that may have produced the original phylogenetic sequence.

Lesions in the genetic material of homoeotic loci producing extreme phenotypic effects are certainly often lethal in homozygous embryos. But what is really surprising is that a simple genetic change such as homozygosity for three mutations *abx*, *bx* and *pbx*, converts the *Drosophila* phenotype into one in which the third thoracic segment is a fully formed wing-bearing segment duplicating the second thoracic segment. Some of these grossly reorganized animals can survive to emergence from the puparium, as do some *bx* mutants with four pairs of legs, although these flies are of poor viability and often show postural and behavioural abnormalities. The important point is that extreme changes in phenotype of this kind do not necessarily disrupt either the larval or adult functions essential to life. Body organization on a unitary compartmentalized basis protects the organism from developmental chaos that would otherwise result from mutations; the alimentary and nervous systems, musculature and so on are maintained as well as possible, consistent with the scale of the disruption experienced.

Another situation in which a single gene may cause substantial and concerted phenotypic change in many characters is exemplified by sex-change phenomena seen in many animals, including mammals. In *Drosophila*, the *transformer* mutation converts the female phenotype into that of the male with respect to genitalia, body pigment patterns, sex-combs and behaviour, although such individuals have small testes and are sterile.

Several recent authors have drawn attention to the idea that rapid evolution in the early Metazoa was favoured by a higher degree of compartmentalization of gene action than that seen in later metazoans. It is difficult to see how any higher degree of focusing of gene action could have existed in more primitive groups, to be later lost. A more plausible view appears to be that animals of the present day have in fact retained their ancestral patterns of compartmentalization, but have increasingly superimposed on them a higher order of organization, both of complex behaviour patterns and of tagmata as functional morphological and physiological units. The apparently higher rate of appearance of larger scale morphological variants in Palaeozoic metazoans would thus reflect restrictions imposed on developmental flexibility in more evolutionarily advanced groups, rather than a higher degree of compartmentalization of gene action in more primitive groups. A corollary of this argument is that the viable homoeotic mutants seem in present day groups, such as the higher arthropods, may represent a subset of the *viable* phenotypes which could have arisen in the absence of these constraints.

Compartmentalization of gene action in morphogenesis as a means of permitting phenotypic flexibility, without chaotic phenotypic disorganization, may also have a logical counterpart in the evolution of integrated behaviour patterns. The whole set of egg-laying behaviours in *Aplysia*, for example, occurs as a systematic repertoire related specifically to egg laying because of the unitary control leading to multiple neuropeptide synthesis and release. A genetic change in any one part of the complex gene producing the polyprotein precursor molecule may lead to behavioural change, but that change will be manifested within the context in time and space of the normal behavioural series to which it relates (unless the regulatory element controlling the gene complex is involved). Perhaps such behavioural compartmentalization plays a part in the rapid evolution of ethological barriers acting at the pre-mating level to reinforce reproductive isolation during the speciation process.

5

ORIGIN OF GENETIC VARIATION

This chapter is based on the premise that natural selection acted to minimize the deleterious effects of (expressed) genetic errors during the early phases of evolution, when many familiar features of the genetic apparatus were 'fixed'. According to this thesis many key features of the information-transmitting system of cells may be viewed as adaptive responses to the heavy error load placed on early genomes by their inability to correct errors made during copying or introduced from the environment. These features include:

1. diploidy and reciprocal recombination in eukaryotes;
2. genome segmentation among RNA viruses and;
3. the processing system which excises introns from RNA precursors.

The title of this chapter is something of a misnomer. I am less concerned with the origin and nature of genetic variation than with its significance for evolution. Let me start by showing how sloppy use of language unconsciously shapes our ideas, often to the detriment of the facts. Compare the words 'variation' and 'error'. In the context of genetics, both usually refer to mutations in DNA (or RNA), that is, they refer to the same thing. But the mental images they evoke are quite different: variation somehow implies 'useful change'—it links with the notion that mutation is a 'good' thing because it fuels evolution. By contrast error implies 'malfunction'—it links with the idea that mutation is a 'bad' thing because it causes information to decay. This confusion of meanings must be avoided. The hard fact is that although mutations may cause genomes to 'vary' they almost always do so in a *destructive* sense. To use an overworked analogy, when a typist copies a sheet of instructions almost any accidental mistake results in a deterioration of the quality of the information. The chances that a misspelled word or a transposed sentence will result in a more meaningful set of instructions are remote.

Part of the problem is that for two generations most textbooks of biology have enshrined the idea that sexual reproduction and reciprocal recombination have been positively selected because they generate diversity. The implication is that evolution needs more diversity than stochastic processes provide. Although most of us now consciously reject these arguments as 'group selectionist' their influence on our thinking remains deep-rooted and pervasive.

These qualitative statements can be put on a quantitative basis. The error rates in most ancient genes can be estimated by measuring the mutation frequency of polynucleotide copying in the absence of enzymes. The documented values are in the range of 10^{-1} to 10^{-3} substitutions per base per generation. This is to be contrasted with the estimated mutation rate of 10^{-9} to 10^{-11} in modern DNA replication. Simple comparisons like these show that, during the course of evolution, the level of noise in the gene copier mechanism has been slashed by a factor of 100,000,000 or so.

This drop in error levels demonstrates that, at least during much of evolution, natural selection has worked powerfully to decrease the frequency of mutation or to minimize its harmful effects. The best known mechanism of error reduction is correction. That is, secondary channels exist which allow a variety of error-detecting and error-correcting systems to repair premutational lesions in genetic systems. Such corrective processes operate not only during DNA synthesis but afterwards, in the resting state (postreplicative mismatch repair).

Protective Effect of Genetic Redundancy

The present concern is not with error correction but error compensation. The most widespread form of compensation is *redundancy*. The selective advantage, K, of redundancy can be modelled mathematically as:

$$K = [1 - (1 - q^L)^n] / q^L$$

where q is the base copying fidelity, n is the number of copies and L is the length of the genetic molecule. The best known example of this protective effect is seen in the structure of the genetic code where the identity of important amino acids is preserved by representing their cognate codons many times over. Fourfold redundancy in the glycine codons, for example, means that 33% of substitution errors in codons will have no effect whatsoever on the phenotype of the encoded polypeptide (assuming that base positions within codons do not affect mutational frequency).

The protective effects of redundancy can be modelled non-mathematically in terms of Figure 5.1. Here we see four DNA molecules encoding two information modules A and B. Provided that there is a critical number of whole genomes, a level of error sufficient to score one 'hit" per molecule per generation should not eliminate intact copies of total information. Thus a reiterated genetic system can survive a degree of genetic scrambling which would consign a single-copy DNA to its doom in short order.

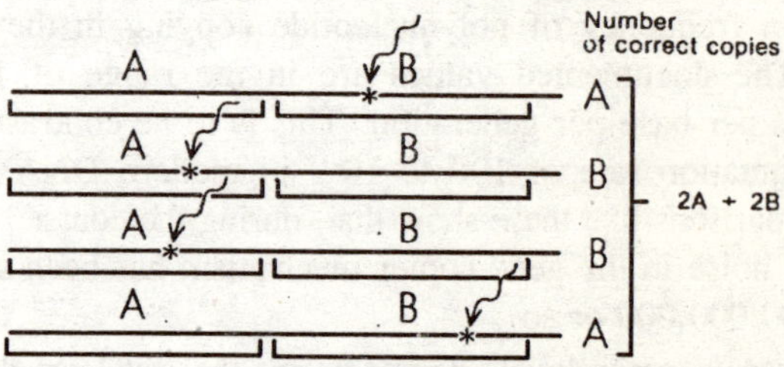

Fig. 5.1. Diagrammatic representation of the protective effect of genetic redundancy.

A striking example of protection due to redundancy may be offered by the unusual prokaryote *Dienococcus radiodurans*. This bacterium is extra- ordinarily resistant to ionizing radiation. This radio-resistance may be linked to the presence of 4—10 'genome equivalents' in each cell of this organism. These DNAs appear to be functionally separate and capable of independent segregation. As we have pointed out elsewhere, in this situation chance works for the preservation of genetic identity following exposure of *D. radiodurans* to ionizing radiation. This is because the *random* nature of mutation ensures that the chances of a given *single* gene being 'hit' in all reiterated copies at *equivalent* positions are remote. The presence of multiple copies of essentially the same information in one 'nucleus' permits recombination between them and this also improves the survival potential of the information since large numbers of multiple crossovers may regenerate some wild-type genomes from damaged ones. I will return to this key point shortly.

Dienococcus radiodurans is adapted to a level of mutation that few other living systems could tolerate. Beyond doubt the most familiar examples of genetic systems protected from error damage by genome redundancy are the *diploid* genomes of eukaryotic cells. It is selectively useful for systems that encode relatively large' amounts of genetic information to present that information to a 'noisy' environment in duplicated form so that mutations in key alleles do not immediately eliminate the organism. The fact that such reiterated systems can store errors as recessive mutations is an 'incidental' consequence of this primary protective strategy. This is not to say that diploidy in modern cells is *maintained* by selection for its protective value. As is often the case in evolution, something that originates in response to one selective pressure may then be subverted to serve quite a separate function. Since diploidy is usually linked to the capacity for sexual reproduction, one can speculate that diploidy arose because it allowed higher cells to alternate between protected (diploid) and unprotected (haploid) phases. Such reversible phases have the capacity to cull out unacceptably damaged genes during cool parts of the seasonal cycle (when the level of beat-induced error is low) and to protect cells during warm intervals (when the level of heat-induced error is high). According to this theory, the familiar alternation of *generations* between haploid and diploid phases had its origins in the equally familiar alternation of *seasons* between cool and hot periods, with the consequent alternation of error rates.

Protective Effect of Genetic Subdivision

Redundancy protects information by *increasing* the mass number of nucleotides per system, i.e. the genetic 'bulk' of the system rises in proportion to the level of protection offered. There is an escalating energy cost to this kind of protection and it is proper to ask if there are alternative mechanisms which allow information to be maintained without incurring such penalties. The answer is 'yes'. It is possible to improve the efficiency of data transfer not by repeating the information but by *subdividing* the information into shorter modules which present a smaller target size to the various error-promoting agents. The principle here is simple: the *shorter* a module of information the *greater* its chances of passing *undamaged* through a noisy channel.

Such divided genomes are today found almost exclusively among the RNA viruses and it can hardly be a coincidence that these tiny genetic objects retain the high error level (3×10^{-4}) characteristic of unrepaired polynucleotide synthesis.

The protective effect offered by genome subdivision can be readily visualized in terms of ultraviolet target theory. Consider an idealized situation in which ultraviolet dosage is calibrated so that it introduces one lethal 'hit' per molecule in a large population of RNAs all of length X. If each RNA is divided into two equal-sized modules, A and

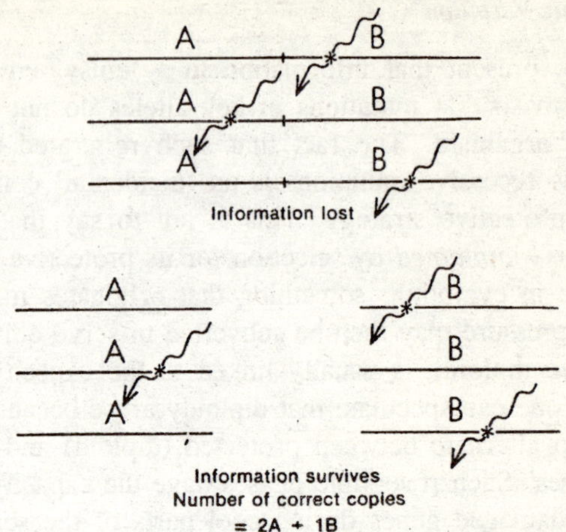

Fig. 5.2. Diagrammatic representation of the protective effect of genetic sub-division.

B, of length X/2 then the same ultraviolet radiation dosage is unlikely to inactivate the whole population. This is because a 'hit' in module A need not affect the integrity of the information in module B, and vice versa. The logic is similar when mutations introduced by an error-prone polymerase are not a result of outside factors.

However, a fundamental distinction must be made between what happens when the modular RNAs are packaged in one capsid (monocompartment viruses) and what happens when each RNA is separately accommodated in a discrete particle (multicompartment viruses). Consider the above example again, and assume that the error rate of the replicase that reproduces the two equal-length modules A

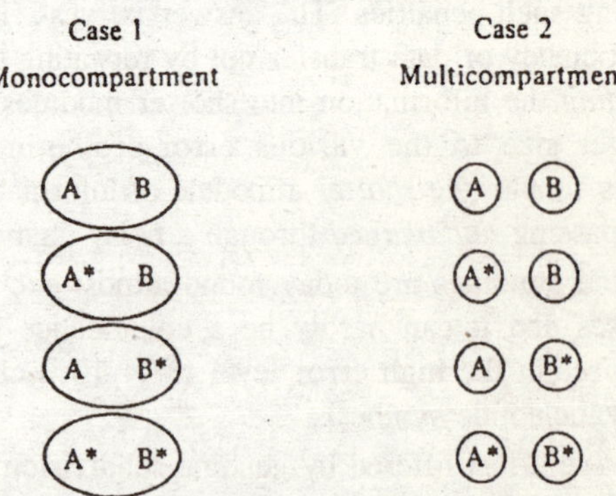

Fig. 5.3. Diagrammatic representation of the protective effect of genetic sub-division in monocompartment and multicompartment viruses.

and B is such that 50% of progeny carry lethal errors. The combinations of modules are then possible, where letters with asterisks indicate lethally damaged modules, and letters without asterisks indicate viable modules. In the first case, each pair of modular RNAs can be accommodated in one capsid so a single particle can initiate infection. However, this strategy limits the number of particles carrying acceptable copies of the genetic information to 25%. In this instance, unless there is some preferential association of non-damaged modules, a segmented genome has no advantage over a non-segmented one. In the second case, 50% of the population of progeny particles carry viable copies of units of the genetic information. This gives the multi-compartment strategy a distinct survival advantage, so long as the transmission of the various encapsidated RNAs remains random and independent. This requirement is met in nature.

The monocompartment viruses appear to have followed an independent evolutionary route. Where multiple RNAs are encapsidated in one coat, sequence-specific interactions must occur if there is to be provision for any kind of discrimination against miscopied information. Does such discrimination exist? Consider the reoviruses. During reovirus maturation *each* particle accumulates the *correct* combination of 10 to 12 *different* modular RNAs. The molecular basis of this specificity is believed to be a set of RNA-RNA or RNA-protein interactions that operate while the RNAs are single stranded.

Lane (1979) has proposed a co-operative process in which the binding for one RNA molecule during assembly alters the nucleation complex to create a binding site for a second RNA and so on. If the binding of a given RNA is faulty, the subsequent RNAs have a lesser chance of entering the nascent particle. Since miscopied RNAs are less likely than faithfully copied RNAs to recognize sequence-specific elements in complementary RNA or RNA binding sites in proteins, their self-amplifying interactions constitute a crude form of molecular proof reading.

Since sequence-specific elements like this are likely to be quite short, one can ask why such interactions—however specific—would select out those molecules whose overall information has been well copied. In answering this question, it is crucial to remember the fundamental difference between a single-stranded and a double-stranded polynucleotide. In double-stranded DNA, the base-pairing capacity of one strand is fully satisfied by its partner, whereas with single-stranded RNA base-pairing capacity is satisfied internally. This means that with single-stranded RNAs *shape* depends on *sequence*. Or, to put it another way, with single-stranded RNA, selection can directly monitor changes in the coding *sequence* through their effects on the three-dimensional *topology* of the molecule. This means that a sequence required to be unpaired (so as to recognize some matching element) must be maintained in its correct configuration by multiple secondary or long-range' tertiary bonds distributed throughout the rest of the molecule. Only if this underlying three-dimensional scaffold is correct can the target sequences adopt the optimal conformation for a faithful interaction either with a complementary sequence on a different RNA or with a stereo-specific RNA binding site in a protein. The target sequence therefore acts as a form of 'quality control' because its availability, i.e. its ability to recognize a matching element, is a sensitive function of the correctness of information elsewhere in the genome.

The reoviruses therefore also conform to my chief thesis in this chapter that genome subdivision is an adaptive response (at least in part) to the high error burden placed on the transmission of information encoded in RNA. Significantly, one can show mathematically that the kind of molecular 'proof reading' envisaged in this model should allow the upper size limit on RNA genomes to be exceeded. It may be no accident that reoviruses, which have the most highly divided genomes of all segmental RNA viruses (10-12 RNA modules), also have the largest genome sizes (12–20 $\times$ 10^6 nucleotides).

'Split' Genes

Sequence-specific interactions among reoviral RNAs provide an intriguing insight into a much wider process. These selective interactions may act as error-screening mechanisms because the reoviral genome is 'split'. But 'split' genes are common in nature. Indeed in the chromosomes of plant and animal cells they are closer to the rule than the exception.

Split genomes and split genes: is there any connexion? I believe there is. Moreover, I believe this connexion may allow us to explain how the exon-intron structure of split genes arose in the first place. The reasoning is as follows.

First, there is now a fairly widespread consensus among biologists that the split gene structure predated the continuous. Perhaps the most compelling evidence for this view is the fact that exons often correspond to modular peptide substructures, which may have discrete functions or roles.

Secondly, a large number of biologists now appear to accept Crick's view that genetic information was originally encoded in single-stranded RNA not double-stranded DNA.

Thirdly, as we have seen, the error rate in the 'first' genes was very high. If we take RNA virus replication as indicative of unrepaired nucleic acid synthesis, we can infer that the error frequency in early genomes was of the order of 10^{-3} to 10^{-4} errors per base per generation. The early copier mechanism was extremely error prone. These postulates together indicate:

(a) early genetic data was encoded in RNA;

(b) such early RNA genomes were 'split';

(c) early replicative processes were very 'noisy'.

This last point is critical. Eigen and Schuster in 1977 worked out a relationship which inversely relates the level of copy error in a self-reproducing system to the maximum size of the information that can be stably transmitted by that system, without experiencing an 'error catastrophe': the *higher* the error rate, the *shorter* the amount of reproducible information. This means that a genetic system without repair capabilities is locked into an 'information freeze'—it cannot evolve towards higher information contents and hence, in an important sense, it cannot evolve at all.

How could genetic systems escape this freeze? Obviously one cannot invoke repair processes such as we see today because these universally use the information specified by the undamaged strand of a duplex molecule to guide restorative processes on its damaged complement. If primitive genetic systems were made of single-stranded RNA it is hard to see how a corrective mechanism that depends on a duplex structure could have evolved or worked.

What options which do not require the *de novo* creation of complex repair mechanisms would be open to such early systems? The most plausible mechanism is genetic redundancy, as this protects the information for the reasons already given, and requires nothing more complicated than the collection of multiple copies of the same information in one 'cell'. Such RNA 'polyploidy' has interesting implications: it immediately allows for recombination between the RNA genomes. Elementary recombination like this need not require protein catalysis in its start-up phases. A model for non-enzymic RNA recombination exists in the self-splicing ribosomal RNA of the ciliate *Tetrahymena*. This transesterification reaction requires strand breakage, strand switching and strand reunion as does DNA recombination, and it occurs in the total absence of proteins. Significantly, the splicing of RNA precursors in the nuclei of higher cells passes through a series of intermediate phases in which branched lariat' structures are produced. There are some tempting similarities between the formation of these branched intermediates and the formation of the postulated intermediates in the auto-excision of RNA from the precursor RNA in *Tetrahymena*.

As Padgett et al. (1984) note of their model for RNA splicing in higher cells: ...this reaction...is consistent with the ribosomal RNA self-catalyzed splicing process seen in *Tetrahymena*. In this latter process, the 3' hydroxyl group from the 5' exon attacks the phosphate group at the 3' splice site in a transesterification reaction to produce the spliced exons and the excised intervening sequence.

I believe then it is entirely reasonable to propose that RNA recombination arose early in evolution for the same reason as DNA recombination did later on. It provided a means of eliminating errors from the pool of mutationally damaged genomes, by regenerating undamaged copies of the genetic information and hence allowing some upward mobility of information content. The selective advantages of recombination between single-stranded RNA molecules are in fact identical to those proposed by Maynard-Smith (1978) for double-stranded DNA recombination. Both function essentially as forms of repair!

If RNA splicing had its origins as a mode of *intermolecular* recombination, what relationship (if any) does it have to the *intramolecular* splicing which occurs today in the nuclei of higher cells? In assessing this question, it is important to bear in mind two points: firstly, the synthesis of RNA from DNA templates retains the high error rate of unrepaired nucleic synthesis; and secondly incorrect splicing has a multiplier or magnifier effect on cell physiology because of its ability to generate out-of-phase messages. In the light of these considerations, it seems likely that the capacity of the modern RNA processing mechanism to discriminate against 'incorrect' sequences is considerable. For example, if errors accumulate in the splice signals or in the sequences that determine topology, then an RNA with multiple exons is unlikely to proceed safely through all the steps needed to remove its introns. This is because RNAs blocked at particular processing steps by mutations in DNA may be degraded in the nucleus. The same is obviously true of errors introduced at equivalent positions during transcriptions. All the processing machinery 'sees' is what is there, whether it came from DNA/DNA or DNA/RNA copying is irrelevant. It follows therefore that the RNA processing machine functions potentially as a kind of molecular 'sieve' or 'filter' ensuring firstly that the average quality of the information reaching the decoding system as message is, in an important sense, *better* than the population of RNA precursors from which it was derived, and secondly that the cell does not waste its energy making nonsense (i.e. out-of-phase) proteins, some of which could damage its internal physiology. The only proviso is that for this system to work, cells must have some mechanism for *separating* primary RNA transcripts and messenger RNAs in a *selective* fashion. It is tempting to ask whether the *nucleus* that characterizes cells with a splicing process was not originally selected for just this reason, as a form of biochemical 'quarantine' to keep in the unproofed primary RNA copies of key genes and let out only those message molecules that had passed the fitness test posed by the multiple-step processing machinery.

According to this thesis, RNA splicing has been retained in the nuclei of cells that encode large' amounts of information because transcriptional copying never evolved efficient repair mechanisms. In DNA, however, the development of a formidable battery of error-detecting and error-correcting processes has seen the error rate fall from about 10^{-3} to 10^{-11} per base per generation. In modern genomes therefore the contribution of genetic noise to the measured rate(s) of mutation is very small. Yet the genomes of higher plants and animals are extraordinarily plastic. Mammalian DNA in particular is something of a genetic museum littered with the 'fossil footprints' of past events due to the insertion and spread of 'selfish' DNA molecules.

Discussion

I believe these two processes—the evolution of DNA editing mechanisms and the development of 'selfish' modular sequences—go hand in hand. Error-correcting mechanisms like the 'cut and patch' process carried out by the *Escherichia coli* DNA polymerase I combine the features of recombination (strand cleavage and ligation) and replication (fill-in synthesis). These kinds of processes could lead rather easily to the kind of 'replicative recombination' that allows semi-autonomous sequences like transposons to multiply in the genome.

This speculation can be put on a more secure footing by going back to Eigen and Schuster's inverse relationship between error rate and genome size. A genetic system operating near the upper limit of its information capacity (as determined by its error frequency) could not tolerate the disruption nor carry the energy burden that the insertion and transposition of large numbers of self-amplifying foreign sequences would bring about. As the error rate falls with improved repair, the maximum information-carrying capacity rises. As genome sizes expand, the cell's energy 'bill' for DNA synthesis also increases. However, beyond a certain point the cell may not, as it were, 'notice' the energy cost of replicating redundant sequences which do not serve any selectively useful function. Such sequences may thus be freed from the selective constraints that dominated previous evolution.

According to this scenario, mutation in modern cells is due less to copy error than to the existence of a multiplicity of mobile genetic elements which have the ability to integrate promiscuously into chromosomal DNA. Evolution in higher plant and animals cells has thus been powered not so much by mutational rewriting of the genetic script as by changes in the 'ecology' of key genes. This idea is consistent with the views that evolution in higher cells is chiefly due to the re-wiring of genetic regulatory circuits and the frequent rearrangements of executive and housekeeping genes.

6

Function of Gene

In general, function in living organisms depends on transforming material and energy outside the organism into processes that take place within it. These living processes must have originally existed on a fairly simple level, not much different from some of the processes we have seen in coacervates and microspheres. For a long time they may have depended at least partly on heat to provide some of the reactions that produce various cellular constituents. However, thermal energy, which may vary in place and in time, would hardly have been reliable, consistent enough for cellular needs. At best, heat from the environment or from rapid oxidations would have provided only an explosive, uncontrolled release of energy. More valuable to the cell was the development of chemical energy providers, such as adenosine triphosphate (ATP), which can release small but significant amounts of phosphate bond energy that the enzyme apparatus of the cell can both control and localize to specific reactions.

The shift to chemical systems of energy therefore meant the elaboration of *organic catalysts* (enzymes), which could restrict chemical reactions to the most opportune times and places. Accompanying this shift, without doubt, must have been a remarkable increase in the efficiency with which particular reactions could take place. For example, inorganic ferric ion (Fe^{3+}) shows some catalytic activity in a variety of reactions, including the decomposition of hydrogen peroxide into water and oxygen. However, when such ions incorporate into porphyrin molecules to form *heme*, the molecules are about a thousand times more effective than Fe^{3+} alone. If the protein component of the enzyme catalase then adds to the heme unit, catalytic efficiency increases by a further factor of 1 billion. How did such proteins evolve?

(a) Aqueous ferric ion **(b)** Heme

(c) Catalase enzyme (heme + protein)

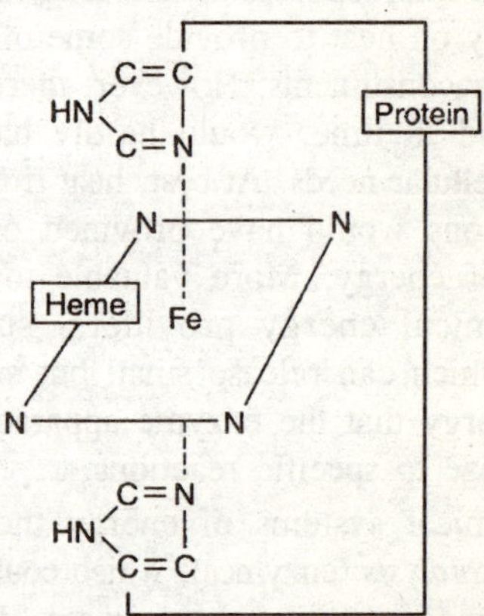

Fig. 6.1. Change in catalytic activity for the reaction $2H_2O_2 \rightarrow 2H_2O + O_2$ when the iron atom is used by itself (a), or in different molecular combinations (b, c).

Proteins or Nucleic Acids First?

At present, the amino acid sequences of the enzymes that serve as catalysts in the cell derive entirely from the nucleotide sequences in ribonucleic acid (RNA), which in turn derive from the nucleotide sequences of deoxyribonucleic acid (DNA). The fact that the entire chain of information transfer-replication (DNA → DNA), transcription (DNA → RNA), translation (RNA → protein)-itself depends on appropriate enzymes poses the serious question of how these functional and informational systems could possibly have evolved independently of each other.

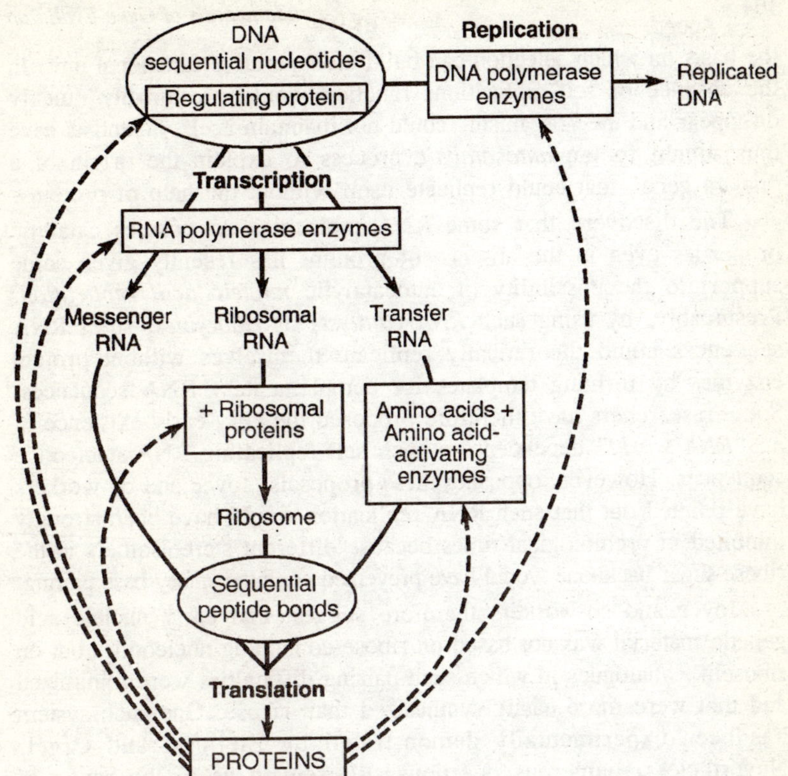

Fig. 6.2. Schematic diagram showing the mutual dependence of information carried by nucleotide sequences and function governed by proteins.

One answer to this problem has been that nucleic acids arose first, and their self-replicating power then enabled selection to develop protein systems that would support further *self-replication*. The geneticist Hermann Muller long ago suggested that since the basic appearance, or phenotype, of an organism derives essentially from its genetic material, or genotype, this relationship must also have existed in the past. That is, the genotype was probably first in the evolutionary sequence. *Protein synthesis* might then have evolved through specific amino acids directly interacting with specific nucleotide sequences, or perhaps through indirect placement of amino acids into such sequences by intermediary *adapter molecules* that brought amino acids to the nucleotide chain.

Supporters of the view that nucleic acid replication arose first have generally argued that only a self-replicating system can provide the basis on which selection can build a cooperative functional unit. In the absence of self-replication, function would presumably quickly disappear and the "organism" could not maintain itself. Scientists have thus sought for an *autocatalytic* process to explain the origin of a "naked gene" that could replicate itself without the help of proteins.

The discovery that some RNA molecules do possess catalytic properties even in the absence of proteins has recently given some support to the possibility of autocatalytic *nucleic acid replication*. Presumably, by using such *RNA catalysts* or *ribozymes*, short RNA sequences could theoretically replicate themselves without protein enzymes by forming templates for complementary RNA sequences. Some researchers have therefore proposed the very early existence of an "*RNA world*" dependent on such self-replicating RNA nucleotide sequences. However, opposing these proposals, Joyce and co-workers have pointed out that such RNA replication would have been strongly inhibited in prebiological times because different stereoisomers in the ribose-sugar backbone would have prevented complementary base pairing.

Joyce and co-workers therefore suggest that early nucleic acid genetic material was not based on ribose-containing nucleotides but on riboselike analogues in which such pairing difficulties were minimized and that were more easily synthesized than ribose. One such system has been experimentally demonstrated by Zielinski and Orgel. Nevertheless, numerous questions still remain as to the range of reactions that RNA could catalyze (its present range is quite limited) and how such systems, even if they occurred, would have made the transition to RNA synthesis and replication.

Instead of nucleic acids, Cairns-Smith suggests that the primitive self-replicating unit may have consisted of organic, claylike silicate crystals or layered clay mixtures (e.g., montmorillonite). Crystals can grow by adding subunits into their highly ordered structures, and thus we can view them as having some self-replicatory powers. According to Cairns-Smith, nucleic acids such as RNA could have incorporated into such an assembly, followed by the formation of peptides and the evolution of a protein-synthesizing system.

Although we cannot exclude such events, we find them difficult to envisage, and some biologists have therefore emphasized more the possibility that either proteins themselves or protein-nucleic acid combinations were the first self-replicating systems. Black, for example, suggests that some early peptides might have served as templates for the aggregation of nucleotides that may then have bonded together to form a precise mold for the replication of these same peptides. This view has the advantage of offering a mechanism by which the replication of both peptide chains and nucleotide chains would have been interdependent from the start. Also, as one chain lengthens during further evolution, so does the other, each gradually improving its replicatory role until some of the present features of nucleic acid replication and protein synthesis evolve. Unfortunately, as with naked genes, we still find it difficult to imagine the spontaneous origin of such a precisely organized nucleic acid template.

Various theories therefore propose that protein systems probably developed diverse functional properties before they coupled to nucleic acid replicative systems. Foremost among the evidence for this view is the likelihood that proteins always had more functionally different forms than nucleic acids. Part of this functional variety arises from the almost inexhaustible array of permuted amino acid sequences that proteins can achieve.

For example, since 20 different amino acid alternatives exist for each position in a polypeptide, a sequence of only 5 amino acids has 20^5 or more than 3 million possible arrangements. By contrast, a sequence of 5 nucleotides in a nucleic acid has only 4^5 or 1,024 possibilities. Moreover, many amino acids are quite different in structure enabling them to interact in many different ways both with each other and with molecules such as water, metal ions, and various monomers and polymers. This confers an astronomical variety of possible three-dimensional configurations on a protein, in contrast to the relatively more rigid shapes assumed by many nucleic acids.

Chemical experiments under presumed prebiological conditions also show the difficulty of producing polymerized nucleic acids spontaneously, whereas long-chained polypeptides are produced in such experiments with relative ease. A number of authors therefore claim that a protein catalytic system must have developed before a nucleic acid replicative system. Many others, however, feel this view has a serious shortcoming, because if proteins arose first and were used by primitive cells or particles for functional purposes, how could they have replicated without a nucleic acid translational system? Could proteins alone have synthesized proteins?

At present, it is difficult to conceive of protein synthesis independent of nucleic acids since no apparent complementary relationships exist between amino acids as do between nucleotides. That is, the precise *stereochemical fit* that occurs between the base

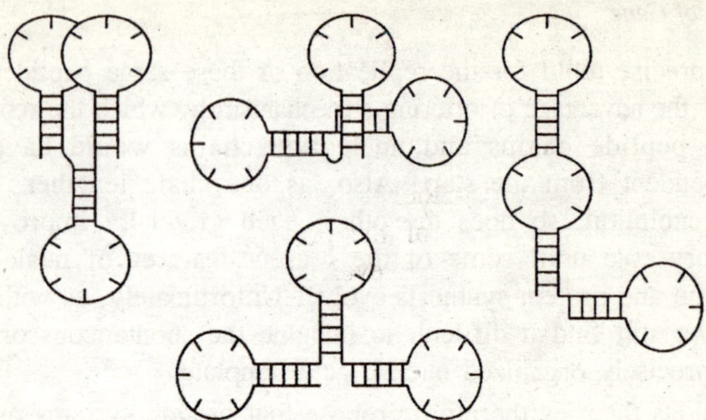

Fig. 6.3. Some triple stem-loop structures that RNA molecules can assume.

pairs of complementary nucleotide chains (adenine-thymine, adenine-uracil, cytosine-guanine) and that accounts for the replicative, transcriptional, and translational properties of nucleic acids nowhere echos a similar complementary stereochemical fit between amino acids in polypeptide chains. Nevertheless, a process does exist in which proteins make proteins.

As Lipmann (1971) and others showed, spore-forming *Bacillus brevis* bacteria produces at least two antibiotics, gramicidin S and tyrocidin, that are formed exclusively by enzymes in the absence of messenger RNA (mRNA). Both these antibiotic molecules are circular *oligopeptides* ten amino acids long and are synthesized by the sequential addition of amino acids. Should one amino acid be omitted, peptide synthesis ceases, indicating that the enzyme involved functions as a template for the amino acid sequence. That is, an unfilled position on the template prevents bonding between amino acids on either side.

Of special interest is the form in which peptides elongate among these antibiotics. Single amino acids bind to sulfhydryl (—SH) groups on the enzyme before they join the peptide chain and are then connected by the sequential removal of their sulfur (thiol) groups and the formation of peptide bonds. That is, the chain maintains a thiol at its "head" end to furnish the connection for sequential growth. Lipmann called this process *head—growth polymerization* and pointed out its striking similarity to the polymerization of carbon groups during fatty acid synthesis (also polymerized by use of sulfhydryl bonds) and to the polymerization of amino acids during ribosomal peptide synthesis (polymerized by phosphate bounds).

Lipmann therefore suggested that these similarities indicate a common underlying polymerization process that may have arisen early during chemical evolution. Of course, the enzymes now involved in antibiotic peptide synthesis are themselves synthesized via information transferred from genetic material, but the fact that proteins can produce proteins in these systems points to the possibility that repeatable copies of short-chained but functional peptides, 15 to 20 amino acids long, may have been produced in the past in the absence of nucleic acids.

Evolution of Protein Synthesis

Whatever the early composition of the templates used in condensing amino acids, the polymerization process itself must have been one of the first functions for which selection occurred, since only in peptide form do amino acids attain their catalytic properties. These early templates, however, were probably inefficient, producing peptide chains with only vague similarities to each other.

Thus, as time went on, selection for improved polymerization must have led to the production of more efficient templates, which were perhaps themselves polymerized by more efficient *polymerization enzymes*. Thus we can say that the circular, autocatalytic tautology of life-to make more of those substances that can interact with the environment to make more such substances-was firmly established during this early period.

Given a template for the synthesis of peptides and a polymerization enzyme, various sequences of events might have followed. The value of considering any of these is simply to show that enough molecular information has been accumulating to permit the development of different step-by-step hypotheses that can be used to solve the origin-of-life puzzle. If we continue with the notion that the first templates were probably not pure nucleic acids but perhaps some combination of peptides and other materials, one possible sequence is as follows:

1. Once peptide polymerization established itself, selection for its improvement must have led to an increased number of polymerase enzymes per protocellular unit, as well as improved template efficiency of these polymerases and their ability to increase the rapidity of amino acid condensation.
2. Improvements in the mechanism that brought amino acids to the template probably accompanied improvement in polymerase function. That is, adaptor molecules evolved that could bind to amino acids in the medium, and then could bind these amino acids to the polymerase enzyme or the polymerizing template. These adaptor molecules were probably not very large, and their

evolution might simply have meant new uses for some of the templates that had previously produced polymerase enzymes.

3. Further improvement probably involved selection for small, nucleotide-like coenzymes used in binding and releasing phosphate energy groups, similar perhaps to present coenzymes such as nicotinamide adenine dinucleotide (NAD) and flavine adenine dinucleotide (FAD). Such nucleotides were not necessarily RNA or DNA but could have consisted of combinations of various sugars, nucleic acid bases, and phosphates.
4. Both the template and the adaptor molecules may have incorporated nucleotide sequences, which could then recognize each other by complementary base pairing via hydrogen bonding similar to the type presently seen between adenine and uracil or guanine and cytosine. These new adaptors could therefore bring their amino acid to the nucleotide portions of the polymerization complex rather than to the peptide portions. Protein synthesis would thus gain the advantage of precise stereochemical pairing between complementary nucleotide bases as compared to the more inefficient amino acid-polypeptide interactions previously used. Such events would therefore lead directly to improved placement of different amino acids into specific positions on polypeptide chains and to a much increased number of amino acids that could incorporate into such chains.
5. If we consider the polymerization enzyme complex to be a primitive ribosome of sorts, some of its nucleotide components thus began to act as a template specifying the sequence of amino acid incorporation by base pairing with adaptor molecules. At the same time, these nucleotide sequences themselves could replicate with considerable precision by base pairing with available nucleotides.
6. Replicating nucleotides led to forming master nucleic acid molecules that could be stored as genetic material, yet that could also serve in translating nucleotides to amino acids or produce complementary messenger strands for this purpose. Three separate functional classes of nucleotide sequences thus eventually arose: storage, messenger, and translational (ribosomal and transfer)—all probably RNA, since this nucleic acid still fills two of these purposes in all organisms (messenger and adaptor) and also fills genetic storage purposes in some viruses.
7. Since RNA was probably used primitively for both information storage and protein translation, difficulties in separating these two

functions must have offered advantages to organisms that could use a different nucleic acid, DNA, for storage purposes. The enzymes that translate RNA into protein do not function with DNA, thus restricting the more uniformly structured, double-helix DNA exclusively to the storage of information and to transcribing one of its strands to form mRNA. The fact that deoxyribonucleotides are synthesized from ribonucleotides in cellular pathways supports the notion that DNA arose later in cellular evolution than RNA.

The evolution of protein synthesis may thus have had its start in a primitive polymerase enzyme that could replicate inefficiently on a template of polypeptides and other materials. Evolution then proceeded to a self-contained ribosome that contained its own stored genetic information and also mRNA committed exclusively to translating a limited number of polypeptides. In later stages these functions sequestered to different parts of the cell, with mRNA moving from its new site of transcription, where genetic information was now stored, to the ribosome for translation. The evolution of new kinds of ribosomes, no longer committed to the production of particular proteins, enabled the same ribosome to translate different mRNAs. This transferred the burden of regulating which proteins to make from the ribosome to the transcriptional process. That is, the particular proteins to produce could now be selected by regulating which mRNA molecules to transcribe from the stored genetic material. By these means, some of the basic modern features of protein synthesis may have come into being.

Evolution of the Genetic Code

Information transfer between nucleic acids and proteins follows a genetic code that determines the placement in a protein of a particular type of amino acid from the placement of a particular trinucleotide sequence in mRNA.

As for any other biological trait, researchers believe that this code must have evolved from a more primitive form, although they have so far discovered no ancestral codes. Attempts at an evolutionary reconstruction of the code have therefore generally relied on a detailed analysis of the features that characterize the present code. These features are as follows:

1. Messenger RNA molecules consist of only four kinds of nucleotide bases, adenine (A), guanine (G), uracil (U), and cytosine (C). These compose chains of varying lengths and varying sequence.
2. An mRNA codon that specifies a particular amino acid is a triplet consisting of a chain of three nucleotides.

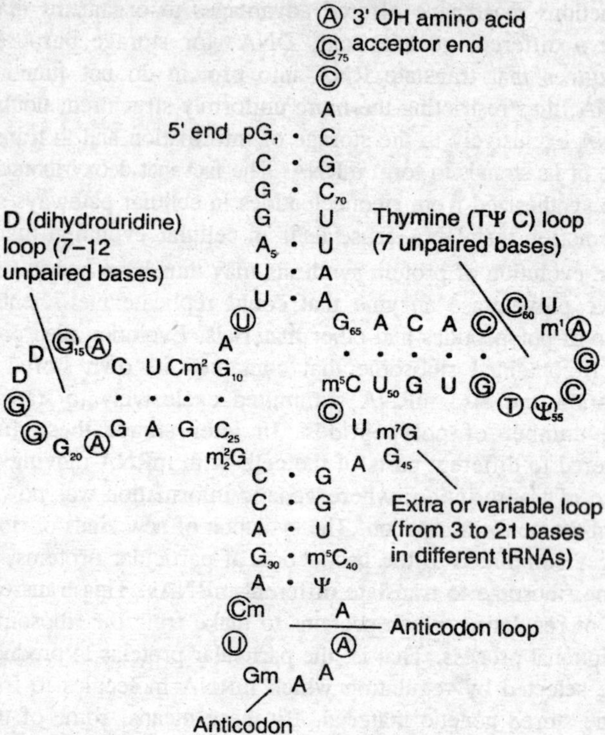

Fig. 6.4. Sequence of the 76 nucleotides in phenylalanine tRNA of yeast showing in the commonly portrayed two-dimensional cloverleaf form.

3. The code is commaless and nonoverlapping: each codon is translated in a continuous sequence, three successive nucleotides at a time, from one end of an mRNA reading frame to the other.
4. The codon sequence complements an anticodon sequence on the adaptor or transfer (tRNA) molecule that carries a particular amino acid to the mRNA codon.
5. All living organisms share the coding dictionary, which is thus universal, with some codon differences in mycoplasmas (bacteria lacking polysaccharide cell walls) and ciliate protistans. Mitochondrial organelles also show a few codon differences from those cellular nuclei use.
6. Ambiguities have not been found in the code; that is, the same codon does not specify two or more different amino acids.
7. With the exception of methionine and tryptophan, all amino acids are each designated by more than one codon.

8. The pattern of degeneracy of the code is mostly in the third codon position. For example, eight amino acids, including valine, threonine, and alanine, use quartets of codons, each member of a quartet varying only at the third position.
9. When an amino acid uses only a duet (two) of the codons in a quartet, the third codon positions in this duet are both pyrimidines (U and C) or both purines (A and G), never one pyrimidine and one purine.

Explanations for some features of code degeneracy have not been difficult to find. In part, degeneracy derives from the presence of more than one kind of tRNA for a single amino acid. The tRNAs used for leucine, for example, may have the anticodons AAU, AAC, and GAG. Crick explained that the degeneracy pattern mostly attaches to the third codon position because of *wobble pairing* between certain bases of the tRNA anticodon and the mRNA codon in this position.

Crick's wobble hypothesis, since proved correct, was that anticodons bearing inosine (I) at this third position can pair with either U, A, or C[6], anticodons bearing G can pair with either of the pyrimidines U and C; and anticodons bearing U can pair with either of the purines A and G. Third codon position degeneracy thus points to the importance of the first two codon positions in specifying amino acids; these positions suffice to code for the eight amino acids that use codon quartets.

The striking universality of the code—the fact that no other code appears in any independent living organism—makes it reasonable to ask, Why this particular code? Since at least 10^{70} possible different codes use 64 codons to code for 21 entities (20 amino acids + chain termination), either the exclusive use of this particular code must have derived from accidental causes, or some relationship between amino acids and their codons (or anticodons) must exclude large numbers of other possible codes—or perhaps both factors operated.

Some authors have suggested that the amino acids originally associated with their codons or anticodons through stereochemical fitting or shared other complementary properties such as hydrophilic and hydrophobic qualities. For example, some kind of pairing may have occurred between an amino acid such as phenylalanine and a codon such as UUC or its anticodon AAG.

However, other authors maintain that, despite considerable search, there are few if any examples of preferential affinity between amino acids and their codons or anticodons. Furthermore, if we consistently applied such a hypothesis, it would be grossly inadequate in explaining why serine uses two such markedly dissimilar sets of codons as the UCN quartet (where N designates any base) and the AGU-AGC duet.

Theorists therefore have offered an alternative hypothesis, suggesting that the universality of the code derives from the survival of only one of the possible codes that may actually have existed in the past. That is, early amino acid-codon relationships arose largely by chance rather than by restricted stereochemical pairing and may therefore have produced a number of different primordial genetic codes, each used by different groups. As time went on, however, only one group carrying a particular code succeeded enough to continue evolving, and the others became extinct. The code thus reached its present form.

Further evolution of the code was then restricted as a *frozen accident*, because protein synthesis in its mature form precisely positions each particular amino acid in all the various long-chain polypeptides in which it is found. So any change in the genetic code for even a single amino acid would significantly change many different proteins. For example, if the code for phenylalanine changed to include the cysteine codons UGU and UGC, tRNA molecules carrying phenylalanine would then insert into all polypeptide positions formerly occupied by cysteine. Since these two amino acids differ considerably in structure and function, the effect of such a sudden change would undoubtedly be lethal.

However, before "freezing," the genetic code must itself have evolved to accompany some of the changes taking place in protein synthesis. That is, early proteins were probably much shorter than present proteins, composed of fewer kinds of amino acids, and produced by a much less accurate translation mechanism. These proteins may therefore have had only minimal functional specificity, and those which a single mRNA molecule produced could best be described as statistically alike rather than exactly alike. Thus, there would likely have been room for changes between some codons of early amino acids, and not all the present amino acids would have incorporated into the primeval code(s).

A very early hypothesis on the evolution of the code was that it derived from a prior code that used fewer than three bases per codon. That the first two nucleotide bases in each codon still primarily specify the amino acids using quartets seemed to support this idea. Thus it seemed possible to suggest that an even earlier code may have been *singlet*, with four mononucleotide codons (A, U, G, C) specifying perhaps four kinds of amino acids. This code evolved later into the *doublet* form with 16 dinucleotide codons (AA, AU, AG, ...) specifying a maximum of 15 amino acids and 1 chain termination codon. Only after undergoing this doublet experience did the code finally achieve its present trinucleotide form of 64 codons.

Although the idea is superficially attractive, evolution of the code in this fashion seems extremely doubtful, because each change in codon size would change the meaning of practically all former codons. For example, a mRNA sequence translated by a doublet code, AU CG UU GU AG CG . . ., would produce entirely different numbers and kinds of amino acids when translated by a triplet code, AUC GUU GUA GCG . . . . In the face of a radical transition of this kind an organism could hardly retain the function of most, if not all, of its genetic material.

Researchers therefore believe the genetic code to have been triplet even during its beginnings, or perhaps doublet with single *nucleotide spacers*. Mechanical considerations support this view, since anything less than a triplet codon would probably not provide a stable pairing relationship between a tRNA anticodon and an mRNA codon. At the same time, quadruplet or quintuplet codons are probably too "sticky," given selection for an optimum turnover rate that would allow rapid dissociation between codons and anticodons. Researchers have also suggested that the trinucleotide width of a triplet anticodon provides a minimal space, enabling tRNA molecules to lie close enough together for peptide bonding between their amino acids.

Given a small group of amino acids coded by such triplets, further evolution of the code would have probably proceeded under three selective conditions:

- Nucleotide substitutions caused by errors in replication (mutations) should produce as few amino acid changes as possible.
- The number of different codons per amino acid should generally be proportional to the frequency in which the amino acid occurs in proteins.
- Errors occurring during the mRNA-tRNA translation process should lead to as few drastic protein changes as possible.

In respect to the first condition, selection would tend to expand the number of different codons used by an amino acid so that random base changes would still produce the same amino acid. Increasing the number of different codons for any particular amino acid would also be advantageous in diminishing the number of stop or nonsense codons that do not code for any amino acid at all, thereby ensuring that most random mutations do not terminate protein production. In contrast, there would be a limit on the number of different codons that a single amino acid could use since the coding dictionary must accommodate a variety of amino acids. Unfortunately, the extent to which these selective forces operated in the past is now hard to determine.

In respect to the numbers of codon assignments, a proportional relationship does seem to exist between the relative number of codons possessed by most amino acids and their frequencies in proteins. However, the source for this relationship is still not clear. It may indicate s selective relationship (more còdons are selected for the use of those amino acids that occur more frequently) or an accidental relationship (the overall composition of proteins derives from the frequency of amino acid codons), or perhaps both factors prevailed.

The third level of codon evolution, selection for minimizing translational errors, may have considerably affected codon assignments. For example, the prevailing degeneracy at the third codon position is precisely what we would expect if this position were the one most involved in translational errors. That is, the various quartets (e.g., GUU, GUC, GUA, GUG) and duets (e.g., UUU, UUC) arise from the selective advantage of assigning to the same amino acid codons that could most easily be mistaken for one another. The next most error-prone translational event occurs at the first codon position, and

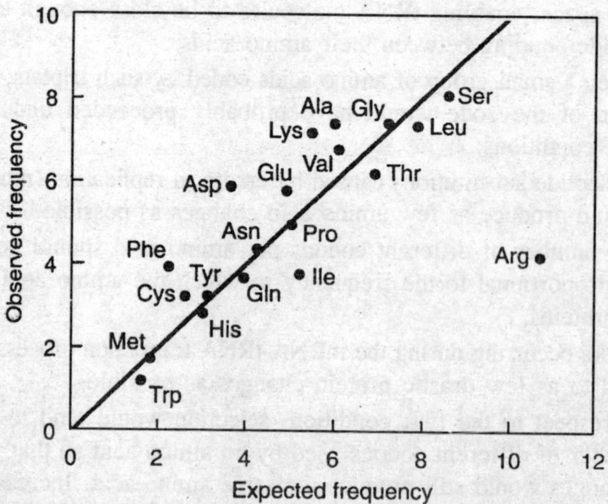

Fig. 6.5. Comparison between observed and expected frequencies of amino acids at 5,492 positions in 53 different vertebrate polypeptides.

the code again shows some degeneracy at this position [e.g., UUA (leucine) → CUA (leucine)], or is so constructed that an error may occasionally enable the substitution of an amino acid with related function [e.g., UUA (leucine) → GUA (valine)]. In general, however, the first codon position is considerably less error prone than the third, apparently because of modifications of the 40th tRNA nucleotide immediately adjacent to this position. These modifications include the addition of methyl or even bulkier groups to nucleotide 40, thereby preventing wobble at the first codon position. For example, one such modified nucleotide is threonyl-6-adenine, which prevents the wobble pairing of U in the adjacent position with G in the first codon position.

Researchers have suggested that the second codon position, the least error prone during translation, may at one time have separated entire classes of amino acids with unique functions. This system would offer a selective advantage by ensuring that amino acids are rarely substituted between classes. A possible remnant of such a grouping may be the assignment of U to the second codon position for leucine, isoleucine, and valine, all of which are *hydrophobic amino acids* that exist mostly in the interior of proteins, and assignment of A to the second codon position for glutamic acid, histidine, aspartic acid, and other *hydrophilic amino acids* that commonly exist on the protein surface.

In summary, it is possible to claim that the triplet genetic code may have initially coded for relatively few amino acids, or, based primarily on the second codon position, may have been able to distinguish only between general classes of amino acids. As time went on, the first codon position also came into use for amino acid positioning because of the modifications that could be made to the immediately adjacent tRNA nucleotide. With the advent of two accurately translated codon positions, the genetic code could then expand its repertoire of amino acids. Thus, until translation accuracy was firmly achieved, some shuffling of codons may have taken place between different amino acids, or entirely new amino acids may have entered the process, using some or all of the codons of older amino acids. The retention of third codon position degeneracy or wobble may have economized on the number of tRNA molecules needed to translate amino acids such as valine, threonine, alanine, and others.

Accordingly, Jukes (1985) has proposed that an early stage in the evolution of the genetic code. Each *codon quartet*, or *family*, of four codons specified perhaps one amino acid, and each amino acid was represented by only one kind of tRNA molecule whose anticodon could pair with all four codons in the family. (This type of code commonly appears in present-day mitochondrial organelles that, because of their small size and limited function, economize in the kinds of both tRNA and proteins they produce.) Such extreme wobbles thus limited this code to no more than 15 or 16 amino acids. One family or partial family of codons terminated translation since there were no amino acid-bearing tRNA molecules whose anticodons could pair with these stop codons. In subsequent evolution, tRNA gene duplications enabled new anticodons to evolve, some of which could now mutate so that new amino acids activated them.

For example, the gene for an early tRNA molecule with anticodon AAU that specifies phenylalanine could form an additional tRNA gene by duplication, which then mutates to produce anticodon AAG. The AAG anticodon becomes restricted to pairing with codons UUU and UUC, whereas the AAU anticodon pairs only with UUA and UUG. Although both tRNAs originally specify phenylalanine, the gene producing the tRNA with anticodon AAU may now mutate so that its tRNA product now binds to leucine rather than to phenylalanine. Thus UUA and UUG codons would now specify leucine. Jukes points out that since such aminoacylation changes have arisen in the genetic codes of mitochondria, which produce relatively few proteins, they could also have occurred in early primitive organisms.

By such means the kinds of tRNA molecules could increase and new amino acids could add to the code. Apparently, when 20 different amino acids had incorporated into the code, these ancestral organisms were producing a large-enough number of proteins so that codon changes necessary to include any further amino acids would lead to widespread protein malfunction and widespread lethality. At that point the code "froze," and all organisms derived from this primitive ancestor shared this common code. The universality of the genetic code, that it is common to all organisms with the exception of a few codons in mitochondria, indicates that only the bearers of this particular code successfully survived the early evolutionary period.

7

PHYLOGENETICS

Living organisms are classified into groups based on observed similarities and differences. A general principle of classification systems is that the more closely related species *b,* the more likely they shared a recent common ancestor. In this way, similarities and differences between organisms can be used to infer *phylogenies* (evolutionary relationships). The branch of science that deals with resolving the evolutionary relationships among organisms is *phylogenetics.*

Phylogenetics can be studied in three ways. In *phenetics*, species are grouped with others they resemble phenotypically and all characters are taken into account. In cladistics, species are grouped only with those that share *derived* characters, that is, characters that were not present in their distant ancestors. The third approach *evolutionary systematics*, incorporates both phenetic and cladistic principles. Cladistics is accepted as the best method available for phylogenetic analysis because it accepts and employs current evolutionary theory, that is, that speciation occurs by bifurcation (*cladogenesis*).

Many different criteria can be used for phylogenetic analysis, including morphological characteristics, biochemical properties and, most recently, the analysis of macromolecular sequences (nucleic acid and protein sequences). Macromolecular sequences are particularly useful for comparison because they provide a large and unbiased data set, which extends across all known organisms, allowing the comparison of both closely related and distantly related taxa. Most importantly, however, the relatedness between sequences can be quantified objectively using sequence alignment algorithms. This is where bioinformatics plays an important role in phylogenetics.

The simple principle behind the phylogenetic analysis of sequences is that the greater the similarity between two sequences, the fewer mutations are required to convert one sequence into the other, and thus the more recently they shared a common ancestor. However, it is important to note that any evolutionary relationships inferred from such analysis sometimes assume a constant rate of mutation and the absence of differential selection in the sequences chosen for comparison. These conditions are seldom met!

Graphs and Trees

The clearest way to visualize the evolutionary relationships among organisms is to use a graph. In mathematics, a *graph* is a simple diagram used to show relationships between entities, such as numbers, objects or places. Entities are represented by *nodes* and relationships between them are shown as *links* or *edges* (connecting lines). These graphs can be used for a variety of purposes. For example, G1 might represent a cyclic chemical compound with atoms as the nodes and chemical bonds as the links, and G2 might represent a street map showing one-way streets. Note that in G2 the links have direction (i.e. they are represented by arrows rather than simple lines). This is a *directed graph* or *digraph*.

G3 represents a special type of graph known as a *tree*. To be classed as a tree, a graph must have *n* nodes, n-1 links and no circuits. A graph contains a circuit if it is possible to move from one node back to itself along a sequence of links where no link is used more than once. G1 and G2 are not trees because each contains a circuit of nodes (A–E–C).

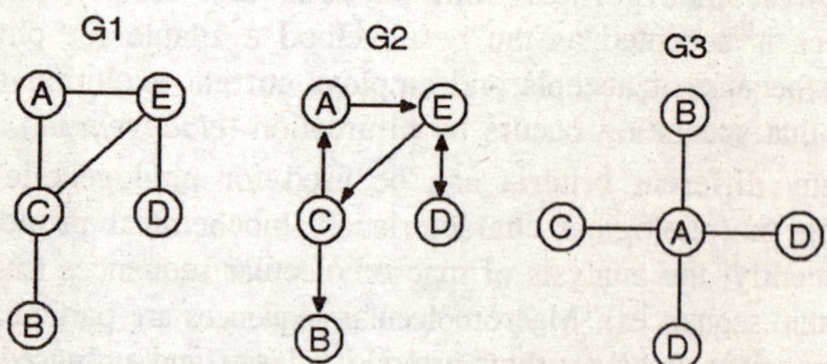

Fig. 7.1. Three examples of simple graphs.

Phylogenetic Trees and Cladograms

Phylogenetic trees (also called *dendrograms*) are used to show evolutionary relationships. The nodes represent different organisms and links are used to show lines of descent. As an example, we consider

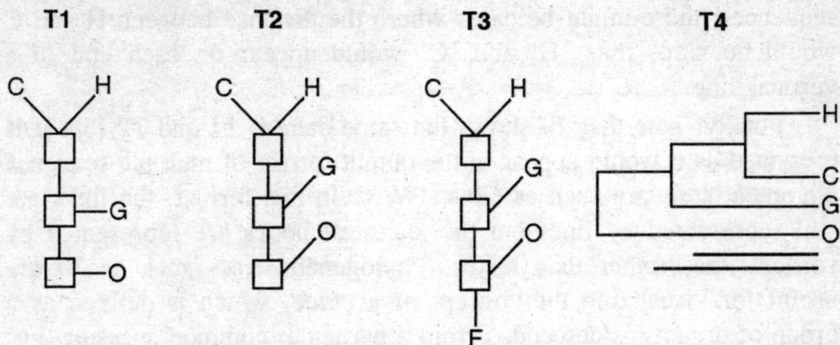

Fig. 7.2. Trees showing the relationship between four species, C, G, H and O (chimpanzee, gorilla, human and orangutan)

the phylogenetic relationship between the entities C (chimpanzee), G (gorilla), H (human) and O (orangutan).

The first point to note about these trees is that there are two types of node. The *ancestral nodes* (represented by boxes) give rise to branches. These may link to other ancestral nodes, or they may link to *terminal nodes* (shown as letters), which are also known as *leaves* or *tips*. Leaves represent known species and mark the end of the evolutionary pathway. Ancestral nodes mayor may not correspond to a known species (e.g. the last common ancestor of humans and chimpanzees is unknown but we can infer its existence).

The second point to note is that *T1* and *T2* are *unrooted trees*, whereas *T3* is a *rooted tree*. *T1* and *T2* are identical except that *T2* is drawn in a conventional style with angled branches to look more like a real tree. These are described as unrooted trees because neither of them shows the position of the last common ancestor of all the species. In *T3,* the position of this ancestor is indicated by the node F.

The third point to note is that each tree is *binary*, that is, no ancestral nodes have more than two branches. Thus, the evolution of species is represented as a series of bifurcations, which fits in with cladistic theory. For this reason, the trees may also be termed *cladograms*.

The fourth point to note is that the length of the branches mayor may not be significant. In *T1, T2* and *T3,* all the branches are of the same length, whereas in *T4* the branches are of different lengths. The lengths of the branches may be used to indicate the actual evolutionary distances between taxa. A cladogram which conveys a sense of evolutionary time using branch lengths may be called a *phylogram*. Note that, if *T4* represented differences between macromolecular sequences, there might be cases where the distance between H and C would be zero, thus 'H' and 'C' would appear on each end of a vertical line.

Finally, note that *T4* shows the same data as *T1* and *T2,* but it is presented as it would appear in the output format of multiple sequence alignment software such as ClustalW/X. In this format, the links are still represented by lines but the ancestral nodes are represented by vertical lines rather than boxes. Phylogenetic trees such as *T4* are useful for visualizing the concept of a *clade*, which is defined as a group of organisms descended from a particular common ancestor (i.e. an ancestor and all its descendants). The groups of organism included within a clade are defined arbitrarily. If, for example, the distance represented by the upper of the two lines beneath *T4* was thought to be significantly close, Hand C would be said to be in the same clade. If, however, the criterion was the length of the lower line, H, C and G would be placed in the same clade. We explore this point in more detail below.

In actual cladogram, the sequence similarities among all the I2-helix integral membrane proteins known in 1996 with redundant sequences removed. In unrooted trees, there is a point referred to as the *point of trichotomy* that corresponds to the place where the ancestor would be located in a rooted tree. It is important to remember that in such a tree, the relative positions of the clades (e.g. PL, PG and PT) are arbitrary.

Classification and Ontologies

For partly historical reasons, biological science abounds with examples of hierarchical or tree-like classifications. The current version of the Linnaean system of classification into species, genera, families, orders, classes and kingdoms was to first to be developed, but the idea has been recruited into the nomenclature of the Enzyme Commission (EC numbers for enzymes) and protein structural classification systems such as SCOP and CATH.

The phylograms suggest there is a method for defining a classifier objectively. That is, by using some criterion or distance, species are represented by organisms that form a clade, genera by those that form a larger clade, families and orders by those that form still larger clades, and so on. The problem that arises is that the criteria are not constant over all living organisms. For example, if the great apes were bacteria they would be regarded as minor variations within the same species, whereas they actually represent different families of mammals. Any attempt to overcome this problem by imposing extra rules leads to inconsistency. For example, one might say chimpanzees and humans are at least distinct species because they cannot interbreed. However, inter-species and even inter-generic crosses are commonplace among ornamental plants and among fish, so this rule cannot be rigorously applied. Among the bacteria (eubacteria), lateral gene transfer occurs widely, but does this mean that all bacteria belong to the same species?

An important point arising from the above is that classifications have to b not only objective and logical but also *useful*. It could be argued that all bacteria logically belong to the same species, but this is not useful when we want to find, the causes of bubonic plague, leprosy and typhoid, or when we want to know which organisms to use in the manufacture of cheese and vinegar. The above exemplifies tile problems of static ontology. *Ontology* is a term used in artificial intelligence systems to describe the relationships between entities within a given area of interest. In a *static ontology*, there is a formal and explicit specification of the relationships between entities, rather like the formal and inflexible classification systems discussed above. The future of classification may well lie in alternative ontologies, which are more flexible. A *dynamic ontology*, for example, allows the relationship between entities to be progressively refined and updated, that is, the rules can be relaxed to fit around the problem rather than forcing descriptions onto the entities themselves.

Building Phylogenetic Trees

Similarity and Distance Tables

Phylogenetic trees can be constructed from either *similarity tables* or *distance tables*, which show the resemblance among organisms for a given set of characters. The characters chosen for comparison may be morphological or biochemical in nature, or may be DNA, RNA or protein sequences. If sequences are used, they are compared initially using multiple sequence alignment tools such as the ClustalW/X programs.

The numbers in the similarity table show the percentage of matches thus the diagonal consists of 100% values. Such data form the basis of *Adansonian analysis* or *numerical taxonomy*. The numbers in the distance table show *percentage differences* or *distances*, thus the diagonal consists of 0% values. Although either table is suitable for phylogenetic tree building using essentially the same methods, macromolecular comparisons are usually recorded as differences, so we should use

(a)	a	b	c	d	e
a	100	65	50	50	50
b	65	100	50	50	50
c	50	50	100	97	65
d	50	50	97	100	65
e	50	50	65	65	100

(b)	a	b	c	d	e
a	0	6	11	11	11
b	6	0	11	11	11
c	11	11	0	2	6
d	11	11	2	0	6
e	11	11	6	6	0

Fig. 7.3. Hypothetical (a) similarity table and (b) distance table for five organisms, a-e.

distance tables in the discussion below. A common measure of difference between macromolecular sequences is 100-S, where S is the percentage of identical monomers when the sequences have been optimally aligned.

Distance Matrix Methods

Some of the most commonly used methods for tree building in phylogenetic analysis involve *agglomerative hierarchical clustering* based on *distance matrices*. The essential basis of this type of algorithm is that the taxa represented in a distance table are transformed into a series of *nested partitions* by merging two taxa together in each step until only on cluster remains.

This process can be formally expressed as follows:

- Begin by creating *n singleton clusters*, that is, clusters containing one taxon from the distance table. For this purpose, a distance table is regarded as an $n \times n$ matrix, with *n* representing the number of taxa being compared.
- Then, determine the differences between each possible pair of clusters. This is formally expressed as a *distance function*, dis(i,j), where *i* and *j* represent any given pair of clusters.
- Next, select the pair of clusters for which dis(i,j) is minimal, that is, select the two most similar clusters.
- Then merge these data into a new cluster (*ij*). This would mean defining a new cluster (*cd*) as the union of *c* and *d*. In the resulting phylogenetic tree, a new ancestral node would be defined with *c* and *d* as branches.
- The number of clusters in the matrix is now reduced by one. Repeat the analysis until there is only one cluster left. This is a *recursive process*, which must be carried out n–1 times.

There are a number of alternative distance matrix algorithms, which differ in the way that the distance between a new (merged) cluster and the remaining taxa in the matrix are calculated for the purpose of recursive searching. The four popular variations of the method are single linkage, complete linkage, average linkage and the centroid method. In the *single linkage* method, the distance between the merged cluster (*ij*) and any candidate taxon *k* is minimized, that is, the smaller of dis(*ik*) and dis(*jk*) is chosen. In the *complete linkage* method, the distance is maximized, that is, the larger of dis(*ik*) and dis(*jk*) is chosen. There are two variations of the *average linkage* method, in which the distance between the merged cluster (*ij*) and candidate taxon *k* is taken as the arithmetic mean of dis(*ik*) and dis(*jk*). In the *unweighted pair group method using arithmetic mean* (UPGMA), the distance is calculated as a simple average because each candidate is weighted equally. In the *weighted pair group method using arithmetic mean* (WPGMA), the clusters are weighted according to their size, that is, so that the candidate taxon *k* is equivalent in weighting to all previous taxa in the cluster. The calculations for these algorithms are therefore slightly more complex than are those for the single and complete linkage methods. Similarly, there are also weighted and unweighted variations of the centroid method, in which the centroid value is used rather than the arithmetic mean. These variations are known as the *unweighted* and *weighted pair group methods using centroid value* (UPGMC, WPGMC).

The UPGMA method discussed above is popular because of its simplicity, but it makes trees with two important assumed properties that are particularly important when trees are made from macromolecular sequence data. First, it is assumed that evolution occurs at the same rate on all tree branches (this is known as the assumption of a *molecular clock*), and second it is assumed that distances in the trees are additive. The additive assumption is that the distance between any two leaves is the sum of distances on edges connecting them. As a consequence of these assumptions, the method can create incorrect trees. For example, two sequences might be very similar not because they have a direct common ancestor, but simply because they are evolving very slowly by comparison with other sequences being analyzed. UPGMA would produce a tree in which they had a direct common ancestor.

Neighbour joining (NJ) is a clustering method related to UPGMA that is able to solve some of the problems discussed above. In particular, it does not make the assumption of additivity. It is also quite fast computationally, and so is almost always a better choice of method, and is used in the ClustalW/X programs to estimate trees from multiple sequence alignments. To start building an NJ tree all taxa are placed in a star, that is, individually joined to a single central node or hub through *n* spokes. From this star, the two taxa with the greatest similarity are chosen and are connected to a new internal node, these taxa are then known as *neighbours*. The process is repeated until the whole star is resolved into a tree. For an unrooted tree there will be *n*-3 internal branches and the process must be repeated *n*-3 times. ClustalW/X also offers bootstrapping to estimate the robustness of the generated trees.

Maximum Parsimony Methods

In distance matrix methods, all possible sequence alignments are carried out to determine the most closely related sequences, and phylogenetic trees are constructed on the basis of these distance measurements. As an alternative *maximum parsimony methods* can be used, in which trees are constructed on the basis of the *minimum number of mutations* required to convert one sequece into another. In proteins, this is achieved by multiple sequence alignment followed by the identification and analysis of corresponding positions in each sequence. For each aligned residue, the minimum number of base substitutions required to convert one amino acid into another is calculated. The final tree is generated by grouping those sequences that can be interconverted with the smallest number of overall changes. This method is very attractive intellectually but, like the maximum likelihood method below, can be expensive in computer time, so NJ is often to be preferred.

Maximum Likelihood Methods

Maximum likelihood methods also involve multiple sequence alignment and, the analysis of changes at each position of the sequence. However, the difference between maximum likelihood and maximum parsimony is that the former incorporates an *expected model* of sequence changes, which weights the probability of any residue being converted into any other. This model can be set by the experimenter. For each possible tree, the likelihood of different sequence changes at each position is calculated, and these values are multiplied provide an overall likelihood for each tree. The most reliable tree is that with the maximum likelihood.

Adding a Root

Some methods automatically generate rooted trees. For instance, the clustering methods described above place the root between the final two clusters to be joined before the algorithm terminates. On the other hand, NJ produces unrooted trees; there are two main ways in which roots can be added to unrooted trees. The easiest is to use an *outgroup*. For instance, if the tree had been generated from a mixture of mammalian and bacterial sequences then it is clear that the root should lie between these two groups so that the first divergence in the tree is to divide bacteria and mammals. In this case one of the distantly related groups (e.g. the bacterial sequences) is known as the *outgroup*. In the case where there is no obvious outgroup in the sequences, a root can be added half way between two most distantly related sequences, essentially assuming a molecular clock.

Limitations of Phylogenetic Algorithms

All clustering methods suffer from three major limitations: incorrect sequence alignments, failure to account for variation of evolution rates at different sites within a sequence, and failure to account for sequences evolving at different rates in different taxa. These limitations often generate incorrect trees through a process known as *long branch attraction*, in which rapidly evolving sequences are grouped even if their relationship is very distant.

The problems can be addressed in several ways. The first is to make sure that sequence alignments are robust. If many diverse sequences are included in the analysis and the alignment contains many gaps, this could be a source of error. It is much better to eliminate outliers before commencing the analysis because each clustering step is definitive and cannot be undone later. Caution should be exercised in particular if a newly built tree disagrees with others generated through the analysis of different genes or proteins.

Improvements and modifications to the clustering algorithms have also made tree building much more accurate. For example, the *Farris transformed distance method* is a useful preliminary to UPGMA and the *Fitch-Margolish method* is a robust modification of WPGMA. The *paralinear distances algorithm* (LogDet) addresses the problem of unequal evolution rates and is now incorporated into many phylogenetic software packages.

It is important to understand the limitations of computational power in phylogenetic analysis. Essentially, the objective of any tree-building experiment is to select the correct tree from many incorrect trees. Assuming all other limitations have been overcome, how many trees is it necessary to build to get the correct one? This depends on the number of data points being analyzed. For example, if there are three sequences, there are three possible rooted trees and one possible unrooted tree. If there are five sequences, there are 105 possible rooted trees and 15 possible unrooted trees. If there are seven sequences, there are 10395 possible rooted trees and 954 possible unrooted trees. *Exhaustive tree search* methods, where all possible trees are created and tested, can therefore only be used for up to about 10 sequences. If more sequences need to be compared, alternative methods must be used. For example, *branch-and-bound analysis* ignores families of trees that cannot possibly give a better answer than a tree that already been found. *Heuristic analysis* samples trees randomly and can be used for many sequences, but the best tree may be missed.

Phylogenetic Software

A large number of software packages are available, some free over the Internet, for the phylogenetic analysis of macromolecular sequences. Some popular programs, such as PAUP (*phylogenetic analysis using parsimony*) and PHYLIP (*phylogenetic inference package*) are versatile and allow distance matrix, parsimony and maximum likelihood method analysis to be carried out. Such packages are frequently updated with the most recent modifications and corrections to the phylogenetic algorithms. Other packages, such as *MacClade* are useful for tree manipulation. A comprehensive resource for phylogenetic software can be found at the following URL: http://evolution.genetics.washington.edul/phylip/software.htrnl

How Reliable are Phylogenetic Trees?

There is no guaranteed way to verify that a phylogenetic tree represents the true path of evolutionary change. However, there are ways in which to test the reliability of phylogenetic predictions. First, if different methods of tree construction give the same result, this is good evidence that the tree is reliable. Second, the data can be resampled to test their statistical significance. In a technique called *bootstrapping*, data are randomly sampled from any position within a multiple sequence alignment, and are built into new artificial alignments, which are then tested by tree building. Since the sampling is random, some positions may be sampled more than once and others not at all. Ideally, the trees built by boot-strapping should always match the original tree, and this would be defined as '100% bootstrap support'. In reality, bootstrap support of 70% or more for any given branch of a tree is taken to provide 95% confidence that the branch is correct. *Jack-knifing* is a similar process in which about 50% of the original data are resampled and used to make a new matrix, from which phylogenetic relationships are reconstructed.

EVOLUTION OF MACROMOLECULAR SEQUENCES

Molecular Phylogeny

The DNA of organisms in different lines of descent accumulates mutations over evolutionary time leading to divergence in *macromolecular sequences* (DNA, RNA and protein sequences). Phylogenetic trees based on differences among macromolecular sequences are known as *molecular phylogenies*. Generally, the greater the divergence between two sequences the more ancient their *last common ancestor* (LCA), so evolutionary trees can be reconstructed on this principle. Molecular phylogenies are very informative compared to those based on traditional anatomical or morphological characters. This is because they are' wider in scope (e.g. it is possible to compare flowering plants and mammals using protein sequences, but not using morphological characters). There are also many different sequences to choose from (Le. a wide range of characters), and data handling is consistent and objective.

Different macromolecular sequences evolve at different rates, even sequences in different regions of the same molecule. This is generally due to variation in *selective constraints*. Residues in an RNA or protein that have a critical structural or functional role in the molecule can accommodate mutations less easily than those in other regions. The rate at which a particular sequence evolves therefore depends largely on the proportion of residues whose substitution would adversely affect normal structure and function. As discussed below, this allows both closely related and distantly related organisms to be studied using the same methods.

Choice of Macromolecular Sequences

The choice of macromolecular sequence for evolutionary studies is dependent on the biological diversity under investigation. The problem is illustrated, which shows two examples of phylogenetic trees each representing three groups of organisms. The trees are qualitative in that the distances shown are not proportionally correct, but the relative relationships between these groups are in accord with current knowledge. The main difference between the trees is the implied time span from the root to the leaves.

In each case, different problems have to be addressed. For the primates, the challenge is to find molecules that evolve quickly enough and therefore differ sufficiently among these closely related species. Conversely, if we wish to study the evolutionary relationship between

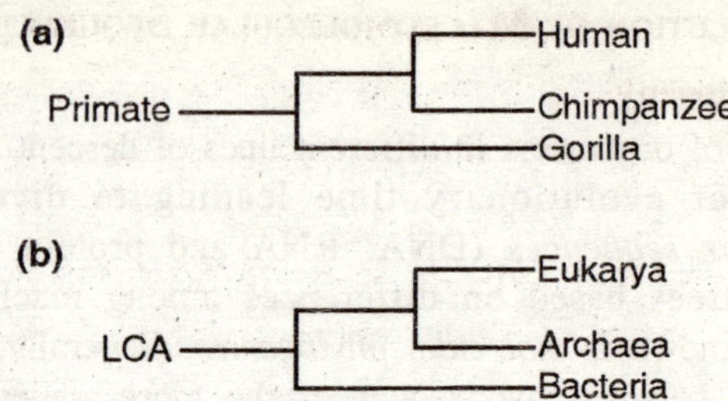

Fig. 7.4. Rooted trees for (a) three great apes with an unspecified primate ancestor, and (b) the three major forms of life on this planet.

the three groups of life highly conserved macromolecules that tolerate few mutations are required.

A useful macromolecular sequence for the study of primates is *mitochondrial DNA* (*mtDNA*). As a consequence of respiratory metabolism, there is a higher concentration of active oxygen species (such as superoxide and the hydroxyl radical) in the mitochondria than in the nucleus and consequently a higher chance of oxidative chemical lesions in mitochondrial DNA. Furthermore, the mitochondrial DNA polymerase is more error-prone than the nuclear enzyme. Therefore, mtDNA evolves more quickly than nuclear DNA due to an increased *intrinsic mutation rate*. There is a short noncoding region in primate mtDNA where selective constraints are low, since point mutations tend not to affect mitochondrial function. This particular sequence evolves at a suitable rate to study primate phylogeny. The tree is consistent with the alignment and clustering of this region and, to a lesser but significant extent, with such analyses of coding genes in mtDNA.

For the study of more divergent taxa, we need a ubiquitous molecule that is highly conserved in all living organisms (including animals, plants, fungi, bacteria, extremophiles and parasites). Such a molecule would be functionally constrained at almost all residues. It would therefore have a low tolerance for mutations and would evolve very slowly. *Ribosomal RNA* (*rRNA*) is such a molecule. The abundant secondary structure of rRNA insures that the rate of evolutionary change is slow, since compensating base changes are required in double helical regions. The tree is consistent with the alignment and clustering of this molecule, and the conclusions are compatible with those of other macromolecular studies, as summarized below:

- The division between the Archaea (archaebatteria) and the Eukarya (eukaryotes) is less deep than that between their common node and the Bacteria (eubacteria). Note also that rRNA trees have proven useful not only for such evolutionary studies; they also provide an objective and useful tool in the difficult problem of eubacterial taxonomy.

- Mitochondria and chloroplasts are derived from prokaryotes. The relationship between the chloroplasts from certain plants and individual cyanobacteria (blue green bacteria) is particularly close.
- Ribosomal RNA sequences are not going to help us to define the LCA of all life. The search for this organism (which is unlikely to occur in the fossil record) will rely on protein sequences and reconstructions of ancient metabolism.

Having commented on good examples of molecules suited to phylogenetic analysis, it is important that the reader understands that there are some molecules that are not well suited. One example is the confusion of orthologs (homologous genes with identical functions in different organisms) and paralogs (homologous genes with different but perhaps related functions). Each of two trees generated from haemoglobin sequences taken from human, chimpanzee and horse. The left-hand tree contains both α and β-globins chains from all species and is easily understood. For each chain, the phylogeny follows the

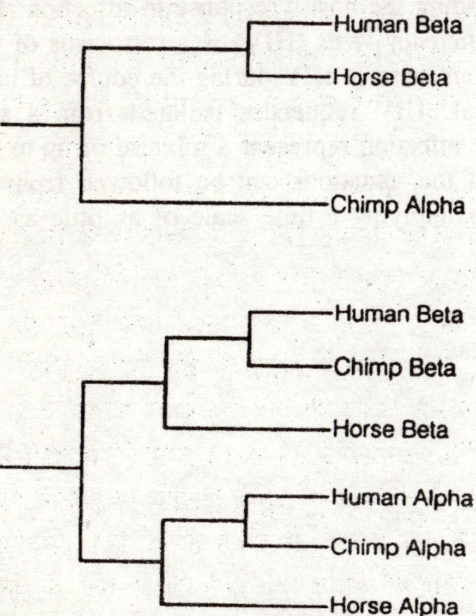

Fig. 7.5. Two trees generated from hemoglobin sequences from human, chimpanzee and horse.

accepted organism phylogeny, with chimpanzee more closely related to human than horse. The fact that the α globins are more closely related to each other than they are to the β globins is evidence that these genes were created by a duplication that happened before the divergence of human, chimpanzee and horse as separate species. The different sequences in the tree have thus been created by two processes, speciation and duplication, and in this case interpretation is easy. The right hand tree is created from the chimpanzee α chain and the β chain of the two other organisms. This does not agree with the accepted organism phylogeny. In fact, if we did not know about the α and β globins chains and had mistakenly thought that the three sequences were orthologs, we might have been tempted to deduce that the horse was our closer relative! The haemoglobin chains are well understood, but other gene families are often not. Very confusing trees can result from sequences that are mixtures of orthologs and paralogs.

Rapidly Evolving Macromolecular Sequences

The phylogenetic trees extend over millions or billions of years. However, some genetic systems evolve at a much faster rate. RNA viruses are genetically unstable because their replication mechanism is highly error-prone. Certain RNA viruses have adopted rapid evolution as a means of evading the host's response to infection. For example, *human immunodeficiency virus* (HIV) is a retrovirus of the lentivirus subgroup. Mutations arise in HIV during the course of its infection of a single individual. HIV sequences isolated from a single patient following a single infection represent a mixture of up to 30 genotypes and the course of the mutations can be followed from phylogenetic trees, in this case implying a time scale of as little as 10 years.

8

GENES IN POPULATION

All evolutionary changes start with changes within populations. The study of genetic changes that occur in populations belongs to the domain of population genetics. This chapter reviews some basic principles of population genetics that are essential for understanding molecular evolution. A basic problem in population genetics is to determine how the frequency of a mutant gene will change with time under the influence of various evolutionary forces. Another basic problem is how genetic variability is maintained in natural populations. In addition, from the long-term point of view, it is important to determine the probability that a new mutant will completely replace existing variants in the population and to estimate how fast the replacement will take place. Unlike morphological changes, many molecular change are likely to have only a small effect on the phenotype of an organism, so the frequencies of molecular variants are subject to strong chance effects. Therefore, chance elements should b taken into account when dealing with molecular evolution. The role of chance effects in evolution is controversial, however, and will be discussed in this chapter.

CHANGES IN ALLELE FREQUENCIES

The chromosomal or genomic location of a gene is called a *locus*, and alternative forms of the gene at a given locus are called *alleles*. In a population, more than one allele may be present at a locus, and their relative proportions are referred to as the *allele frequencies* or *gene frequencies*. For example, assume that there are two alleles with n_1 and n_2 copies at a certain locus, in a haploid population of size N. Then, their allele frequencies are equal to n_1/N and n_2/N, respectively. Note that $n_1+n_2=N$, and $n_1/N+n_2/N=1$.

Evolution is a process of change in the genetic makeup of populations, with the most basic component being change in allele frequencies with time. In fact, from the evolutionary point of view, a new mutation becomes significant only if its frequency increases with time and ultimately reaches 1; the mutant gene is ten said to have become *fixed* in the population. Without increasing its frequency, a mutation will have but a passing effect on the evolutionary history of the species; the only exception is that it is maintained in the population for a long time by balancing selection. For a mutant allele to increase in frequency, factors other than mutation must come into play. These factors include natural selection, random genetic drift, recombination, and migration.

To understand the process of evolution, we must study how the above factors govern the changes of allele frequencies. In this book, we discuss only natural selection and random genetic drift. In classical evolutionary studies involving morphological traits, natural selection has been considered as the major driving force of evolution. In contrast, random genetic drift is thought to have played an important role in evolution at molecular level.

There are two mathematical approaches to studying genetic changes in populations: deterministic and stochastic. The *deterministic model* is simpler. It assumes that changes in the frequencies of alleles in a population from generation to generation occur in a unique manner and can be unambiguously predicted from knowledge of initial conditions. Strictly speaking, this approach applies only when two conditions are met: (1) the population is infinite in size and (2) the environment either remains constant with time or changes according to deterministic rules. These conditions are obviously never met in nature, and therefore a purely deterministic approach may not be sufficient to describe the temporal changes in allele frequencies in populations. Random or unpredictable fluctuations in allele frequencies must also be taken into account.

Dealing with random fluctuations requires a different mathematical approach. *Stochastic models* assume that changes in allele frequencies occur in a probabilistic manner, that is, from knowledge of the conditions in one generation we cannot predict unambiguously the allele frequencies in the next generation, but can only determine the probabilities with which certain allele frequencies will be attained. Obviously, stochastic models are preferable over deterministic ones, since they are based on more realistic assumptions. However, deterministic models are much easier to treat mathematically and, under certain circumstances, they yield sufficiently accurate approximations. In the following, we shall deal with natural selection in a deterministic fashion.

Natural Selection

Natural selection is defined as the differential reproduction of genetically distinct individuals or genotypes within a population. Differential reproduction is caused by differences among individuals in such factors as mortality, fertility, fecundity, mating success, and the viability of offspring. Natural selection is predicated on the availability of genetic variation among individuals in characters related to reproduction. It cannot occur in a population that consists of individuals that do not differ from one another in such traits. Selection leads to changes in allele frequencies over time. However, a mere change in allele frequencies from generation to generation does not necessarily indicate that natural selection is at work. Other processes, such as random genetic drift, can bring about temporal changes in allele frequencies as well.

The *fitness* of a genotype, commonly denoted as w, is a measure of the individual's ability to survive and reproduce. Since the size of a population is usually constrained by the carrying capacity of the environment in which the population resides, the evolutionary success of an individual is determined not by its *absolute fitness*, but by its *relative fitness* in comparison with the other genotypes in the population. In nature, the fitness of a genotype is not expected to remain constant for all generations and under all environmental circumstances. However, by assigning a constant value of fitness to each genotype, we are able to formulate simple theories that are useful for understanding the dynamics of change in the genetic structure of populations brought about by natural selection. In the simplest class of models, we assume that the fitness of an organism is determined solely by its genetic makeup. We also assume that all loci contribute independently to the fitness of the individual, so that each locus can be treated separately.

Most new mutants arising in a population reduce the fitness of their carriers. Such mutations will be selected against and most will be eventually removed from the population. This type of selection is called *negative* or *purifying selection*. Occasionally, a new mutation may be as fit as the best allele in the population. Such a mutation is selectively *neutral*, and its fate is determined not by selection but by chance events. Rarely, a mutant that confers a selective advantage on its carriers may arise. Such a mutation will be subjected to *positive selection*. If the new mutant is advantageous only in heterozygotes but not in homozygotes, the resulting selective regime will be *overdominant selection*.

In the following, we shall consider the case of one locus with two alleles, A_1 and A_2. Each allele can be assigned as intrinsic fitness value; it can be advantageous, deleterious, or neutral. However, this assignment is only applicable to haploid organisms. In diploid organisms the fitness is determined by the interaction between the two alleles at the locus. With two alleles at a locus, there are three possible diploid genotypes: A_1A_1, A_1A_2, and A_2A_2, and their fitnesses can be denoted by w_{11}, w_{12}, and w_{22}, respectively.

Given that the frequency of allele A_1 in a population is p, and the frequency of the complementary allele, A_2, is q=1–p, we can show that under random mating, the frequencies of A_1A_1, A_1A_2 and A_2A_2 are p^2, 2pq, and q^2, respectively. A population is which such genotypic ratios are maintained is said to be at *Hardy-Weinberg equilibrium*.

In the general case, the three genotypes are assigned the following fitness values and initial frequencies:

Genotype:	A_1A_1	A_1A_2	A_2A_2
Fitness:	w_{11}	w_{12}	w_{22}
Frequency:	p^2	2pq	q^2

Let us now consider the dynamics of allele frequency changes following selection. Given the frequencies of the three genotypes and their fitnesses as above, the relative frequencies of the three genotypes in the next generation will become p^2w_{11}, $2pqw_{12}$, and q^2w_{22}, for A_1A_1, and A_1A_2 and A_2A_2, respectively. Therefore, the frequency of allele A_2 in the next generation will become:

$$q' = \frac{pqw_{12} + q^2w_{22}}{p^2w_{11} + 2pqw_{12} + q^2w_{22}} \quad \text{...(1)}$$

The extent of change in the frequency of allele A_2 per generation is denoted as Δq=q'–1. We can show that:

$$\Delta q = \frac{pq[p(w_{12} - w_{11}) + q(w_{22} - w_{12})]}{p^2w_{11} + 2pqw_{12} + q^2w_{22}} \quad \text{...(2)}$$

In the following, we shall assume that A_1 is the original or "old" allele in the population. We shall then consider the dynamics of change in allele frequencies following the appearance of a new allele, A_2.

For mathematical convenience, we shall assign a fitness value of 1 to the A_1A_2 genotype. The fitness of the newly created genotypes, A_1A_2 and A_2A_2, will depend on the mode of interaction between A_1 and A_2. For example, if A_2 is completely dominant to A_1, then w_{11}, w_{12} and w_{22} can be written as 1,1+s, and 1+s, respectively, where *s* is the difference between the fitness of an A_2-carrying genotype and the fitness of A_1A_1. A positive value of *s* means an increase in fitness in comparison with A_1A_1, while a negative value means a decrease in fitness. If A_2 is completely recessive, the fitnesses of the three genotypes become 1, 1, and 1+s, respectively.

Two common modes of interaction will be considered: (1) codominance, or genetic selection, and (2) overdominance. Codominance represents a case of directional selection and is mathematically the simplest mode of interaction, while overdominance represents a type of balancing selection.

Codominance

In the *codominant mode* of selection, or *genetic selection*, the two homozygotes have different fitness values, while the fitness of the heterozygote is the mean of the fitnesses of the two homozygous genotypes. The relative fitness values for the three genotypes can be written as:

Genotype:	A_1A_1	A_1A_2	A_2A_2
Fitness:	1	1+s	1+2s

For Equation (1) we obtain the following change in the frequency of allele A_2 per generation under codominance:

$$\Delta q = \frac{spq}{1+2spq+2sq^2} \qquad \ldots(3)$$

By iteration, this equation can be used to compute the frequency (q) of A_2 at any generation. However, the following approximation leads to a much more convenient formula. If *s* is small, as is usually the case, the denominator of Equation (3) is approximately equal to 1 and the equation becomes approximately

$$\Delta q = spq$$

This difference equation can be approximated by the following differential equation

$$\frac{dq}{dt} = spq = sq(1-q) \qquad \ldots(4)$$

The solution of this equation is given by

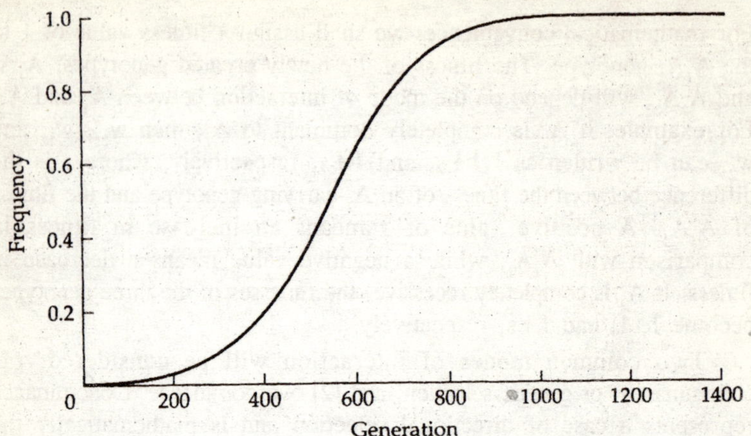

Fig. 8.1. Frequency of a codominant advantageous allele with s=0.01 following its appearance as a result of mutation in generation 0.

$$q_t = \frac{1}{\left(\frac{1-q_0}{q_0}\right)e^{-st}} \quad ...(5)$$

where q_0 and q_t are the frequencies of A_2 at time 0 and t, respectively.

Fig. 8.1 illustrates the increase in the frequency of allele A_2 for $s=0.01$. Clearly, the frequency of A_2 always increases with time. For this reason, genic selection is a type of *directional selection*. Note, however, that at low frequencies, selection for a codominant allele is not very efficient (i.e., the change in allele frequencies is slow). The reason is that, at low frequencies of A_2, the proportion of A_2 alleles residing in heterozygotes is large. For example, when the frequency of A_2 is 0.5, 50% of A_2 alleles will be carried by heterozygotes, whereas when the frequency of A_2 is 0.01, 99% of all such alleles reside in heterozygotes. Because heterozygotes, which contain both alleles, have a smaller selective advantage than do A_2A_2 homozygotes (i.e., *s* versus 2s), the overall change in allele frequencies at low values of q is small.

In Equation (5) the frequency q_t is expressed as a function of time t. Alternatively, t can be expressed as a function of the frequency q as follows:

$$t = \frac{1}{s}\ln\frac{q_t(1-q_0)}{q_0(1-q_t)} \quad ...(6)$$

From this equation, one can calculate the number of generations required for the frequency of A_2 to change from one value (q_0) to another (q_t).

Overdominance

In the *overdominant mode of selection*, the heterozygote has the highest fitness. Thus:

Genotype:	A_1A_1	A_1A_2	A_2A_2
Fitness	1	$1+s_1$	$1+s_2$

In this case, $s_1>0$ and $s_1>s_2$. Depending on whether the fitness of A_2A_2 is greater than, equal to, or less than that of A_1A_1, s_2 can be positive, zero, or negative. The change in allele frequencies is expressed as:

$$\Delta q = \frac{-pq(2s_1q - s_2q - s_1)}{1+2s_1pq+s_2q^2} \qquad ...(7)$$

Fig. 8.2 illustrates the changes in the frequency of an allele subject to overdominant selection. In contrast to the codominant selection regime, in which one of the alleles is eventually eliminated from the population, under overdominant selection the population sooner or later will reach an equilibrium in which the two alleles coexist. After equilibrium is reached, on further change in allele frequencies will be

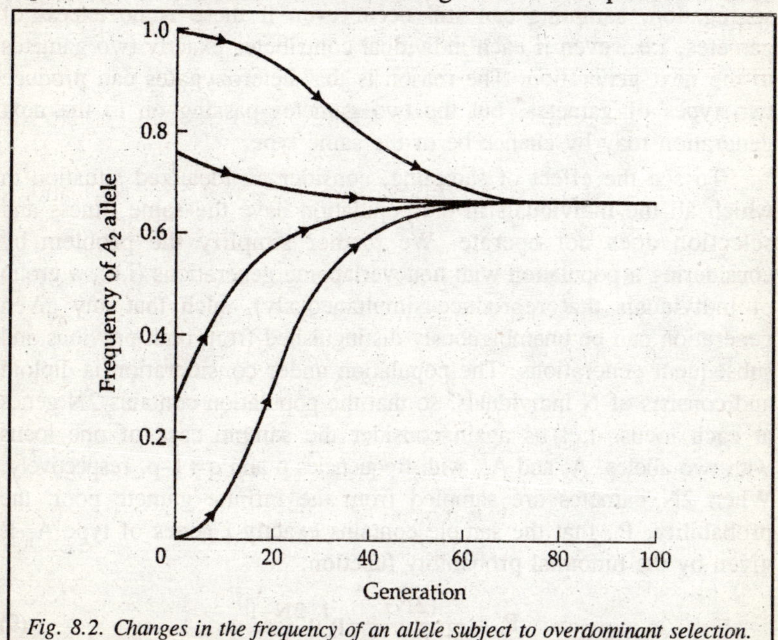

Fig. 8.2. Changes in the frequency of an allele subject to overdominant selection.

observed (i.e., $\Delta q=0$). Thus, overdominant selection belongs to a class of selection regimes called *balancing* or *stabilizing selection.*

The frequency of allele A_2 at equilibrium is obtained by solving Equation (7) for $\Delta q=0$:

$$\hat{q} = \frac{s_1}{2s_1 - s_2} \qquad ...(8)$$

When $s_2=0$ (i.e., both homozygotes have identical fitness values), the equilibrium frequencies of both alleles will be 50%.

Random Genetic Drift

As noted above, natural selection is not the only factor that can cause changes in allele frequency. Allele frequency changes can also occur by chance, though in this case the changes are not directional but random. An important factor in producing random fluctuations in allele frequencies is the random sampling of gametes in the process of reproduction. Sampling occurs because, in the vast majority of cases in nature, the number of gametes available in any generation is much larger than the number of adult individuals produced in the next generation. In other words, only a minute fraction of gametes succeed in developing into adults. In a diploid population under Mendelian segregation, sampling can still occur even if there is no excess of gametes, i.e., even if each individual contributes exactly two gametes to the next generation. The reason is that heterozygotes can produce two types of gametes, but the two gametes passing on to the next generation may by chance be of the same type.

To see the effect of sampling, consider an idealized situation in which all the individuals in the population have the same fitness and selection does not operate. We further simplify the problem by considering a population with nonoverlapping generations (i.e., a group of individuals that reproduce simultaneously), such that any given generation can be unambiguously distinguished from both previous and subsequent generations. The population under consideration is diploid and consists of N individuals, so that the population contains 2N genes at each locus. Let us again consider the sample case of one locus with two alleles, A_1 and A_2, with frequencies p and $q=1-p$, respectively. When 2N gametes are sampled from the infinite gamete pool, the probability, P_i, that the sample contains exactly *i* genes of type A_1 is given by the binomial probability function:

$$P_i = \frac{(2N)}{i!(2N-i)!} p^i q^{2N-i} \qquad ...(9)$$

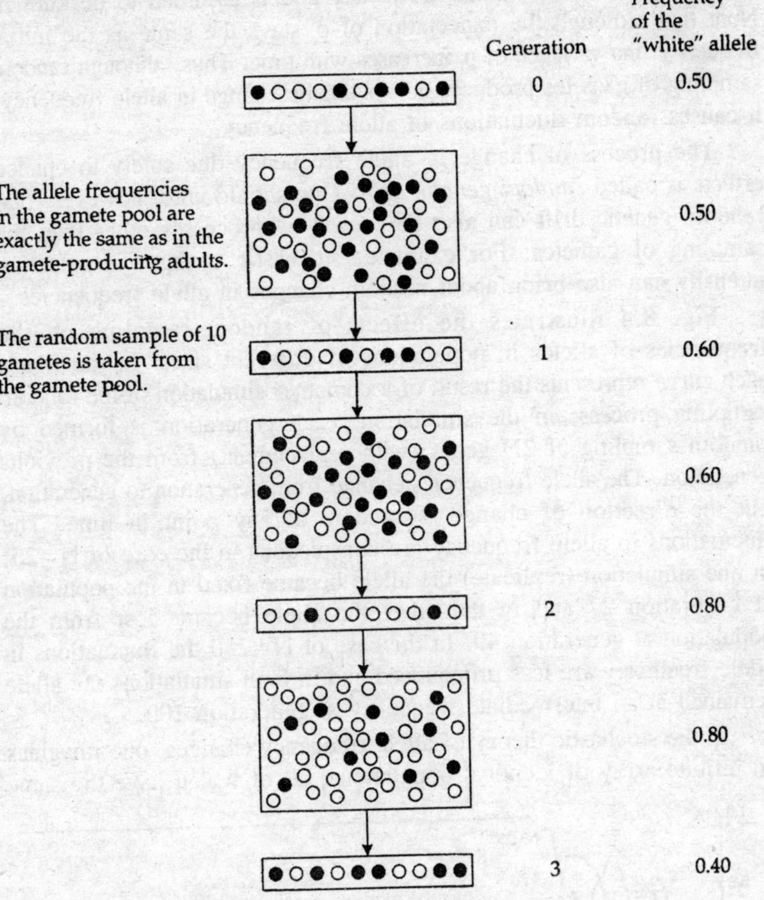

Fig. 8.3. Random sampling of gametes.

Since P_i is always greater than 0 for populations in which $0<p<1$, the allele frequencies may change from generation to generation without the aid of selection.

The frequency of the A_1 at generation *t*, denoted by p_t, is a random variable. The mean and variance of p_t are given by

$$E(p_t) = p_0 \quad \text{...(10)}$$

$$V(p_t) = p_0(1-p_0)\left[1-\left(1-\frac{1}{2N}\right)^t\right] \quad \text{...(11)}$$

$$\approx p_0(1-p_0)(1-e^{-t/(2N)})$$

where p_0 denotes the initial frequency and is assumed to be known. Note that although the expectation of p_t stays the same as the initial frequency, the variance of p_t increases with time. Thus, although random sampling of gametes produces no systematic change in allele frequency, it causes random fluctuations in allele frequency.

The process of change in allele frequency due solely to chance effects is called *random genetic drift*. One should note, however, that random genetic drift can also be caused by processes other than the sampling of gametes. For example, stochastic changes in selection intensity can also bring about random changes in allele frequencies.

Fig. 8.4 illustrates the effects of random sampling on the frequencies of alleles in populations of different sizes. In the figure each curve represents the result of a computer simulation of the random sampling process; in the simulation, each generation is formed by random sampling of 2N genes (with replacement) from the previous generation. The allele frequencies change from generation to generation, but the direction of change is random at any point in time. The fluctuations in allele frequency are conspicuous in the case of N=25; in one simulation (replicate) the allele became fixed in the population at generation 27 and in the other the allele became lost from the population at generation 49. In the case of N=250 the fluctuations in allele frequency are less pronounced and in both simulations the allele remained at an intermediate frequency at generation 100.

In the stochastic theory of allele frequency changes, one imagines an infinite array of identical populations, all of which have the same

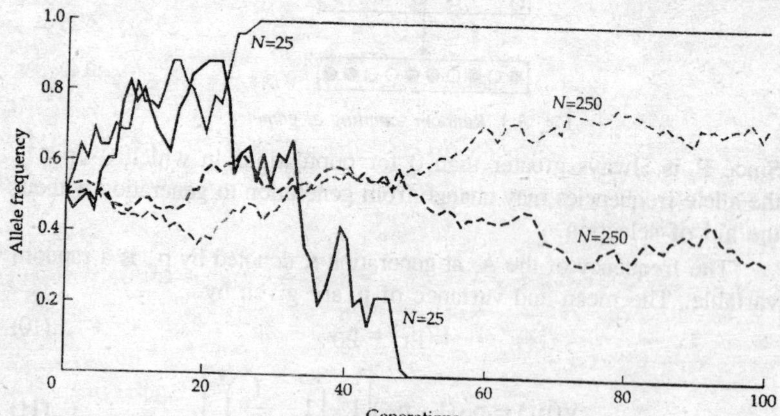

Fig. 8.4. Changes in frequencies of alleles subject to random genetic drift in populations of different sizes (N).

population size, are subject to the same sampling procedure, and start with the same initial frequency. At any time point *t* the allele frequencies in different populations represent the distribution of allele frequencies at time *t*. For example, the probability that the frequency of allele A_1 is in a certain frequency interval is equal to the proportion of populations in which the frequency of allele A_1 is in the interval. In computer simulation, of course, one cannot stimulate an infinite array of populations. However, if the size of the array of simulated populations is large, the frequency distribution obtained will be close to the true distribution. For example, the computer simulation was conducted with an array of 1000 populations. (In practice, it is difficult to simulate 1000 populations at the same time, and so instead one population is simulated at one time and the simulation is repeated 1000 times. Each repeat is known as one replicate.) In the simulation each population consists of N=10 diploid individuals, or 20 genes, and starts with the frequencies 0.3 and 0.7 for the A_1 and A_2 alleles, respectively (a small N is used to save computer time). Since the number of genes is only 20, the exact distribution can be obtained by the Markov chain method. It is noted that the simulated distributions in different generations are very close to the exact distributions.

At generation 1, the distribution of the frequency of A_1 follows a binomial distribution, with p=0.3 and 2N=20. Since the variance is small, i.e., only 0.3(1-0.3)/20=0.01, the distribution is concentrated around the mean (0.3). At generation 5, the distribution becomes flatter. The A_1 allele has become lost in 10.8% of the populations and has become fixed in 0.1% of the populations (these percentages correspond to the heights of the distribution at frequencies 0 and 1, respectively). At generation 20, the distribution becomes fairly uniform for frequencies between 0 and 1, and the probabilities of loss and fixation become 47.4% and 10.4%, respectively. At generation 100, the A_1 allele has become lost in 69.7% of the populations and fixed in 29.7% of the population; that is, it has become either lost or fixed in 99.4% (almost 100%) of the populations. If *t* increases further, the A_1 allele will eventually become lost in 70% of the populations and fixed in 30% of the populations; since the initial frequency of A_1 is 0.3 and since there is no factor favouring either of the two alleles, the probability that A_1 will become fixed in the population is equal to the initial frequency of A_1

The time required for most of the populations to become fixed for the A_1 or A_2 allele can be roughly estimated from Equation (11). In this equation the only term that changes with time is $e^{-t/(2N)}$, which, for N=10, reduces to only 0.007 as *t* increases to 100. Thus, for $t \geq 100$, $V(p_t)$ is close to $p_0(1-p_0)$, which is the maximum value for $V(p_t)$ and is obtained when all populations have become fixed for either A_1 or A_2. This means that when $t \geq 100$, almost all populations should have already become fixed for either A_1 or A_2. This is in agreement with the simulation result.

Effective Population Size

The above mathematical formulation of random genetic drift assumed an idealized population in which all individuals contribute gametes to the next generation with equal probability, generations are nonoverlapping and population size is constant over time. In practice, it is unlikely that all these conditions hold. Furthermore, as will be discussed below, there are often other factors that complicate the mathematical treatment. To simplify the mathematical formulation, Wright (1931) introduced the concept of *effective population size*, Ne, which is the size of an idealized population that would have the same effect of random sampling on gene frequency as that in the actual population. Consider a population with actual size N and assume that the frequency of allele A_1 at the present generation is p. If any of the above conditions is violated, then the variance of the frequency of allele A_1 (p′) in the next generation is expected to be larger than the following binomial variance

$$V(p') = p(1 - p)/(2N) \qquad ...(12)$$

which can be obtained from Equation (11) by putting t=1. The concept of effective population size is to define N_e in such a way that the actual variance is given by

$$V(p') = p(1 - p)/(2N_e) \qquad ...(13)$$

In general, N_e is smaller, sometimes much smaller, than N, the actual population size. Various factors can contribute to this difference. For example, in a population with overlapping generations, at any given time, part of the population will consist of individuals in either their pre- or post-reproductive stage. Due to this developmental stratification, the effective size can be considerably smaller than the census size, N. For example, according to Nei and Imaizumi (1966), in humans N_e is only slightly larger than N/3.

Reduction in the effective population size in comparison to the census size can also occur if the number of males involved in reproduction is different from the number of females. This disparity is especially pronounced in polygamous species, such as social mammals and territorial birds, or in species in which a nonreproducing caste exists (e.g., bees, ants, and termites). If a population consists of N_m males and N_f females, N_e is given by:

$$N_e = \frac{4N_m N_f}{N_m + N_f} \qquad ...(14)$$

Note that, unless the number of females equals the number of males, N_e will always be smaller than N. As an extreme example, let us assume that in a population of size N, all the females (N/2) and only one male take part in the reproductive process. By using Equation (24), we see that $N_e = 2N/(1 + N/2)$. If N is considerably larger than 1, N_e becomes 4, regardless of the census population size.

The effective population size can also be much reduced due to long-term variations in the population size, which in turn are caused by such factors as environmental catastrophes, cyclical modes or reproduction, and local extinction and recolonization events. For example, the *long-term effective population size* in a species for a period of *n* generations is given by:

$$N_e = n/(1/N_1 + 1/N_2 + \ldots + 1/N_n) \qquad ...15)$$

where N_i is the population size of the *i*th generation. In other words, N_e equals the harmonic mean of the N_i values, and consequently it is closer to the smallest value of N_i than to the largest one. Similarly, if a population goes through a bottleneck, the effective population size is greatly reduced.

GENE SUBSTITUTION

Gene substitution is defined as the process whereby a mutant allele completely replaces the predominant or *wild type* allele in a population. In this process, a mutant allele arises in a population, usually as a single copy, and becomes *fixed* after a certain number of generations. The time it takes for a new allele to become fixed is called the *fixation time*. Not all new mutants, however, reach fixation. In fact, the majority of them are lost after a few generations. Thus, we also need to address the issue of *fixation probability* and discuss the factors affecting the chance that a new mutant allele will reach fixation in a population. New mutations arise continuously within populations. Consequently, gene substitutions occur in succession, with one allele replacing another and being itself replaced in time by a new allele. Thus, we can speak of the *rate of gene substitution*, i.e., the number of substitutions or fixations per unit time.

Fixation Probability

The probability that a particular allele will become fixed in a population depends on (1) its initial frequency, (2) its selective advantage or disadvantage, s, and (3) the effective population size, N_e. In the following, we shall consider the case of genic selection and assume that the relative fitness of the three genotypes A_1A_1, A_1A_2, and A_2A_2 are 1, 1 + s, and 1 + 2s, respectively.

Kimura (1962) showed that the probability of fixation of A_2 is given by

$$P = (1 - e^{-4N_e sq}) / (1 - e^{4N_e s}) \quad \text{...(16)}$$

where q is the initial frequency of allele A_2. Since $e^{-x} \approx 1 - x$ when x is small, Equation (16) reduces to $P \approx q$ as s approaches 0. Thus, for a neutral allele, the fixation probability equals its frequency in the population. For example, the initial frequency of allele A_1 is 30% and so it will eventually become fixed in 30% of the cases and become lost in 70% of the cases. This is intuitively understandable because in the case of neutral alleles, fixation occurs by random genetic drift, which favours neither allele.

We note that a new mutant arising as a single copy in a diploid population of size N has an initial frequency of 1/(2N). The probability of fixation of an individual mutant allele, P, is thus obtained by replacing q with 1/(2N) in Equation (16). When s ≠ 0,

$$P = [1 - e^{(2N_e s)/N}] / (1 - e^{-4N_e s}) \quad \text{...(17)}$$

For a neutral mutation, Equation (17) becomes

$$P = 1/(2N) \quad \text{...(18)}$$

If the population size is equal to the effective population size, Equation 2.17 reduces to

$$P = (1 - e^{2s}) / (1 - e^{-4Ns}) \quad \text{...(19)}$$

If the absolute value of s is small, we obtain

$$P = 2s / (1 - e^{-4Ns}) \quad \text{...(20)}$$

For positive values of s and large values of N, Equation (20) reduces to

$$P = 2s \quad \text{...(21)}$$

Thus, if an advantageous mutation arises in a large population and its selective advantage over the rest of the alleles is small, say >50%, the probability of its fixation is approximately twice its selective advantage. For example, if a new mutation with $s=0.01$ arises in a population, the probability of its eventual fixation 2%.

Let us now consider a numerical example. A new mutant arises in a population of 1000 individuals. For simplicity, we assume that $N=N_e$. The probability that this allele will become fixed in the population is $1/(2N)=0.05\%$ if it is neutral, 2% if it confers a selective advantage of 0.01, and 0.004% if it has a selective disadvantage of 0.001 (the last two cases are computed from Equation 19). These results are quite noteworthy, for they mean that an advantageous mutation does not always become fixed in the population. In fact, 98% of all the mutations with a selective advantage of $s=0.01$ will be lost by chance. On the other hand, even slightly deleterious mutations have a finite probability of becoming fixed in a population, albeit a small one. The mere fact that a deleterious allele may become fixed in a population illustrates in a powerful way the importance of chance effects in determining the fate of mutations during evolution. If the population size becomes larger, the chance effect, of course, becomes smaller. For instance, in the above example if the population size is $N=N_e=10{,}000$ instead of 1,000, then the fixation probabilities become 0.005%, 2%, and $\approx 10^{-20}$, respectively. While the fixation probability for the advantageous mutation remains approximately the same, that for the deleterious allele has become very small when N_e increases from 1000 to 10,000. Therefore, in a large population, it is almost impossible for a deleterious mutation to become fixed in the population and the chance for a neutral mutation to become fixed in the population is very small; however, see below for the rate of substitution for neutral mutation.

Fixation Time

The time required for the fixation or loss of an allele depends on the frequency of the allele and the size of the population. The mean time to fixation or loss becomes shorter as the frequency of the allele approaches 1 or 0, respectively.

In terms of evolution, we are more interested in the chance of fixation of new mutations. Thus, in the following we shall deal with the mean fixation time of those mutants that will eventually become fixed in the population. This variable is called the *conditional fixation time*. In the case of a new mutation $[q=1/(2N)]$, the mean conditional fixation time, $\bar{t}$, was calculated by Kimura and Ohta (1969). For a neutral mutation, it is approximated by

$$\bar{t} = 4N \text{ generations} \qquad \text{...(22)}$$

and, for a mutation with a selective advantage of s, it is approximated by

$$\bar{t} = (2 / s)\ln(2N) \text{ generations} \qquad \ldots(23)$$

To illustrate the difference between different types of mutation, let us assume a mammalian species with an effective population size of about 10^6 and a mean generation time of 2 years. Under these conditions, it will take a neutral mutation, on average, $4 \times 10^6 \times 2 = 8$ million years to become fixed in the population. In comparison, a mutation, with a selective advantage of 1% will become fixed in the same population in only about 5800 years. Interestingly, the conditional fixation time for a deleterious allele with a selective disadvantage s is exactly the same as that for an advantageous allele with a selective advantage s. This is intuitively understandable given the high probability of loss for a deleterious allele. That is, for a deleterious allele to become fixed in a population, fixation must occur very quickly.

Advantageous mutations are either rapidly lost or rapidly fixed in the population. In contrast, the frequency changes for neutral alleles are slow, and the fixation time is much longer than for advantageous mutants.

(a) Advantageous mutations

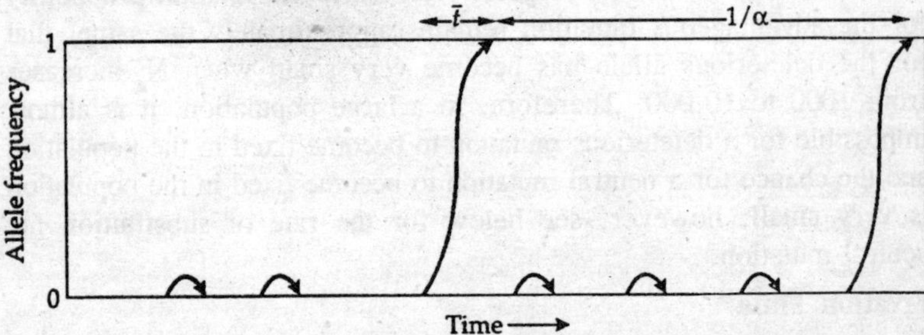

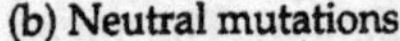
(b) Neutral mutations

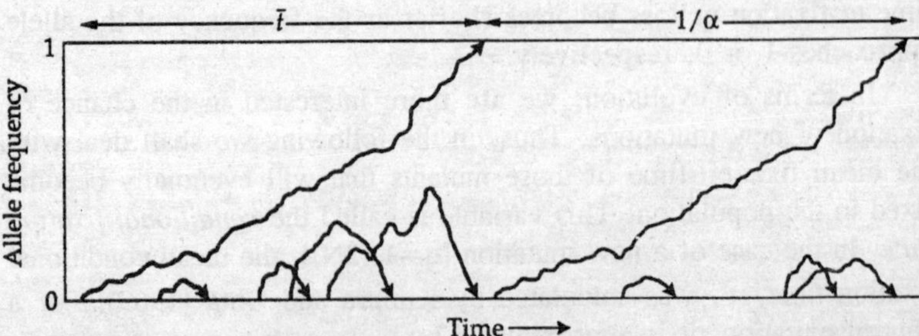

Fig. 8.5. Dynamics of gene substitution for (a) advantageous and (b) neutral mutations.

Rate of Gene Substitution

Let us now consider the *rate of substitution*, defined as the number of mutants reaching fixation per unit time. First, consider neutral mutations. If neutral mutations occur at a rate of *u* per gene per generation, then the number of mutants arising at a locus in a diploid population of size N is 2Nu per generation. Since the probability of fixation for each of these mutations is P = 1/(2N) and since the rate of substitution is K = 2NuP, we have

$$K = u \qquad \text{...(24)}$$

Thus, for neutral mutations, the rate of substitution is equal to the rate of mutation, a remarkably simple result. This result can be intuitively understood by noting that, in a large population, the number of mutations arising every generation is high, but the fixation probability of each mutation is low. In comparison, in a small population, the number of mutations arising every generation is low, but the fixation probability of each mutation is high. As a consequence, the rate of substitution for neutral mutations is independent of population size.

For advantageous mutations, the rate of substitution can also be obtained by multiplying the number of mutations arising every generation (i.e., 2Nu) by the probability of fixation for such alleles as given in Equation (21). For genic selection with $s>0$, we obtain

$$K = 4Nsu \qquad \text{...(25)}$$

In other words, the rate of substitution for the case of genic selection depends on the population size (N) and the selective advantage (s), as well as on the rate of mutation (u).

Extinction of an Allele under Mutation Pressure

At the DNA (protein) level an allele represents a nucleotide (protein) sequence. The allele will become a different allele if a mutation occurs at any nucleotide site (amino acid residue). Once it mutates to a different allele, the chance for the new allele to mutate back to the original allele is very small because the mutation must occur at the mutated site and in a specific direction. To see this, consider a gene consisting of 300 nucleotides, a fairly small gene. Suppose that a mutation occurs at the tenth nucleotide in allele A_1, changing the nucleotide from T to C, so that a new allele, A_2, is created. If a new mutation occurs in A_2, the probability for it to mutate back to A_1 is less than 1/300 because the mutation has a probability of 1/300 to occur at the tenth nucleotide, and even if it occurs at that site, the nucleotide may change to A or G rather than back to T. Therefore, practically every allele in a population is subject to irreversible mutation and will eventually become extinct from the population.

The question then is, "How long will it take for an allele to become extinct from a population under irreversible mutation?" The problem can be formulated as follows. Let A be the allele under consideration and put all other possible allelic forms into one single class and denote it by *a*. Let *u* be the mutation rate per gene per generation. Then, A mutates to *a* irreversibly at the rate of *u* per generation. For simplicity, let us consider only genic selection and assume that the relative fitnesses of genotypes AA, Aa, and aa are 1, 1+s, and 1+2s, respectively. Thus, $s>0$ ($s<0$) means that all new mutations have a selective advantage (disadvantage) of s over A, whereas $s=0$ means that new mutations are all neutral. Let the initial frequency of A be p and effective population size be N_e. Li and Nei (1977) developed a general formula for the mean extinction time of an allele T(p), which is not presented here because it is rather complicated in the presence of selection. For neutral mutation, if $p=1$, the formula becomes

$$T(1) = \sum_{i=1}^{\infty} \frac{4N_e}{i(\theta + i - 1)} \qquad \text{...(26)}$$

where $\theta = 4N_e u$. If $\theta \leq 0.1$, Equation 26 becomes approximately

$$T(1) = 4N_e\left(1 + \frac{1}{\theta}\right) = 4N_e + \frac{1}{u} \qquad \text{...(27)}$$

Table 8.1 shows the mean extinction times for neutral mutations ($S=4N_e s=0$), advantageous mutations ($S>0$), and disadvantageous mutations ($s<0$). The case of $p=1$ is of especial interest because if we consider all alleles currently existing in the population as a single allele and denote it by A, then $p=1$ and T(1) is the expected time until all presently existing alleles become extinct from the population.

First, let us consider neutral mutations. For $\theta \leq 0.1$, the mean extinction time is roughly inversely proportional to $q=4N_e u$ or, in other words, it is roughly proportional to 1/u if N_e is fixed; note that in Table 8.1, time is measured in units of $4N_e$ generations. The dependence of the mean extinction time on u is easily seen from Equation (27), which shows that if $4N_e<1/u$, the mean extinction time is roughly proportional to 1/u. For a given θ, the mean extinction time decreases with the initial frequency p. For example, for $\theta \leq 0.1$,

Table 8.1. Mean extinction time of an allele under mutation pressure

		θ			
s	*p*	*1*	*0.1*	*0.01*	*0.001*
0	1	1.65	10.94	101.0	1001.0
	0.5	1.06	5.80	50.8	500.8
10	1	0.54	1.53	10.4	99.1
	0.5	0.29	0.32	0.39	0.99
100	1	0.10	0.20	1.05	9.61
	0.5	0.05	0.05	0.05	0.05
-5	1	9.93	251	—	—
-50	1	2×10^{18}	7×10^{20}	—	—

the mean extinction time for p=0.5 is only about half of that for p=1. This can be understood by noting that when $\theta \leq 0.1$, mutations occur rarely, so that for p=0.5, roughly 50% of the cases the allele will become lost in a relatively short time (less than $4N_e$ generations) and in the other 50% of the cases the allele will become fixed or nearly fixed in the population, and the mean extinction time is then similar to that for p=1.

Next, let us consider advantageous mutations. Clearly, if all new mutations are advantageous, the mean extinction time for A is greatly reduced. For example, for p=1 and θ=0.01, the mean extinction times (in units of $4N_e$ generations) are 101.0 for S=0, 10.4 for S=10. Note also that for the case of p=0.5 and S=100, the mean extinction time is independent of mutation rate. This is because the selection intensity is strong so that allele A is quickly replaced by the mutant alleles already existing in the population.

Finally, if all mutations are disadvantageous, the mean extinction time is greatly increased. For example, if p=1, θ=1, and S=-50, the mean extinction time is $2 \times 10^{18} \times 4N_e$ generations, which is extremely long. Note that if N_e=12,500, then S=-50 implies s=-0.001, which is fairly small. Thus, if the effective population size is fairly large, disadvantageous mutations have practically no chance of becoming fixed in the population. The assumption that all mutations are disadvantageous implies that the present allele is the optimal allele and that it is subject to purifying selection arising from functional or structural requirements of the sequence. The above result implies that if such requirements are stringent, then the optimal allele can persist in the population for an extremely along time. As we shall see, a protein sequence with stringent structural requirements can indeed persist for millions of years without change.

Genetic Polymorphism

A locus is said to be *polymorphic* if two or more alleles coexist in the population. However, if one of the alleles has a very high frequency, say, 99% or more, then none of the other alleles is likely to be observed in a sample unless the sample size is large. Thus, for practical purposes, a locus is commonly defined as polymorphic if the frequency of the most common allele is less than 99%. This definition is obviously arbitrary and other criteria have been used in the literature.

One of the simplest ways to measure the extent of polymorphism in a population is to compute the proportion of polymorphic loci (P) by dividing the number of polymorphic loci by the total number of loci sampled. For example, if 5 of the 20 loci studied are polymorphic, then P=5/20=25%. This measure, however, is dependent on the number of individuals studied. A more appropriate measure of genetic variability within populations is *gene diversity*. This measure does not depend on an arbitrary delineation of polymorphism, can be computed directly from knowledge of the gene frequencies, and is less affected by sampling effects. Gene diversity at a locus is defined as:

$$h = 1 - \sum_{i=1}^{m} x_i^2 \qquad \ldots(28)$$

where x_i is the frequency of allele i and m is the number of alleles observed at the locus. For any given locus, h is the probability that two alleles chosen at random from the population are different from each other. In a randomly mating population, h is also the *expected heterozygosity*, i,e., the expected frequency of heterozygotes in the population for a locus with allele frequencies x_i, $i=1, \ldots, m$. The average of h values over all the loci studied can be used as a measure of genetic variability in the population.

The introduction of electrophoresis into population genetics in the early 1960s provided a convenient, powerful tool for studying protein polymorphism in natural populations. It was then discovered that natural populations such as humans and *Drosophila* contain a large amount of genetic variability. In fact, early surveys revealed that the proportion of polymorphic loci is about 30% in mammalian species and can be more than 50% in *Drosophila* species. The question was then how such high genetic variabilities are maintained in natural populations.

Table 8.2. Surveys of protein polymorphism in a number of organisms

Species	*Number of populations*	*Number of loci*	*P*	*H*
Homo sapiens	1	71	0.28	0.067
Mus musculus musculus	4	41	0.29	0.091
M. m. brevirostris	1	40	0.30	0.110
M. m. domesticus	2	41	0.20	0.056
Peromyscus polionotus	7 (regions)	32	0.23	0.057
Drosophila pseudoobscura	10	24	0.43	0.128
D. persimilis	1	24	0.25	0.106
D. obscura	3 (regions)	30	0.53	0.108
D. subobscura	6	31	0.47	0.076
D. willistoni	10	20	0.81	0.175
D. melanogaster	1	19	0.42	0.119
D. simulans	1	18	0.61	0.160

Since the 1960s there has been much interest in developing mathematical models for studying genetic variability. A commonly used one is the infinite-allele model, which assumes that every mutation creates a new allele or an allele that is not currently existing in the population. Under this model and the assumption of selective neutrality, the expected heterozygosity at equilibrium is given by

$$h = \frac{4N_eu}{1+4N_eu} \qquad \text{...(29)}$$

where N_e is the effective population size and u is the mutation rate per gene per generation. This simple formula has been heavily used as a reference point for the neutral expectation. There are of course many factors, such as selection and migration, that can cause deviations from this expectation. For example, overdominant selection can greatly increase the heterozygosity, whereas purifying selection can reduce it.

It was questioned whether the infinite-allele model is suitable for analyzing electrophoretic data, because the electrophoretic mobility of a protein may change in a stepwise manner so that recurrent and backward mutation can occur. In other words, a mutation may not create a new allele. To make the model more realistic, Ohta and Kimura (1973) proposed the stepwise-mutation model. In this model the possible allelic states are visualized as the integers on a line, and a mutation causes the state of the allele to move one step either to the right or to the left. Under this model and the assumption of neutrality, the expected heterozygosity at equilibrium is given by

$$h = 1 - 1\sqrt{1 + 8N_e u} \qquad ...(30)$$

The h value under the step-mutation model can be substantially lower than that under the infinite-allele model. This model has been extended to include the possibility of two-step changes. Although electrophoresis is now only occasionally used in polymorphism study, the stepwise model is introduced here, because it will be used when we discuss the population genetics of tandem repeats of DNA.

Neo-Darwinian theory and Natural Mutation Hypothesis

Darwin proposed his theory of evolution by natural selection without knowledge of the sources of variation in populations. After Mendel's laws were rediscovered and genetic variation was shown to be generated by mutation, Darwinism and Mendelism were used as the fremework of what came to be called the synthetic theory of evolution, or neo-Darwinism. According to this theory, mutation is recognized as the ultimate source of genetic variation, but natural selection is given the dominant or "creative" role in shaping the genetic makeup of populations and in the process of gene substitution.

In time, neo-Darwinism became a dogma in evolutionary biology, and selection came to be considered the only force capable of driving the evolutionary process, while other factors such as mutation and random drift were thought of as minor contributors at best. This particular brand of neo-Darwinism was called *selectionism*.

According to the selectionist or neo-Darwinian perception of the evolutionary process, gene substitutions occur as a consequence of selection for advantageous mutations. Polymorphism, on the other hand, is maintained by balancing selection. Thus, neo-Darwinists regard substitution and polymorphism as two separate phenomena driven by different evolutionary forces. Gene substitution is the end result of a positive adaptive process, whereby a new allele takes over future generations of the population if and only if it improves the fitness of the organism, while polymorphism is maintained when the coexistence of two or more alleles at a locus is advantageous for the organism or the population. Neo-Darwinian theories maintain that most genetic polymorphisms in nature are stable.

The 1960s witnessed a revolution in population genetics. The introduction of electrophoresis into population genetics studies soon led to the discovery of the existence of large amounts of genetic variability in natural populations such as human and *Drosophila* populations. The availability of protein sequence data removed the species boundary in population genetics studies and for the first time provided adequate empirical data for examining theories pertaining to the process of gene substitution. In 1968, Kimura postulated that the majority of molecular changes in evolution are due to the random fixation of neutral or nearly neutral mutations; it was also independently proposed by King and Jukes (1969). This hypothesis, now known as the *neutral theory of molecular evolution*, contends that at the molecular level the majority of evolutionary changes and much of the variability within species are caused neither by positive selection of advantageous alleles nor by balancing selection, but by random genetic drift of mutant alleles that are selectively neutral or nearly so. Neutrality, in the sense of the theory, does not imply strict equality in fitness for all alleles. It only means that the fate of alleles is determined largely by random genetic drift. It only means that the fate of alleles is determined largely by random genetic drift. In other words, selection may operate, but its intensity is too weak to offset the influences of chance effects. For this to be true, the absolute value of the selective advantage or disadvantage of an allele must be smaller than $1/(2N_e)$.

According to the neutral theory, the frequency of alleles is determined largely by stochastic rules, and the picture that we obtain at any given time is merely a transient state representing a temporary frame from an ongoing dynamic process. Consequently, polymorphic loci consist of alleles that are either on their way to fixation or on their way to extinction. Viewed from this perspective, all molecular manifestations that are relevant to the evolutionary process should be regarded as the result of a continuous process of a mutational input and a concomitant random extinction or fixation of alleles. Thus, the neutral theory regards substitution and polymorphism as two facets of the same phenomenon. Substitution is a long and gradual process, whereby the frequencies of mutant alleles increase or decrease randomly, until the alleles are ultimately fixed or lost by chance. At any given time, some loci will possess alleles at frequencies that are neither 0% nor 100%. These are the polymorphic loci. According to the neutral theory, most genetic polymorphism in populations is transient in nature.

The essence of the dispute between neutralists and selectionists essentially concerns the distribution of fitness values of mutant alleles. Both schools agree that most new mutations in proteins are deleterious and that these mutations are quickly removed from the population os that they contribute neither to the rate of substitution nor to the amount of polymorphism within populations. The difference concerns the relative proportion of neutral mutations among nondeleterious mutations. While selectionists maintain that very few mutations are selectively neutral, neutralists maintain that most nondeleterious mutations are effectively neutral. Of course, not all selectionists hold the same view of evolution nor do all neutralists. For example, Ohta's (1973, 1974) hypothesis of slightly deleterious mutation emphasizes the importance of slightly deleterious mutations in gene substitution and molecular polymorphism, whereas in Nei's (1987) view such mutations do not play an important role.

The heated controversy over the neutral-mutation hypothesis during the last two decades has had a strong impact on molecular evolution. First, it has led to the general recognition that the effect of random drift cannot be neglected when considering the evolutionary dynamics of molecular changes. Second, the synthesis between molecular biology and population genetics has been greatly strengthened by the introduction of the concept that molecular evolution and genetic polymorphism are but two facets of the same phenomenon. Although the controversy still continues, it is now recognized that any adequate theory of evolution must be consistent with both of these aspects of the evolutionary process at the molecular level.

9

GENE FREQUENCIES

For populations to evolve—that is, to change their gene frequencies—mutation must first introduce the nucleotide differences that give rise to such changes. The mere appearance of new genes (alleles), however, is no guarantee that they will persist or prevail over others. For example, no certainly exists that a newly mutated gene such as *a* (e.g., $A \rightarrow a$) will transmit to the next generation, since its carrier (e.g., *Aa*) may or may not survive and may or may not mate. Even if the *Aa* mutant carrier does mate ($Aa \times AA$), the chance of a transmission declines since a significant proportion (40 percent) of matings in most stable populations produce families with zero surviving offspring (*a* is lost) or only one offspring (*a* has a 50 percent chance of being lost during meiosis). Larger families may also lose the gene since, for example, even when the $Aa \times AA$ family produces two offspring the mutant gene has a 25 percent chance ($.5 \times .5$) of not being transmitted to either of them.

Fisher has calculated that the chances that a newly mutated gene may be eliminated within one generation is more than 33 percent because of its possible random loss in families of such different sizes. By the time 30 generations have passed after its introduction, this probability of elimination has risen to almost 95 percent. To explain the persistence of many mutations and their increase in frequency, we must therefore look somewhere other than the original mutational event.

MUTATION RATES

One factor that can be expected to affect gene frequency is the frequency of mutation. Obviously if gene *A* continually mutates to *a* and the reverse mutation never occurs, the chances improve that *a* will increase in frequency with each generation. Given a long enough period of time and a persistent mutation rate in a population of constant size, *a* can eventually replace *A*. Of course, the mutation rate does not always occur in only one direction. For example, if *u* is the mutation rate of *A* to *a*, the allele *a* may mutate back to *A* with frequency *v*. We can estimate these effects quantitatively by calling the initial frequencies of alleles *A* and *a*, p_0 and q_0, respectively, and noting that a single generation of mutation will produce a frequency of *A* equal to $p_0 + vq_0$ and a frequency of *a* equal to $q_0 + up_0$.

If we now confine our attention to only one of the alleles, *a*, clearly it has gained the fraction up_0 (new *a* alleles) but lost the fraction vq_0 (new *A* alleles). In other words, the change in the frequency of *a*, which we call delta *q* (Δq), can be expressed as $\Delta q = up_0 - vq_0$. Thus, if *p* were relatively large and *q* small, Δq would be large and *q* would increase rapidly; when *q* became larger and *p* became smaller, Δq would diminish. The point at which Δq is zero—that is, the point where there is no further change and *p* and *q* are balanced in relation to their mutation frequencies—we call the *mutational equilibrium* (frequency $a = \hat{q}$ or "q hat"): $\Delta q = 0 = up - vq$, or $up = vq$ at $\hat{q}$. However, since there are only two alleles, *A* and *a*, $p = 1 - q$, which leads to

$$up = vq$$

$$u(1-q) = vq$$

$$u = uq + vq = q(u+v)$$

$$\hat{q} = \frac{u}{u+v}$$

The same procedure applied to the frequency of *A* gives $\hat{p} = v/(u + v)$, so that $\hat{p} / \hat{q} = [v/(u+v)]/[u/(u+v)] = v/u$. Thus when the mutation rates are equal—that is, $u = v$—the *equilibrium gene frequencies* $\hat{p}$ and $\hat{q}$ will be identical. If the mutation rates differ, so will the equilibrium frequencies. For example, if $u = .00005$ and $v = .00003$, the equilibrium frequency $\hat{q}$ equals $5/8 = .625$ and $\hat{p} = 3/8 = .375$. However, the rate at which mutation reaches this equilibrium frequency is usually quite slow, and we can derive it by calculus methods from Δq as

$$(u+v)n + \ln[q_0 - \hat{q}) / (q_n - \hat{q})]$$

where *n* is the number of generations required to reach a frequency q_n when starting with a frequency q_0. For the example just considered, the number of generations necessary for *q* to increase from a frequency of one-eighth to three-eighths is

$$(.00008)_n = \ln\frac{.125-.625}{.0375-.625}$$

$$= \ln 2.00 = .69315$$

$$n = \frac{.69315}{.00008} = 8{,}664 \text{ generations}$$

Thus the approach to equilibrium based on the usually observed mutation rates of 5×10^{-3} or less is very slow, and mutational equilibrium is probably rarely if ever reached, especially since mutation rates are probably not constant. As a rule, therefore, the attainment of mutational equilibrium does not appear to be the sole cause for existing gene frequencies. A more efficient mechanism that can help explain how gene frequencies change is the effect of selection, the "scrutinizing process" that Darwin proposed.

SELECTION

The fact that genotypes can differ in viability and fertility can evidently produce important effects on their frequencies. Obviously if individuals carrying gene *A* are more successful in producing viable and fertile offspring than individuals carrying its allele *a*, then the frequency of the former will tend to increase relative to the latter. The wide variety of mechanisms that affect the reproductive success of a genotype is known collectively as *selection*, and the extent to which a genotype contributes to the offspring of the next generation is commonly known as its *fitness*, *selective value*, or *adaptive value*. That is, we can describe selection as a composite of the forces that limit the reproductive success of a genotype and we can describe fitness as the comparative ability of a genotype to withstand selection. The genetic effect of selection on a particular trait in a population therefore is confined to fitness differences among the different alleles that affect that particular trait. Thus, when the selective process operates, gene frequencies tend to change among generations, unless the population has reached a genetic equilibrium, as described later.

In simplest form, fitness and selection are measured by the number of fertile offspring produced by one genotype compared to those produced by another. For example, if individuals of genotype *A* produce an average of 100 offspring that reach full reproductive maturity while genotype *a* individuals produce only 90 in the same environment, the adaptive value of *a* relative to *A* is reduced by 10 offspring, or the fraction 10/100 =.1. If we designate the adaptive value of a genotype

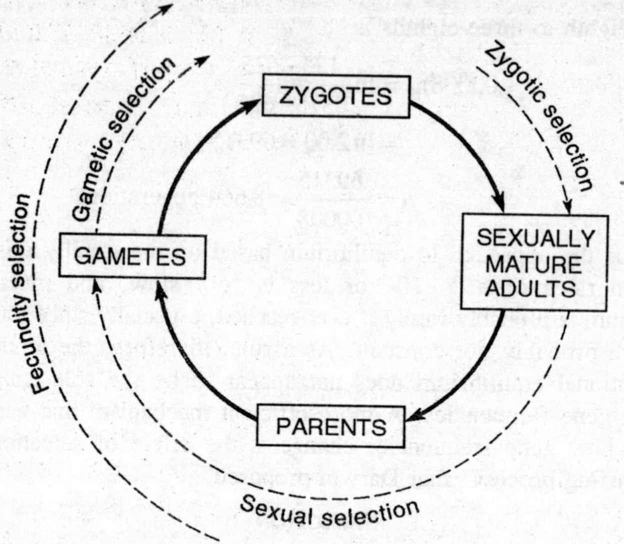

Fig. 9.1. Types of selection acting on the life stages of an organism.

as W and the selective force acting to reduce its adaptive value as s (the *selection coefficient*), then we can say that $W = 1$ and $s = 0$ for A in the preceding example, and $W = .9$ and $s = .1$ for a. The relationship between W and s for a particular genotype is therefore simply $W = 1 - s$;, or $s = 1 - W$.

Selection against a genotype may occur in either the haploid (gametic) or diploid (zygotic) stage or both, depending at which of these stages gene expression influences survival or fertility. In any of these stages, selection may be obvious or subtle, its effects ranging from complete lethality or sterility ($s = 1$) to only slight reductions in adaptive value (e.g., $s = .01$). When selection occurs among haploids, there is of course no difference between dominant and recessive genes, since their carriers phenotypically express both kinds of alleles.

Thus, as we might expect, the effect of selection on haploids is much more rapid and direct than on diploids because *deleterious recessive* alleles cannot be hidden from selection among heterozygotes as they are in diploids. The number of generations necessary to change the frequency of deleterious genes in haploids under a variety of selective conditions. Note that in contrast to diploids haploids completely eliminate a *deleterious lethal* gene ($s = 1$) in one generation, and even lesser selection coefficients result in relatively rapid gene frequency changes.

In most higher animals and plants selection takes place primarily in the diploid or zygotic stage. In diploids however, there are three possible genotypes for a single gene difference (e.g., *AA*, *Aa*, *aa*), so that the effectiveness of selection depends, among other things, on the degree of dominance.

The number of generations necessary to change such deleterious recessive gene frequencies when we project selection over periods of time and for various selection coefficients. Note that the initial change in gene frequency from .99 to .10 is relatively rapid for practically all selection coefficients. Further reductions in gene frequency are considerably slower: to reduce the gene frequency below .01 may lake thousands of generations, even when the selection coefficient is relatively high. The reason for the relative inefficiency of selection against rare recessives is simply that most recessive genes are present in heterozygotes where they are protected from selection: the more rarely a gene appears in a population, the more frequently it occurs in heterozygotes compared to homozygotes.

The selective situation can, of course, reverse so that the dominant allele is selected against and the recessive is favoured. When this occurs, selection will obviously be more effective, because the *deleterious dominant* gene is subject to selection in all genotypes in which it occurs. For example, should a dominant allele become lethal, its frequency falls to zero in a single generation. However, as the selection coefficient against the dominant allele decreases, replacement by the recessive is considerably slower. For the general case, selection against a dominant allele of gene frequency p results in a change of

$$-sp(1-p)^2 / [1-sp(2-p)]$$

Note that if s is small, the denominator is close to 1, and Δp is effectively equal to $-sp(1 - p)^2$. Since $1 - p$ is q and p is $1 - q$, this means that Δp is now $-sq^2(1 - q)$, or Δp is identical to Δq for a deleterious recessive at low selection coefficients. Under these conditions, we may apply the values in reverse order. That is, for a selection coefficient of .10 in favour of a recessive allele, 90,023 generations are necessary to increase its frequency from .0001 to .001, or to reduce the frequency of the dominant allele from .9999 to .9990. Subsequent changes in frequency are more rapid as the favoured recessive homozygotes become more frequent.

When dominance of the advantageous allele is incomplete, heterozygotes will show the effect of a deleterious gene since the heterozygous phenotype is at least partially harmful. If dominance is absent completely and the heterozygote has a phenotype exactly intermediate between the two homozygotes, its selection coefficient will be exactly half that in the deleterious homozygotes. The resultant change in gene frequency in one generation $[-sq\ (1-q)]/[1-2sq]$ is almost identical to that for gametic selection $[-sq(1-q)]/[1-sq]$.

In other words, the absence of dominance uncovers deleterious alleles and makes all of them available for selection, allowing rapid changes in gene frequencies. The effectiveness of selection therefore strongly depends on the degree to which the heterozygote expresses the deleterious gene. Since population geneticistcs believe most recessive genes have some heterozygous expression, selection efficiency for or against them probably falls between the extremes of slow progress for complete dominance and rapid progress for absence of dominance.

Heterozygous Advantage

The examples of selection just considered always go in one direction, toward elimination of the deleterious allele and establishment or *fixation* of the favoured allele. As long as the selection coefficient does not change, an equilibrium between favoured and unfavoured alleles is impossible without new mutations. Various conditions, however, permit the establishment of an equilibrium through which both alleles

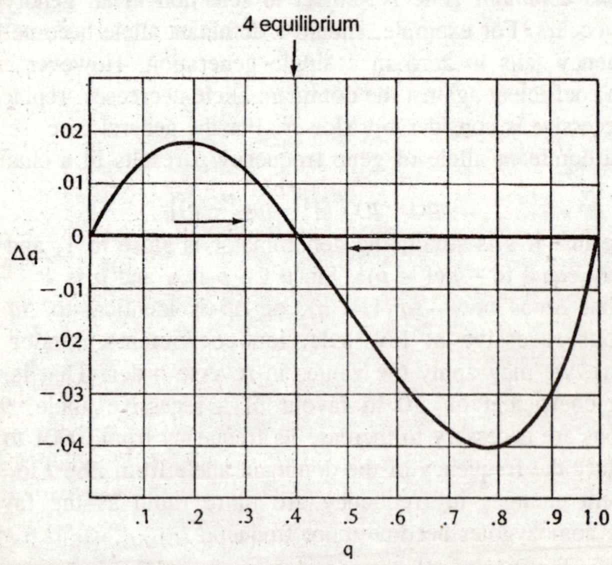

Fig. 9.2. Change in frequency (Δq) of allele a when the genotypic adaptive values are AA = .8, Aa = 1.0, aa = .7, and population size is infinite.

may remain indefinitely within the population. One such condition, *overdominance*, occurs when the heterozygote has superior reproductive fitness to both homozygotes.

In general, if the heterozygote *Aa* has an adaptive value of 1.00 while the fitnesses of the homozygotes *AA* and *aa* reduce by the selective coefficients s and t, respectively, the change in frequency of a in a single generation. When Δq is zero, equilibrium has been reached and gene frequency will not change further. Note that three possible conditions will cause the numerator $[pq(ps - qt)]$ to be equal to zero and therefore Δq to equal zero. Under the first two conditions, when either p or q are zero, both alleles will not be present in the population at the same time, and balance, or equilibrium, will be absent. The third condition occurs when $ps = qt$, so that the numerator of Δq is $pq(0) = 0$. When this happens, the following relationships can be derived:

$$ps = qt$$

Add qs to both sides of preceding equation	Add pt to both sides of preceding equation
$ps + qs = qt + qs$	$ps + pt = qt + pt$
$s(p + q) = q(s + t)$	$p(s + t) = t(p + q)$

(Now, since $p + q = 1$)

$$q = \frac{s}{s+1} \qquad\qquad p = \frac{t}{s+1}$$

It is easy to see that if s and t are constant values, both p and q will reach a stable equilibrium: if q departs from the equilibrium value, selection pressure will force it back. That is, if Δq is positive, the gene frequency q increases, but if Δq is negative, q decreases, the negative or positive sign of Δq depending on whether q is above or below its equilibrium value. For example, when $s = .2$ and $t = .3$, the equilibrium value for q is $s/(s + t) = .2/(.2 + .3) = .4$. Values of q below .4 cause Δq to be positive, which increases q, whereas values of q above .4 cause Δq to be negative, which decreases q. The effect of such heterozygote superiority is to drive the frequencies of the two alleles in the population to a stable equilibrium at $q = .4$.

Selection and Polymorphism

The persistence of different genotypes through heterozygote superiority is an example of *balanced polymorphism*, a term Ford invented to describe the preservation of genetic variability through selection. In general, we consider a gene locus polymorphic if at least two alleles are present, with a frequency of at least 1 percent for the second most frequent allele. Although selection coefficients are difficult to measure in natural populations, such polymorphisms are certainly ubiquitous in practically all populations researchers have examined so far, both on the chromosomal level and on the genic level. One prominent example of polymorphism that overdominance causes is the sickle cell gene in humans, where heterozygotes (Hb^A/Hb^S) survive the malarial parasite more successfully than either normal (Hb^A/Hb^A) or sickle cell homozygotes (Hb^S/Hb^S). This gene (as well as others that appear to offer protection against malaria) persists in notable frequencies in geographical areas where malaria is prevalent.

In laboratory populations, where researchers can more easily control and measure genetic variability, many experiments show balanced polymorphism, apparently by some sort of overdominance. In *Drosophila pseudoobscura*, for example, Dobzhansky and Pavlovsky have shown that the frequencies of the Standard (ST) and Chiricahua (CH) third-chromosome arrangements come to a stable equilibrium when flies carrying these arrangements are placed together in a population cage

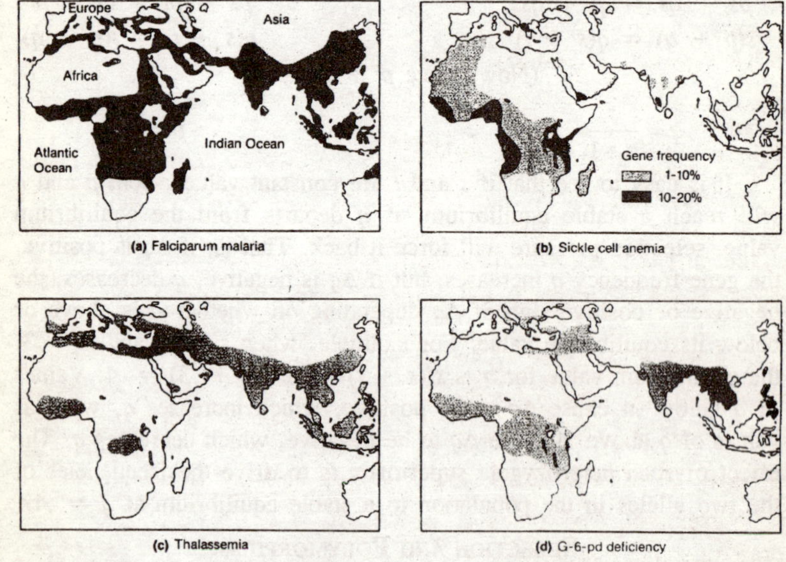

Fig. 9.3. Relationship between the geographic distribution of malaria and genes that confer resistance against the disease.

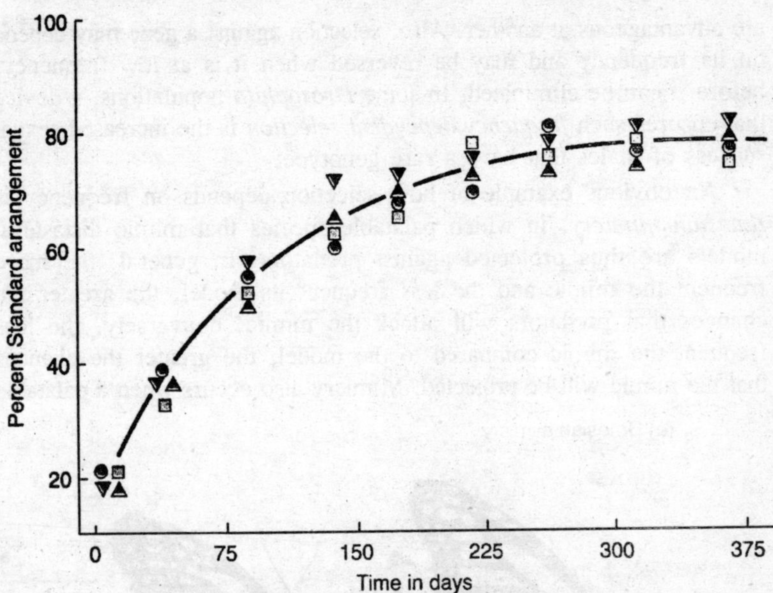

Fig. 9.4. Results of four Drosophila pseudoobscura population cage experiments in which two third-chromosome arrangements are competing, Standard (ST) and Chiricahua (CH).

kept continuously for a year or longer. The superiority of the heterozygote can be seen in the relative adaptive values calculated for the various third-chromosome combinations: ST/ST = 0.90, ST/CH = 1.00, CH/CH = .41.

In fact, we can calculate that even lethal recessive genes may remain in a population if they confer only a small heterozygous advantage. For example, gene *a*, lethal in *aa* homozygous condition but providing a 1 percent advantage to the *Aa* heterozygote compared to the *AA* homozygote, would reach a frequency of approximately 1 percent at equilibrium:

	Genotypes		
	AA	Aa	aa
Adaptive value	.99	1.00	0
Selection coefficient	s=.01	0	t=1

$$\hat{q} \text{ (frequency of } a) = \frac{s}{s+t} = \frac{.01}{1.01} = .0099$$

Other conditions responsible for polymorphism may include a change in selection coefficients so that genes detrimental at one time are advantageous at another. Also, selection against a gene may depend on its frequency and may be reversed when it is at low frequency, before it can be eliminated. In some *Drosophila* populations, a device that ensures such *frequency dependent selection* is the increased sexual success of males that have a rare genotype.

An obvious example of how selection depends on frequency is *Batesian mimicry*, in which palatable species that mimic distasteful models are thus protected against predators. In general, the more frequent the mimic and the less frequent its model, the greater the chances that predators will attack the mimic; conversely, the less frequent the mimic compared to the model, the greater the chances that the mimic will be protected. Mimicry also occurs when a palatable

(a) Batesian mimicry

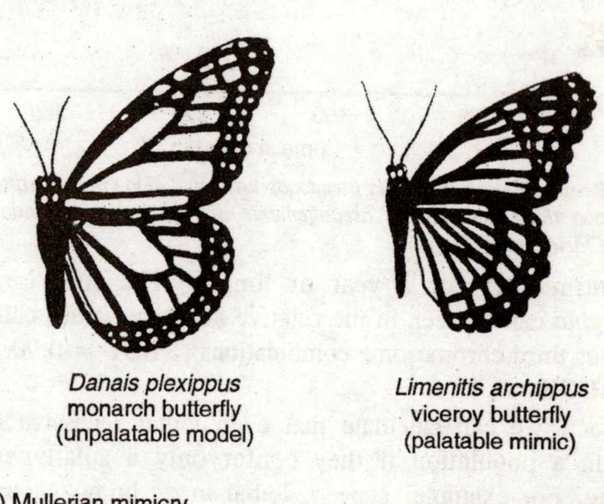

Danais plexippus
monarch butterfly
(unpalatable model)

Limenitis archippus
viceroy butterfly
(palatable mimic)

(b) Mullerian mimicry

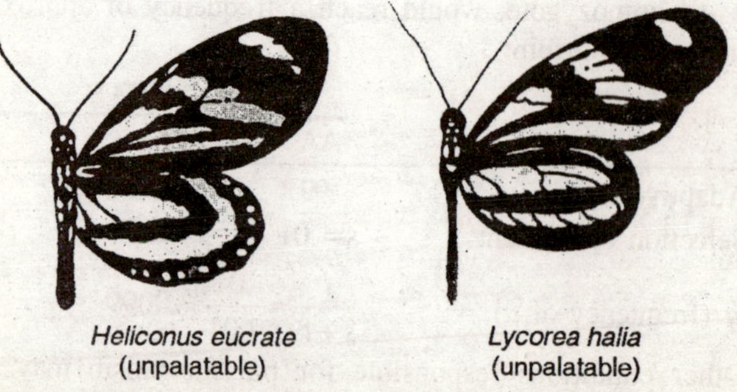

Heliconus eucrate
(unpalatable)

Lycorea halia
(unpalatable)

Fig. 9.5. Mimicry in different species of butterflies.

mimic imitates a conspicuous warning (aposematic) colouration or pattern shared by two or more different unpalatable species. Mimicry between different unpalatable species (*Mullerian mimicry*) benefits all such species by enabling predators to learn a single warning pattern that applies to all these potential but distasteful prey.

In plants, *self-sterility genes* that prevent fertilization between closely related individuals are also frequency-dependent. For example, a haploid pollen grain carrying a self-sterility allele, S^1, will not grow well on a diploid female style carrying the same allele, such as S^1S^2, but can successfully fertilize a plant carrying S^2S^3 or S^3S^4. Thus once an allele becomes common (e.g., S^1), its frequency is reduced by the many sterile mating combinations to which it is now exposed. Rare alleles, in contrast, will successfully fertilize almost every female plant they meet, until they, too, become common. Thus, because of frequency dependence, self-sterility systems of considerable numbers of alleles can become established, reaching as high as 200 alleles or more in red clover.

Polymorphism may also become established when selection coefficients are not constant but vary from one environment to another. A population sufficiently widespread to occupy many environments may therefore maintain a variety of genotypes, each of which is superior in a particular habitat. A prominent example is the polymorphism associated with the phenomenon known as *industrial melanism*. Certain moths and butterflies show increased proportions of dark-coloured, or melanic, forms, usually caused by the increased frequency of a dominant gene in areas where air pollution darkens trees.

In the industrial city of Birmingham, Kettlewell and others demonstrated the selective advantage of such melanic genes by releasing known numbers of both light and melanic forms of the British peppered moth, *Biston betularia*, and recapturing a significantly greater proportion of melanic forms. In other words, these sooty areas offer greater protection to the melanic forms than to the light-coloured forms, since more of the former survived to be recaptured. As Kettlewell showed, the adaptive value of the melanic types probably lies in their ability to remain concealed on darkened tree trunks from bird predators. In nonindustrial areas, in contrast, trees covered with normal gray lichens offer decided advantages to the light-coloured moths. Thus English *B. betularia* populations show various degrees of polymorphism, ranging from high frequencies of the melanic gene in industrial areas to almost zero in rural areas.

Interestingly, passage of clean air legislation in Britain in 1956 has reduced industrial smoke and sulfur dioxide in many formerly polluted areas. This reduction in pollution is now correlated with "reverse evolution": the frequency of melanic forms of *B. betularia* and other insects has declined dramatically.

Levins has pointed out that both the spatial and temporal organization of the environment may significantly affect the extent to which a population will rely on genetic polymorphism as an adaptive strategy. *Coarse-grained environments*, in which different individuals in a population endure different experiences, promote greater genetic polymorphism than do *fine-grained environments*, in which all individuals experience the environmental differences. Hartl and Clark discuss other mechanisms that can maintain polymorphism.

KINDS OF SELECTION

When selection has occurred for particular conditions over long periods of time, we can consider most populations to have achieved phenotypes that are optimally adapted to their surroundings. That is to say, many phenotypes will tend to cluster around some value at which fitness is highest. We can thus expect individuals that depart from these *optimum phenotypes* to show less fitness than those closer to the optimal values. In a classic 1899 study on sparrows that survived a storm, Bumpus showed that measurements taken on eight of nine different characteristics tended to cluster around intermediate phenotypic values, while sparrows killed by the storm showed much greater variability. In Bumpus's terms, "it is quite as dangerous to be conspicuously above a certain standard of organic excellence as it is to be conspicuously below the standard." Many studies on a variety of organisms, including snails, lizards, ducks, chickens, and others have since supported this view.

In humans, measurements of birth weights of newborn babies, among other characteristics, also show selection for optimum values. Most survivors cluster around a birth weight of 8 pounds, and those who depart from this value have less chances for survival. This reduction in frequency of extreme phenotypes has been called *stabilizing*, or *centripetal*, *selection*, because it signifies selection for an intermediate stable value.

However, not all selection is stabilizing, because selection may well favour an extreme phenotype by proceeding in one or the other direction of the phenotypic distribution. Such *directional selection* is commonly practiced by animal and plant breeders, who select for

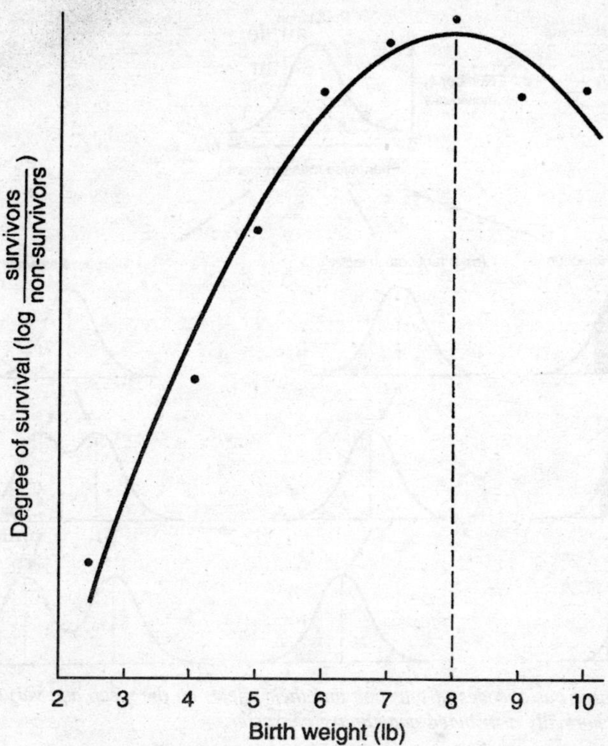

Fig. 9.6. Relationship between birth weight and the degree of survival in female births.

extremes of yield, productivity, resistance to disease, and so forth. The role of directional selection in evolution is especially important when the environment of a population is changing and only extreme phenotypes happen to be adapted for new conditions.

Selection, whether stabilizing or directional, may act in a constant fashion if the selective environment is uniform. However, when conditions are changeable, a population may be subjected to divergent or cyclically changing (oscillating) environments to which different genotypes among its members are most suited. Such selection is therefore *disruptive*, or *centrifugal*, because it establishes genetic differences in a population.

To ultimately affect evolution, selection, however it occurs, must change the frequencies of genes involved in fitness. This means that genetic variability must be present, and *pure lines* that are homozygously uniform for such fitness genes offer no opportunity for selection to produce any noticeable evolutionary change. Fisher

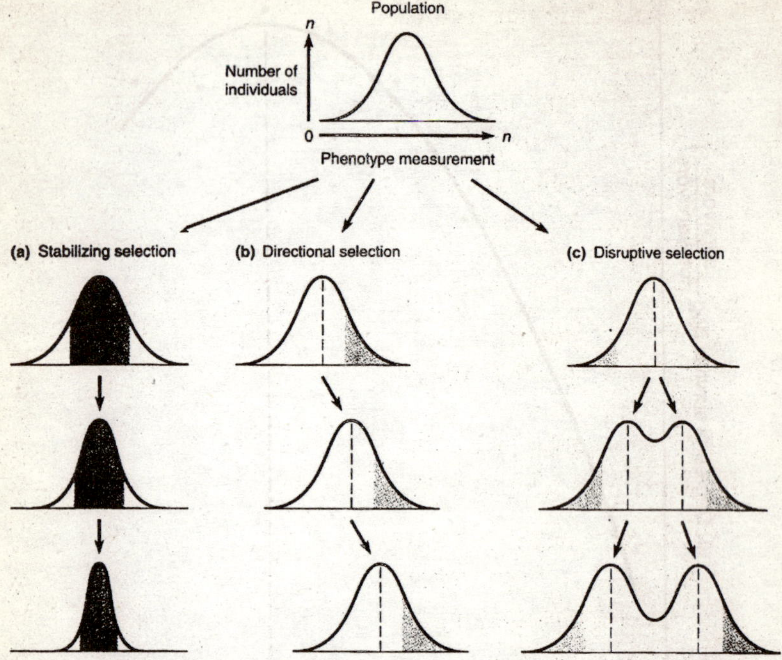

Fig. 9.7. Three basic modes of selection and their effects on the mean and variation of a normally distributed quantitative character.

formulated this principle mathematically as a fundamental theorem that essentially states, "The greater the genetic variability upon which selection for fitness may act, the greater the expected improvement in fitness."

One consequence of Fisher's theorem is that we would expect populations long subjected to selection—and this includes all populations—to have little remaining variability for genes affecting fitness, because selection would have diminished such variability. The continued existence of selection therefore implies that variability itself is favourably selected: continuous changes in the environment affect formerly unselected genes whose presence now offers new opportunities for improving fitness. Such environmental changes obviously include changes in resources, supplies, waste products, and predator and parasite populations.

Van Valen has proposed that species generally compete with each other for resources, so that an advantage, or improvement in fitness, for one species represents a deterioration in the environment of others.

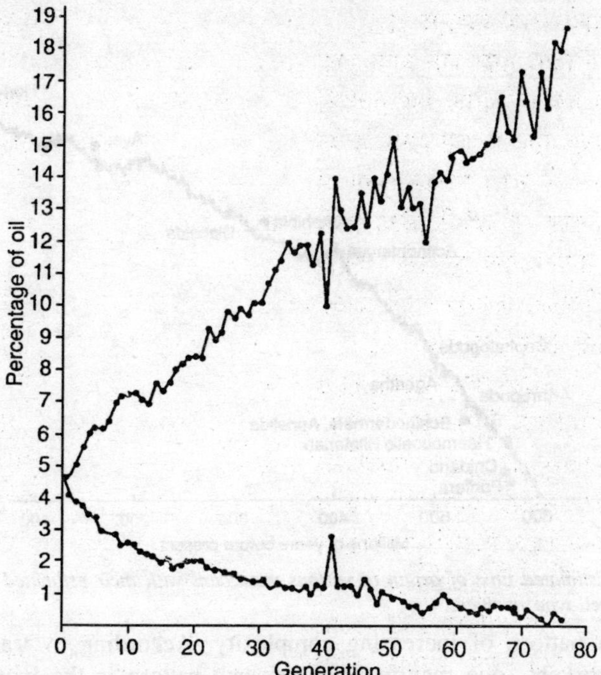

Fig. 9.8. Results of selection for high and low oil content in corn kernels in an experiment.

He points out that species survival is very much in accord with the remark made by the Red Queen whom Alice meets in Lewis Carroll's *Through the booking Glass*: "Here, you see, it takes all the running you can do to keep in the same place." According to the *Red Queen hypothesis*, each species continually faces new selective challenges because of environmental changes often associated with variations in the fitness of its interacting populations, so that to survive each species must constantly cope with and overcome threats to its fitness. Thus nature perpetuates a cyclical process where adaptations in any one organism continually elicit selection for adaptations in others. Or as Darwin stated in *On the Origin of Species*, "If some of these many species become modified and improved, others will have to be improved in a corresponding degree or they will be exterminated."

The long-term consequence of the Red Queen's reign is therefore to increase the competitive fitness of each interacting population, a view that has gained support from experiments with RNA viruses. The adaptations resulting from this process will, of course, vary between organisms, but researchers have suggested that we can nevertheless

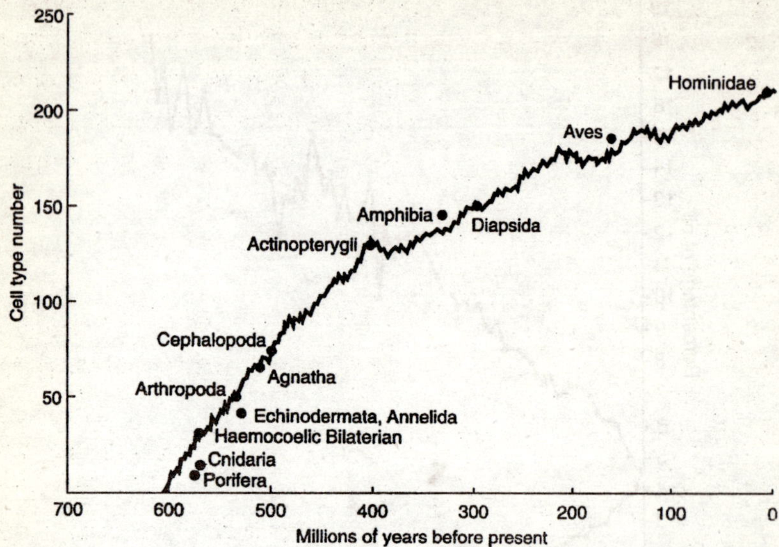

Fig. 9.9. Estimated time of origin of various metazoans with their estimated somatic cell-type numbers.

discern a pattern of increasing complexity. According to Valentine and co-workers, one manifestation of such pattern is the increasing morphological complexity among metazoans, as measured by their estimated number of different somatic cell types. This number has increased at an average rate of about one cell type per 3 million years, starting with the Precambrian-Cambrian period, with no evidence of any downward trend. Note, however, that the increased cell-type numbers that provide new adaptational opportunities do not necessarily imply replacement of all organisms that have fewer cell types. In a sense, therefore, the race for survival goes on at many levels—at lower as well as higher morphological complexities—with survivors at each level, yet with obvious pressure to generate new levels of complexity.

Equilibrium between Mutation and Selection

For convenience we have considered changes in gene frequency to be caused by either mutation or selection acting separately. In nature, however, mutation and selection are simultaneous processes, and both factors influence gene frequency values. Predictions on the basis of one factor alone may therefore be misleading.

For example, even though a recessive gene is detrimental in homozygous condition, it may nevertheless persist in a population because of its mutation frequency. That is, a population reaches a certain equilibrium point at which the number of genes being removed by loss of homozygotes through selection is replaced by the same number of genes introduced into heterozygotes through mutation. We may determine this equilibrium frequency by the following argument.

We have seen that the change in gene frequency per generation for a deleterious recessive *a* with frequency q is equal to a loss of $sq^2(1 - q)/(1 - sq^2)$. If s is small, we can consider the denominator 1, and the loss in frequency is then $sq^2(1 - q)$. The frequency of newly mutated *a* genes, however, is equal to the mutation rate (u) of $A \rightarrow a$ multiplied by the *A* frequency, which is $1 - q$. Thus the loss of *a* genes through selection is exactly balanced by the gain of newly mutated *a* genes when

$$sq^2(1-q) = u(1-q)$$

$$sq^2 = u$$

$$q^2 = \frac{u}{s}$$

$$q = \sqrt{\frac{u}{s}}$$

The equilibrium frequency of a mutant gene in a population is thus a function of both the mutation frequency and the selection coefficient. As you can see in the hydraulic model of this relationship, when the mutation rate increases the equilibrium gene frequency also increases, but the equilibrium frequency decreases when the selection coefficient increases.

For a deleterious dominant allele, similar algebraic manipulations point to an equilibrium frequency of about u/s, a value almost identical to the equilibrium frequency of genes that lack dominance. Since such dominant or partially dominant genes are of considerable disadvantage to heterozygotes, Fisher proposed that their deleterious effect probably diminishes in most organisms by selection of *modifier genes* at other loci that change the degree of dominance. For example, mutant alleles at a particular locus *A* (e.g., A^1, A^2, A^3 ...) may act as partial dominants in the presence of the wild-type allele A^+. Since these mutant alleles are mostly deleterious, modifier genes at other loci (e.g., B^1, B^2 . . ., or C^1, C^2 . . ., etc.) that increase the dominance of A^+ will be selected until the effects of mutations at the *A* locus are relatively recessive.

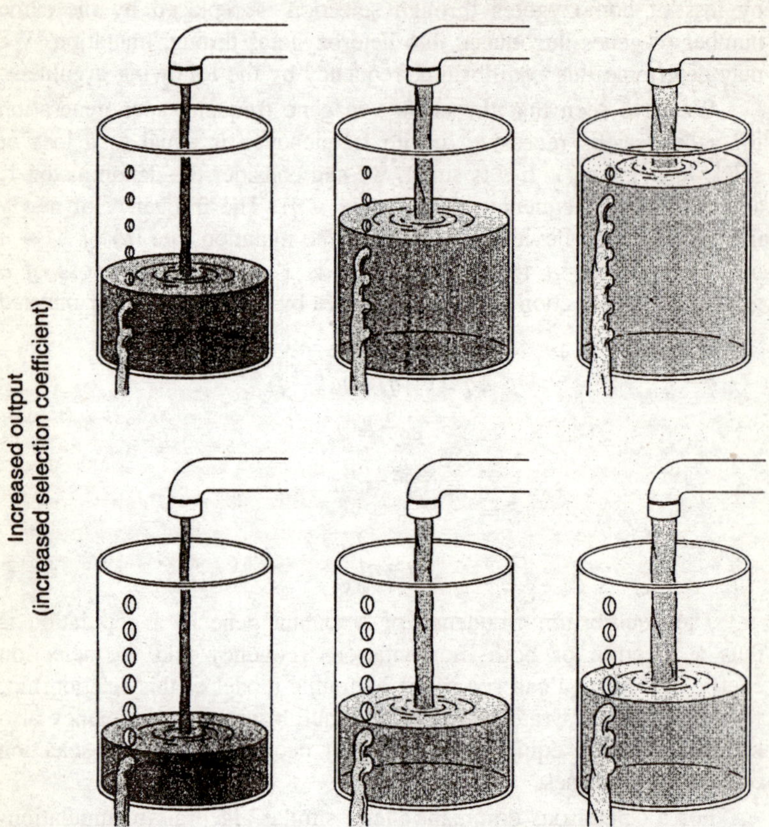

Fig. 9.10. Hydraulic model of mutation-selection equilibrium.

The successful selection for dominant and recessive modifiers that Ford demonstrated in the currant moth *Abraxas grossulariata* provides evidence for Fisher's view. In this moth a single gene, *lutea*, in homozygous condition, produces yellow instead of the normal white ground colour but has an intermediate effect as a heterozygote. After four generations of selecting moths for greater and lesser expression of the *lutea* phenotype in the heterozygote, Ford obtained two distinct strains: in one case *lutea* acted almost as a complete dominant and in the other case almost as a complete recessive. In each strain special modifiers had been chosen, some enhancing and some detracting from the dominance of this particular gene.

Instead of modifiers, Haldane suggested that special wild-type alleles are selected (e.g., A^{X+}, A^{Y+}, A^{Z+}) that act as dominants in the presence of a mutant allele (e.g., A^1, A^2, A^3, . . .), and the entire subject has been considerably discussed. In general, evolution has probably used both kinds of dominance mechanisms mentioned, as Harland and others demonstrated in two species of cotton, *Gossypium barbadense* and *G. hirsutum*. In these plants certain alleles show simple dominance when variants of the same species cross. Interspecific crosses, by contrast, show the effect of many modifying genes on these traits, as well as differences in the degree of dominance of particular alleles. However, despite these dominance-producing mechanisms, many harmful genes are probably still not completely recessive and seem to have some effect in heterozygous condition. Thus, in natural populations the equilibrium frequencies of harmful genes are probably higher than for dominants but lower than for pure recessives.

MIGRATION

Mutation is not the only mechanism by which new genes enter a population. A population may receive alleles by *migration* (also called *gene flow*) from a nearby population that maintains an entirely different gene frequency. When this occurs, two factors are important to the recipient population: (1) the difference in frequencies between the two populations and (2) the proportion of migrant genes that are incorporated each generation. If we designate q_0 as the initial gene frequency in the recipient, or hybrid, population, Q as the frequency of the same allele in the migrant population, and m as the proportion of newly introduced genes each generation, then the gene frequency in the hybrid population will suffer a loss of q_0 equal to mq_0 and a gain of Q equal to mQ. Over n generations of migration, when the gene frequency of the hybrid population becomes q_n one can calculate that the relationship between these factors will reach

$$q_n - Q = (1 - m)^n (q_0 - Q)$$

or

$$(1 - m)^n = \frac{q_n - Q}{q_0 - Q}$$

For populations whore this equation can be applied, we must know four of these factors to calculate the fifth. One such example can be found in human populations where blood group gene frequencies are known for both American blacks and American whites, two populations between which gene exchange has occurred. In general, although some black genes undoubtedly enter the white population, the white population is so large that this introduction probably makes little difference in white gene frequencies. In contrast, the black population is much smaller and has remained isolated from its African origin for two or more centuries. On this basis the white population can be considered as the gene donor or migrant population (Q) and the present black population as the hybrid (q_n).

To obtain the original gene frequency of one of the Rh blood group alleles, R^0, in the black population (q_0), researchers used data of present East African blacks on the assumption that these data may reflect the original gene frequencies of 200 to 500 years ago. Among the East Africans, R^0 showed a frequency of .650, indicating that the frequency of this gene had fallen in American blacks to its present frequency of .446. This fall in frequency could be ascribed to interbreeding with the American white population, where the frequency of R^0 is about .028, much lower than among blacks. According to Glass and Li, this reduction had begun at the time of the initial introduction of blacks into the American colonies 500 years ago and probably continued throughout the 10 generations since. Substituting these values into the preceding formula, we obtain

$$(1-m)^{10} = \frac{q_{10}-Q}{q_0-Q} = \frac{.446-.028}{.630-.028} = .694$$

$$1-m = \sqrt[10]{.694} = .964$$

$$m = .036$$

This value of m means that, excluding all other causes such as mutation, 36 genes per 1,000, or 3.6 percent of genes in the black population, entered from the white population each generation. Since 1 – m represents the proportion of nonintroduced genes, $(1 - m)^{10}$ = .694 is the proportion of genes that have remained of African origin over the 10-generation period. Supported by somewhat similar estimates in more recent studies, blood group gene frequencies generally indicate that the American black population is genetically about 70 to 80 percent African and 20 to 30 percent white, with some differences between Southern and Northern blacks.

Where we lack exact information on gene frequency exchanges between populations, and this includes most populations, considerable discussion and dispute have flourished about the importance of migration. According to Mayr and to Stanley, migration can hinder local evolutionary changes by infusing genes from populations that are not adapted to local conditions. For example, some populations of mammals who live on dark, formerly volcanic lava flows have dark fur when they are isolated from neighbouring populations who live on lighter coloured backgrounds, but do not have dark fur when they receive immigrants from the lighter-coloured surroundings. In contrast, Ehrlich and co-workers describe populations of the butterfly *Euphydras editha*, which show no phenotypic changes whether or not they are subject to migration from phenotypically different populations. In the absence of genetic information, these issues are, so far, difficult to resolve.

RANDOM GENETIC DRIFT

The three forces we have considered up to now—mutation, selection, and migration—share one important quality: they usually act directionally to change gene frequencies progressively from one value to another. Unopposed, these forces can fix one allele and eliminate all others; when balanced, they can lead to equilibrium between two or more alleles. However, in addition to these directional forces, there are also changes that have no predictable constancy from generation to generation. *Random genetic drift*, one of the most important of such *nondirectional forces*, arises from variable sampling of the gene pool each generation. This is apparent if we consider that, in the absence of directional forces to change gene frequencies, there is always a strong likelihood of obtaining a good sample of the genes of the previous generation as long as the number of parents in a population is consistently large. However, since real populations are limited in size, genetic drift will cause gene frequency changes because of *sampling errors*. For example, if only a few parents are chosen to begin a new generation, such a small sample of genes may deviate widely from the gene frequency of the previous generation.

The extent of the deviation for all sizes of populations can be measured mathematically by the standard deviation of a proportion $\sigma = \sqrt{pq / N}$, where p is the frequency of one allele, q of the other, and N the number of genes sampled. For diploid parents, each carrying two alleles, $\sigma = \sqrt{pq / 2N}$, where N is the number of actual parents. For example, if we begin with a large diploid population, where $p = q = .5$, and continue this population each generation by using 5,000 parents, then $\sigma = \sqrt{(.5)(.5) / 10{,}000} = \sqrt{.000025} = .005$. The values of such populations will therefore fluctuate mostly around .5 ± .005, or between .495 and .505. A choice of only two parents as founders will produce a standard deviation of $\sqrt{(.5)(.5) / 4} = \sqrt{.0625} = .25$, or values of .50 ± .25 (from .25 to .75).

In other words, sampling accidents because of smaller population size can easily yield gene frequencies that depart considerably from the initial .5 values in a single generation. If the population remained small and the next generation began with either of these extremes—that is, a gene frequency of .25 or .75 for a particular allele—in the following generation the frequency of that allele may fall to almost zero($.25 \pm \sqrt{(25)(.75)/}$ =.25±.22: a range of .03 to .47) or increased almost to $1(.75 \pm \sqrt{(.75)(.25)/4} = 75 \pm 22$: a range of .53 to .97). Should such small populations continue each generation, the likelihood increases that one or more will eventually reach fixation for one of the alleles. The proportion of such populations that attain fixation—that is, the *rate of fixation*—will eventually reach $1/2N$. Obviously, if N is large, fixation proceeds slowly, but even large populations can show some degree of drift.

This reliance of drift on population number emphasizes the importance of what is called *effective population size* (N_e). It differs from the observed population size because not all members of a population are necessarily parents and because parentage can also be limited by a reduced number of one of the sexes. For example, if out of a total population of 1,000, 3 males mated to 300 females produce

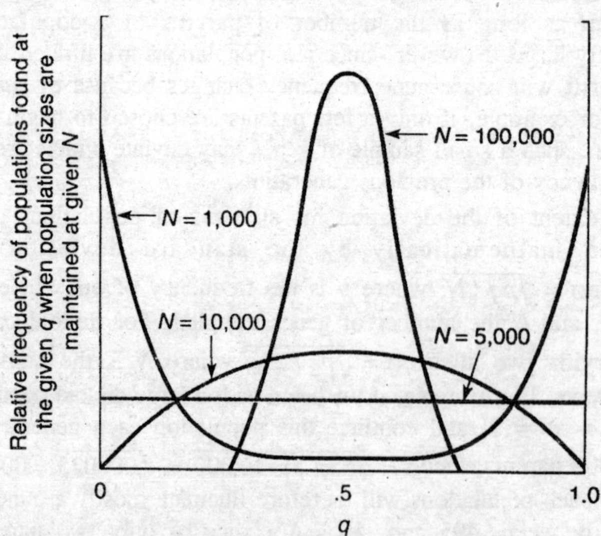

Fig. 9.11. Distribution of equilibrium gene frequencies for populations of different sizes when selection is zero and a small amount of migration occurs in each population (m = 0.0001) from a population whose gene frequency is q = .5.

the next generation, the effective population size is more than 6 but still less than 303. Wright has expressed the relationship as $N_e = 4N_fN_m/(N_f + N_m)$, where N_f is the number of parental females and N_m the number of parental males. In the preceding case N_e would be $4(300)(3)/303 = 11$. Inequalities in numbers of offspring among different parents will also reduce the effective population size.

Wright has therefore proposed that genetic drift may be quite important in changing gene frequencies among populations when their effective sizes are small. Among the observations illustrating this concept is that of Buri, who set up 107 separate lines of *D. melanogaster*, each line carrying two alleles at the *brown* locus (*bw* and bw^{75}) at initially equal frequencies of 50 percent. Buri then continued the lines for 19 generations by randomly selecting 8 males and 8 females as parents from each preceding generation ($N = 16 = 32$ *brown* alleles) and scoring the frequency of the two different *brown* alleles. By the first generation a number of populations already showed departures from the original 50 percent bw^{75} frequency, and genetic drift continued to increase successively so that by generation 19 more than half the 107 populations reached fixation for either the *bw* or bw^{75} alleles. Despite these genetic differences, note that the average frequencies of *brown* alleles when combining all populations remain at

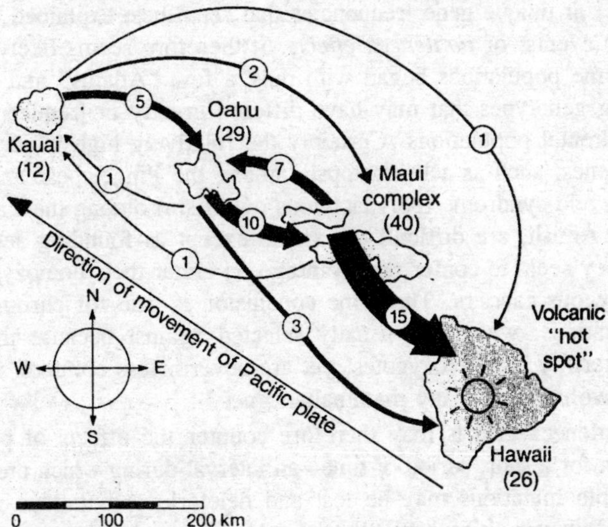

Fig. 9.12. Colonization pattern showing the founder events that Carson proposed to explain the origin of native "picture-winged" Drosophila species found on the Hawaiian islands.

about .5. Genetic drift therefore increases variation between populations, but on the average, not in any particular direction.

Although the persistence of small population size over many generations is an obvious cause of genetic drift, occasional size reductions for only one or a few generations may also have pronounced effects on gene frequencies and future evolution. At the extreme of such reductions, which Mayr called the *founder principle*, a population may occasionally send forth only a few founders to begin a new population. Whatever genes or chromosome arrangements these founders take with them, detrimental or beneficial, all stand a good chance of becoming established in the new population because of this sudden sampling accident.

Thus Carson, by carefully analyzing salivary chromosome banding patterns, has shown that the more than 100 native Hawaiian "picture-winged" *Drosophila* species can derive from founder events in which each island was settled by relatively few individuals whose descendants evolved into different species. For example, the 40 species unique to the Maui island complex derive from only 12 founders—10 from Oahu and 2 from Kauai—with each single founder providing unique chromosome arrangements that we can trace in the descendant species.

There are by now numerous examples in many organisms, including humans, of unique gene frequencies that seem best explained by such founder events, or *bottleneck effects*. It therefore seems likely that at least some populations began with only a few "Adams" and "Eves" carrying genotypes that may have differed greatly in frequency from their parental populations. Certainly the relatively high incidences of some genes, such as achromatopsia among the Pingelapese and Ellis-van Creveld syndrome (polydactylous dwarfism) among the Lancaster County Amish, are difficult to explain except as founding accidents, since they seem to confer no advantage on either their homozygous or heterozygous carriers. The same conclusion is true for chromosomal translocations, which are usually selected against because they can cause sterility in heterozygotes, yet are nevertheless common features in the evolution of many mammalian lines.

Bottleneck effects may therefore counter the effects of previous selection for a short period of time—an interval during which previously favourable mutations may be lost and deleterious mutations may be fixed. However, it is difficult to imagine that any genetic trait that affects the armament with which organisms face their environment can long continue to escape selective environmental pressures. Through nonselective genetic changes, a bottleneck may cast the evolution of a population in a new direction, but, from all we understand of evolution, this direction could not remain nonadaptive without extinction.

In general, therefore, the consequences of founding events on gene frequencies are largely unpredictable compared to variability estimates (i.e., σ) that can be arrived at when population size is constant. Other factors that make the variability of gene frequencies unpredictable are unique historical events such as a change in the direction or intensity of selection because of a radical change in environment, an unusually favourable mutation, a rare hybridization event with another variety, or an unusual swamping of a population by mass immigration. The effects of these factors on evolutionary changes may be quite important, although genetic data for such events are still difficult to obtain.

10

Population Genetics

The genotypes at a locus may all have the same fitness. In small populations, gene frequencies then evolve by random genetic drift. The chapter begins by explaining why drift happens and what it means, and presents examples of random sampling effects. We then see how drift can ultimately fix one allele, and in a sense undermine the Hardy-Weinberg equilibrium. We then add the effects of mutation, which introduces new variation; the variation observed in a population will be a balance between the drift to homozygosity and mutation that creates heterozygosity.

Imagine a population of 10 individuals, of which 3 have genotype *AA*, 4 have *Aa*, and 3 *aa*. The population contains 10 *A* genes and 10 *a* genes; the gene frequencies of each gene are 0.5. We also imagine that natural selection is not operating—that is, all genotypes have the same fitness. What will the gene frequencies be in the next generation? The most likely answer is 0.5 *A* and 0.5 *a*. However, this answer is only the most likely; it is not a certainty. The gene frequencies may, by chance, change somewhat from the previous generation. This situation is possible because the genes that form a new generation are a *random sample* from the parental generation.

Random sampling starts at conception. In every species, each individual produces many more gametes than will ever fertilize, or be fertilized, to form new organisms. The successful gametes that do form offspring represent only a sample of the many gametes that the parents produce. If a parent is homozygous, the sampling makes no difference to the genes received by the offspring because all of a homozygote's gametes contain the same gene. However, sampling does matter if the parent is a heterozygote, such as *Aa*. A heterozygote will then produce a large number of gametes, of which approximately one-half will be *A* and the other half *a*. (The proportions may not be exactly one-half, however. Reproductive cells could die at any stage leading to gamete formation, or after they have become gametes; also, in the female, a randomly picked three-fourths of the products of meiosis are lost as polar bodies.) If that parent produces 10 offspring, it is most likely that 5 will inherit an *A* gene and 5 will inherit an *a*. Because the gametes that formed the offspring were sampled from a much larger pool of gametes, the proportions could possibly differ from those figures. For example, 6 could inherit *A* genes and only 4 *a*, or 3 *A* and 7 *a*.

In what sense is the sampling of gametes random? Let us consider the first two offspring produced by an *Aa* parent. When it produces its first offspring, one gamete is sampled from its total gamete supply, and there is a 50% chance it will be an *A* and 50% that it will be an *a*. Suppose this gamete happens to be an *A*. Sampling is then seen as random because it is no more likely that the next gamete to be sampled will be an a gene simply because the last one sampled was an *A*—that is, the chance that the next successful gamete will be an *a* remains 50%. (Coin flipping is random in the same way; if you first flip a head, the chance that the next flip will be a head is still $^1/_2$.)

The alternative to random sampling would be a "balancing" system in which, after an *A* gamete had been successful in reproduction, the next successful gamete would be an *a*. If reproduction occurred in such a manner, the gene frequency contributed by a heterozygote to its offspring would always be exactly $^1/_2A$:$^1/_2a$. Random drift would then be unimportant in evolution. In fact, reproduction does not proceed in this way, so the successful gametes are a random sample from the gamete pool.

The sampling of gametes is only the first stage at which random sampling occurs. It continues at every stage as the adult population of a new generation matures. To see how this sampling works, imagine a line of 100 packhorses walking single file along a hazardous mountain path. Only 50 of the horses make it to the end of the path; the other 50 fall off the path and crash down the ravine. The 50 survivors could potentially be, on average, genetically surer of foot than the rest, in which case the sampling of 50 survivors out of the original 100 would be nonrandom. Natural selection would determine which horses survived and which died. If we examined the genotypic frequencies among the smashed horses at the bottom of the ravine, they would differ from those among the survivors.

Alternatively, death could be accidental, occurring whenever a large rock bounced down the mountainside, and knocked a horse into the ravine. Suppose that the rocks fall at unpredictable times and places and arrive so suddenly that defensive action is impossible; the horses do not vary genetically in their ability to avoid the falling rocks. The loss of genotypes would then be random in the sense defined above. If an *AA* horse has just fallen victim to a rock, then it is no more or less likely that the next victim will have the *AA* genotype. Now if we compared the genotype frequencies in the survivors and nonsurvivors, the two would most likely not differ. The survivors would be a random genetic sample from the original population. They could, however, differ by chance. More *AA* horses might have been unlucky with falling rocks; more *aa* might have been lucky. Some increase in the frequency of the a gene would then be observed in the population.

The sampling of packhorses is imaginary, but analogous sampling may happen at any time in a population, and at any life stage as juveniles develop into adults. Because many more eggs are produced than adults, abundant opportunity exists for sampling as each new generation grows up. Random sampling occurs whenever a smaller number of successful individuals (or gametes) are sampled from a larger pool of potential survivors and the fitnesses of the genotypes are the same.

Genetic Drift

The condition that genotypes *AA*, *Aa*, and *aa* leave, on average, the same number of offspring (they have identical fitness) is called selective *neutrality*. The fitnesses can be expressed in the following form:

genotype	AA	Aa	aa
fitness	1	1	1

In this chapter, we are not concerned with how likely it is that all the genotypes at a locus would have the same fitness, when we will see that the condition is quite common for some kinds of allelic variant. Instead, we will consider the abstract theory. If there are two alleles at a locus, and they have the same fitness, random sampling can cause their relative frequencies in a population to change, for the reason discussed in the previous section. These random changes in gene frequencies between generations are called *genetic drift*, *random*

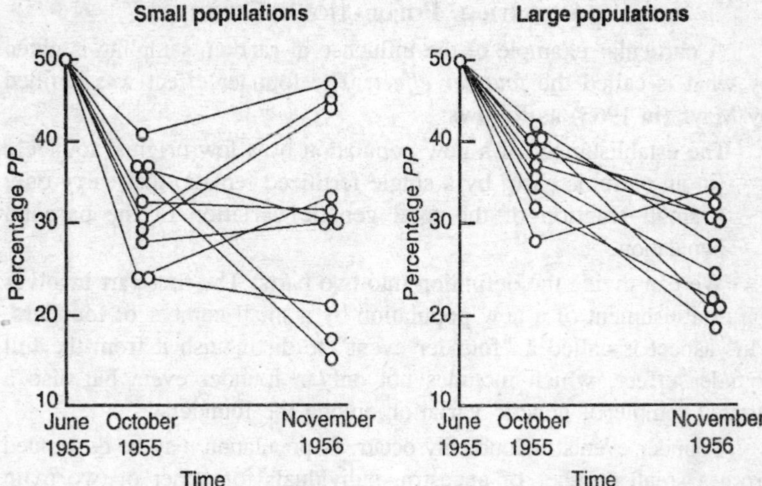

Fig. 10.1. Random sampling is more effective in small populations than in large.

drift, or simply *drift*. Genetic drift has important consequences for the random substitution of genes and the Hardy-Weinberg equilibrium.

The rate of change of gene frequency by random drift depends on the size of the population. Random sampling effects are more important in smaller populations. For example, Dobzhansky and Pavlovsky, working with the fruitfly *Drosophila pseudoobscura*, made 10 populations with 4000 initial members (large populations) and 10 with 20 initial members (small populations), and followed the change in frequency of two chromosomal variants for 18 months. The average effect was the same in both the small and large populations, but the variability was significantly greater among the small populations.

An analogous result could be obtained by flipping 10 sets of 20, or 4000, coins. On average, both cases would include 50% heads, but the chance of flipping 12 heads and 8 tails in the small population is higher than the chance of flipping 2400 heads and 1600 tails in the large population.

If a population is small, it is more likely that a sample will be biased away from the average by a given percentage amount; genetic drift is, therefore, greater in smaller populations. Indeed, because random effects become strong only in small populations, this chapter could have been called "population genetics in small populations." The smaller the population, the more important the effects of random sampling.

Ancestral Population's Genes

A particular example of the influence of random sampling is given by what is called the *founder effect*. The founder effect was defined by Mayr (in 1963) as follows:

> The establishment of a new population by a few original founders (in an extreme case, by a single fertilized female) that carry only a small fraction of the total genetic variation of the parental population.

We can divide the definition into two parts. The first part involves the establishment of a new population by a small number of founders. This aspect is called a "founder event" to distinguish it from the full founder effect, which includes not only a founder event but also a limited sample of genetic variation among the founders.

Founder events undoubtedly occur. A population may be descended from a small number of ancestral individuals for either of two main reasons. A small number of individuals may colonize a place previously uninhabited by their species; the 250 or so individuals making up the modern human population on the island of Tristan da Cunha, for example, are all descended from 20-25 immigrants in the early nineteenth century, and most are descended from the original settlers—one Scot and his family—who arrived in 1817. Alternatively, a population that is established in an area may fluctuate in size; the founder effect then occurs when the population passes through a "bottleneck" in which only a few individuals survive, and later expands again when more favourable times return.

If a small sample of individuals is taken from a larger population, what is the chance that the sample will have reduced genetic variation? We can express the question exactly by asking what the chance is that an allele will be lost. In the special case of two alleles (*A* and *a* with proportions *p* and *q*), if one allele is not included in the founder population, the new population will be genetically monomorphic. The chance that an individual will be homozygous *AA* is simply p^2· The chance that two individuals drawn at random from the population will both be *AA* is $(p^2)^2$, in general, the chance of drawing *N* identical homozygotes is $(p^2)^N$. The founding population could be homozygous either because it consists of *N AA* homozygotes or *N aa* homozygotes, and the total chance of homozygosity is therefore

$$\text{chance of homozygosity} = (p^2)^N + (q^2)^N \qquad ...(1)$$

Figure 10.2 illustrates the relation between the number of individuals in the founder population and the chance that the founder

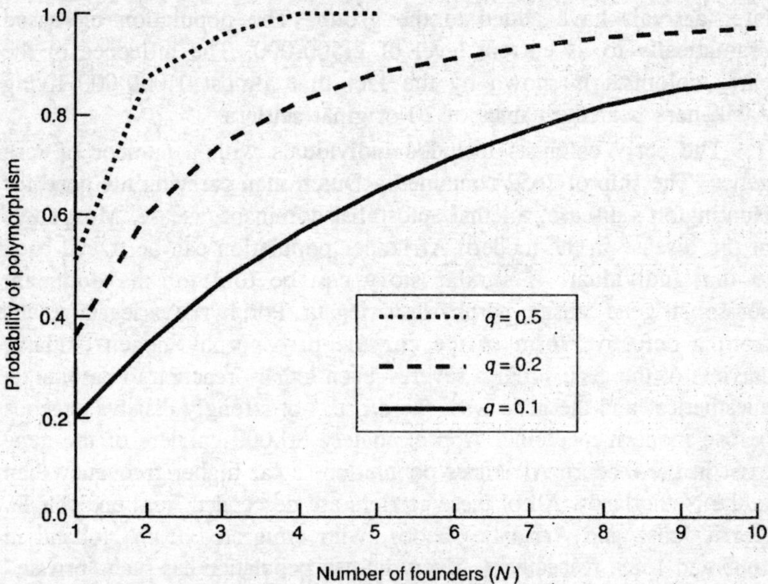

Fig. 10.2. The chance that a founder population will be homozygous depends on the number of founders and the gene frequencies.

population is genetically uniform. The interesting result is that founder events are not effective at producing a genetically monomorphic population. Even if the founder population is very small, with $N < 10$, it will usually possess both alleles. An analogous calculation could be performed for a population with three alleles, in which we asked the chance that one of the three would be lost by the founder effect. The resulting population would not then be monomorphic, but have two (instead of three) alleles. In general, then, founder events—whether by colonizations or population bottlenecks—are unlikely to reduce genetic variation unless the number of founders is tiny.

However, founder events can have other consequences. Although the sample of individuals forming a founder population is likely to include nearly all of the ancestral population's genes, the frequencies of the genes may differ from the parental population. Isolated populations often have exceptionally high frequencies of otherwise rare alleles, and the most likely explanation is that the founding population had a disproportionate number of those rare alleles. The clearest examples of this consequence involve human populations.

Consider the Afrikaner population of South Africa, which is mainly descended from one shipload of immigrants that landed in 1652, although later arrivals have added to the group. The population increased dramatically to its current level of 2,500,000. The influence of the early colonists is shown by the fact that almost 1,000,000 living Afrikaners bear the names of 20 original settlers.

The early colonists included individuals with a number of rare genes. The ship of 1652 contained a Dutch man carrying the gene for Huntington's disease, a lethal autosomal dominant disease. Most cases of the disease in the modern Afrikaner population can be traced back to that individual. A similar story can be told for the dominant autosomal gene causing porphyria variegata. Porphyria variegata results from a defective form of the enzyme protoporphyrinogen oxidase; carriers of the gene suffer a severe—even lethal—reaction to barbiturate anesthetics, and the gene was, therefore, not strongly disadvantageous before modern medicine. Approximately 30,000 carriers of the gene exist in the modern Afrikaner population, a far higher frequency than in the Netherlands. All of these carriers are descended from one couple, Gerrit Jansz and Ariaantje Jacobs, who emigrated from Holland in 1685 and 1688, respectively. Every human population has such "private" polymorphisms, which were caused by the genetic peculiarities of founder individuals in many cases.

Both of the examples involving the Afrikaner population involve medical conditions. The individual carriers of the genes have lower fitness than average, and selection will, therefore, act to reduce the frequency of the gene to zero. For most of history, the porphyria variegata gene may have had a similar fitness to its allele; it may have been a neutral polymorphism until its "environment" came to contain (in selected cases) barbiturates.

In contrast, the gene for Huntington's disease will have been consistently selected against. Thus, its present high frequency suggests that the founder population had an even higher frequency, because selection would have probably decreased its frequency in subsequent generations. Any particular founder sample would not be expected to show a higher than average frequency of the Huntington's disease gene. If enough colonizing groups set out, however, some of them are bound to have peculiar, or even very peculiar, gene frequencies. In the case of Huntington's disease, founders with more copies of the gene than average are known in instances other than the Afrikaner population. For example, 432 carriers of Huntington's disease in Australia are descended from a Miss Cundick who left England with her 13 children. In addition, a French nobleman's grandson, Pierre Dagnet d'Assigne de Bourbon, bequeathed all of the known cases of Huntington's disease on the island of Mauritius.

Random Drift

The frequency of a gene is equally likely to decrease as to increase by random drift. On average, the frequencies of neutral alleles remain unchanged from one generation to the next. In practice, their frequencies drift up and down, and it is possible for a gene to enjoy a run of luck and be carried up to a much higher frequency—in the extreme case, its frequency could, after many generations, increase to 1 (become fixed) by random drift.

During each generation, the frequency of a neutral allele has a chance of increasing, a chance of decreasing, and a chance of staying constant. If it increases in one generation, it again has the same chance of increasing, decreasing, or staying constant in the next generation. Thus, the frequency has a small chance of increasing for two generations in a row (equal to the square of the chance of increasing in any one generation), a still smaller chance of increasing through three generations, and so on. For any one allele, fixation by random drift is very improbable. The probability is finite, however, and if enough neutral alleles, at enough loci, and over enough generations, are randomly drifting in frequency, one of them will eventually be fixed. The same process can occur, regardless of the initial frequency of the allele. It is less likely that a rare allele will reach fixation by random drift than a common allele, because it would take a longer run of good luck. Nevertheless, that scenario is still possible. Indeed, a unique neutral mutation has some chance of eventual fixation. Any one mutation is most likely to be lost, but if enough mutations arise, one will be bound to be fixed eventually.

As a result, a gene can be substituted by random drift. What is the rate of this kind of neutral evolution? We might expect it would occur more rapidly in smaller populations, because most random effects are more powerful in smaller populations. However, an elegant argument demonstrates that the neutral evolution rate exactly equals the neutral mutation rate, and is independent of population size. In a population of size N, there are a total of $2N$ genes at each locus. On average, each gene contributes one copy of itself to the next generation. Because of random sampling, however, some genes will contribute more than one copy and others will contribute none. When we look two generations ahead, those genes that contributed no copies to the first generation cannot contribute copies to the second, third, fourth, or any subsequent

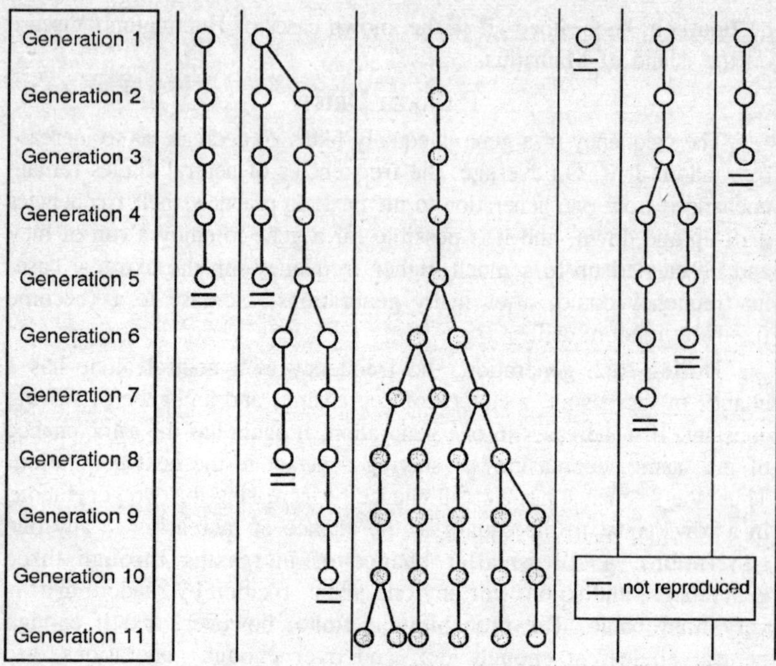

Fig. 10.3. The drift to homozygosity.

generation—once a gene fails to be copied, it is lost forever. In the next generation, some more genes will likewise "drop out" and be unable to contribute to future generations. Each generation, some of the $2N$ original genes are lost in this way.

If we look far enough ahead, we eventually come to a time when all $2N$ genes are descended from just one of the $2N$ genes now. This evolution occurs because some genes will fail to reproduce in every generation. Thus, we must eventually come to a time when all but one of the original genes have dropped out. That one gene will have experienced a long enough run of lucky increases and will have spread through the whole population. That is, it will have been fixed by genetic drift. Because the process is based entirely on luck, each of the $2N$ genes in the original population has an equal chance of being the remaining gene. Any one gene in the population, therefore, has a $1/(2N)$ chance of eventual fixation by random drift (and a $(2N - 1)/(2N)$ chance of being lost by it).

Because the same argument applies to any gene in the population, it also applies to a new, unique, neutral mutation. When the mutation arises, it will be one gene in a population of $2N$ genes at its locus (i.e., its frequency will be $1/(2N)$, and it has the same $1/(2N)$ chance of eventual fixation as does every other gene in the population. The mutation will, therefore, most likely be lost (probability of being lost $= (2N - 1)/(2N) \approx 1$ if N is large), but it does have a small $(1/(2N))$ chance of success. Thus, according to the first stage of the argument, the probability that a neutral mutation will eventually be fixed is $1/(2N)$.

The rate of evolution equals the probability that a mutation is fixed, multiplied by the rate at which mutations appear. We define the rate at which neutral mutations arise as u per gene per generation. (Note that u is the rate at which new selectively neutral mutations arise, not the total mutation rate. The total mutation rate includes selectively favourable and unfavourable mutations as well as neutral mutations.) At each locus, there are $2N$ genes in the population. The total number of neutral mutations arising in the population will be $2Nu$ per generation. The rate of neutral evolution is then $^1/_2N \times 2Nu = u$. The population size cancels out and the rate of neutral evolution is equal to the neutral mutation rate.

Hardy-Weinberg Equilibrium

Let us stay with the case of a single locus, with two selectively neutral alleles A and a. If genetic drift does not occur—if the population is large—the gene frequencies will remain constant from generation to generation and the genotype frequencies will also be constant, in Hardy-Weinberg proportions. In a smaller population, however, the gene frequencies can drift around. The average gene frequencies in one generation will be the same as in the previous generation, and it might be thought that the long-term average gene and genotype frequencies will simply be those of the Hardy-Weinberg equilibrium, but with a bit of "noise" around them. That, however, turns out not to be correct. The long-term result of genetic drift is that one of the alleles will be fixed; the polymorphic Hardy-Weinberg equilibrium is unstable in a small population.

Suppose that a population consists of five individuals, containing five A alleles and five a alleles (obviously a tiny population, but the same point would apply if there were 500 copies of each allele). The genes are randomly sampled to produce the next generation. Imagine that six A alleles and four a alleles are sampled. This sample represents the starting point to produce the next generation. The most likely ratio in the next generation is six A and four a, because no "compensating" process pushes the ratio back toward five and five. Perhaps six *A* and four *a* are drawn in the next generation. The fourth generation might be seven *A* and three *a*, the fifth generation six *A* and four *a*, the sixth generation seven *A* and three *a*, then seven *A* and three *a*, eight *A* and two *a*, eight *A* and two *a*, nine *A* and one *a*, and

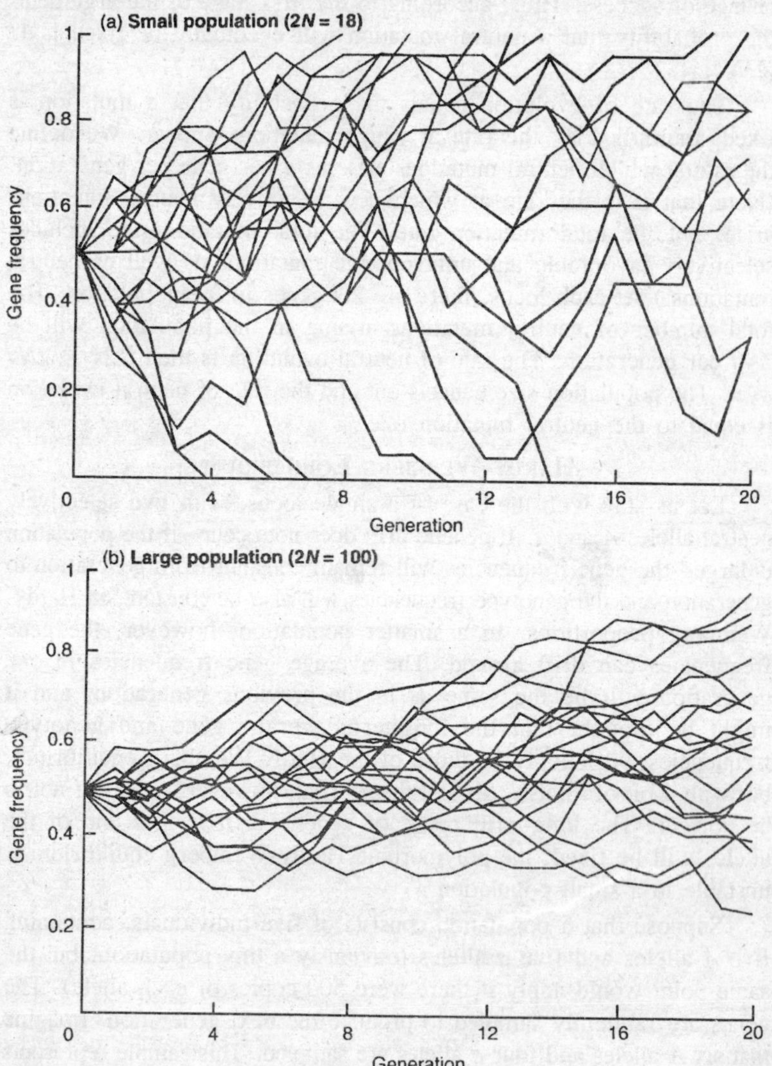

Fig. 10.4. Twenty repeat simulations of genetic drift for a two-allele locus with initial gene frequency 0.5.

then 10 A. The same process could have moved in the opposite direction, or begun by favouring A and then reversing to fix a—random drift is directionless. However, when one of the genes is fixed, the population is homozygous and will remain homozygous.

The Hardy—Weinberg equilibrium is a good approximation for all but small populations, and retains its importance in evolutionary biology. It is also true that, once we allow for random drift, the Hardy—Weinberg ratios are not an equilibrium. The Hardy—Weinberg ratios apply to neutral alleles at a locus and the Hardy—Weinberg result suggests that the genotype (and gene) ratios are stable over time. However, in a small population, gene frequencies drift about, and one of the genes will eventually be fixed. Only at that point will the system be stable. The true equilibrium of the Hardy—Weinberg system in a small population appears at homozygosity.

Some geneticists would say that this conclusion is overdramatized. The Hardy-Weinberg theorem has two features. One feature specifies a certain ratio of genotype frequencies, given the gene frequencies; this ratio results from simple algebra and is not affected by drift. The other feature is the claim that the genotypic ratios, once reached, are stable. This is true only in the absence of drift, because drift alters the gene frequencies. However, some population specialists would deny that this second claim is substantially affected by drift. They would prefer to say that the Hardy-Weinberg equilibrium refers only to a large population, and events in small populations are unrelated to the Hardy-Weinberg equilibrium. The important thing to remember is small populations do not support Hardy-Weinberg equilibrium.

Neutral Drift

Over the long term, pure random drift causes the population to "march" to homozygosity at a locus. The process by which this march occurs has already been considered and illustrated. All loci at which several selectively neutral alleles appear will tend to become fixed for only one gene.

It is not difficult to derive an expression for the rate at which the population becomes homozygous. First, we define the degree of homozygosity. Individuals in the population are either homozygotes or heterozygotes. Let f = the proportion of homozygotes, and $H = 1 - f$ represent the proportion of heterozygotes (f comes from "fixation"). Homozygotes here include all types of homozygotes at a locus. If, for example, there are three alleles A_1 A_2, and A_3 then f is the number of A_1A_1, A_2A_2, and A_3A_3 individuals divided by the population size; H is the sum of all heterozygote types. N will again stand for population size.

How will f change over time? We will derive the result in terms of a special case: a species of hermaphrodite in which an individual can fertilize itself. Individuals in the population discharge their gametes into the water and each gamete has a chance of combining with any other gamete. New individuals are formed by sampling two gametes from the gamete pool. The gamete pool contains 2N gamete types, where "gamete types" should be understood as follows. There are $2N$ genes in a population made up of N diploid individuals. A gamete type consists of all gametes containing a copy of any one of these genes. Thus, if an individual, with two genes, produces 200,000 gametes, there will be, on average, 100,000 copies of each gamete type in the gamete pool.

To calculate how f, the degree of homozygosity, changes through time, we derive an expression for the number of homozygotes in one generation in terms of the number of homozygotes in the previous generation. We must first distinguish between *a* gene-bearing gametes in the gamete pool that are copies of the same parental *a* gene, and those that are derived from different parents. A homozygote may be produced in two ways: when two *a* genes from the same gametic type meet or when two a genes from different gametic types meet; the frequency of homozygotes in the next generation will be the sum of these two homozygote offspring.

The first way of making a homozygote is called "self-fertilization." There are $2N$ gamete types but, because each individual produces many more than two gametes, there is a chance $1/(2N)$ that a gamete will combine with another gamete of the same gamete type as itself. If

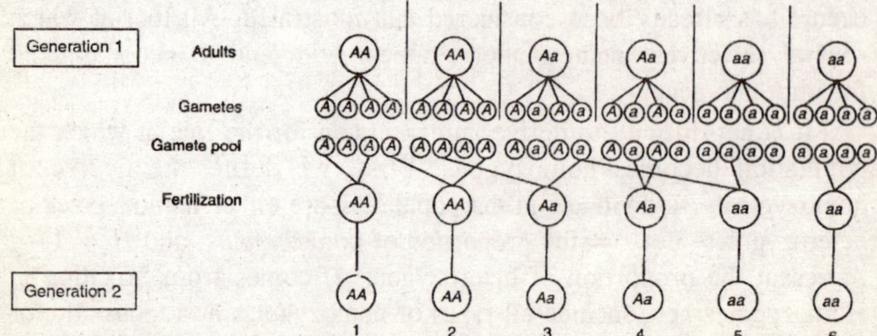

Fig. 10.5. Inbreeding in a small population produces homozygosity.

such a combination occurs, the offspring will be homozygous. (If, as above, each individual makes 200,000 gametes, the gamete pool would contain 200,000*N* gametes. We first sample one gamete from the pool. Of the remaining gametes, practically 100,000 of them—99,999 in fact—are copies of the same gene. The proportion of gametes left in the pool that contains copies of the same gene as the gamete we sampled is 99,999/200,000N, or $1/(2N)$.)

The second way to produce a homozygote is by combining two identical genes that were not copied from the same gene in the parental generation. If the gamete does not combine with another copy of the same gamete type—chance $1 - (1/(2N))$—it will still form a homozygote if it combines with a copy made from the same gene but from another parent. For a gamete with an *a* gene, if the frequency of *a* in the population is *p*, the chance that two *a* genes meet is simply p^2, which is the frequency of *aa* homozygotes in the parental generation. For two types of homozygote, *AA* and *aa*, the chance of forming a homozygote will be $p^2 + q^2 = f$. In general, the chance that two independent genes will combine to form a homozygote is equal to the frequency of homozygotes in the previous generation. The total chance of forming a homozygote by this second method is the chance that a gamete does not combine with another copy of the same parental gene, $1 - (1/(2N))$, multiplied by the chance that two independent genes combine to form a homozygote (f). That is, $f(1 - (1/(2N)))$.

Now we can write the frequency of homozygotes in the next generation in terms of the frequency of homozygotes in the parental generation. This frequency is the sum of the two ways of forming a homozygote. Following the normal notation for f' and f (f' is the frequency of homozygotes one generation later),

$$f' = \frac{1}{2N} + \left(1 - \frac{1}{2N}\right)f \qquad ...(2)$$

We can follow the same march to increasing homozygosity in terms of the decreasing heterozygosity in the population. *Heterozygosity* is a general measure of the genetic variation per locus in a population. A population may contain 1, 2, 3 . . . *n* alleles at a locus, and we can write the frequency of those alleles as $p_1, p_2, p_3 \ldots p_n$. In general, the frequency of the *i*th allele (where *i* can have any value from 1 to *n*, depending on the allele to which it refers) is p_i,. Heterozygosity is then defined as:

$$H = 1 - \Sigma p_i^2$$

This definition can be understood intuitively. In a population of size N, there are $2N$ genes per locus. Imagine drawing out two of those genes at random. The heterozygosity is the chance that the two will be different alleles. If only one allele exists, its frequency is one, and $H = 1 - 1 = 0$. When there are n alleles, for any one of them p_i^2: is the chance of drawing two copies of that allele—p_i^2, is the sum of all chances of drawing two copies of one allele. $1 - p_i^2$ is one minus the total chance of drawing two identical alleles for all alleles, which equals the chance of drawing two different alleles.

If the population is in Hardy-Weinberg equilibrium, then the heterozygosity equals the proportion of heterozygous individuals. H is a more general definition of genetic diversity than the proportion of heterozygotes, however. The chance that two random genes differ measures genetic variation in all populations, regardless of whether they are in Hardy-Weinberg equilibrium. $H = 50\%$, for example, in a population consisting of half *AA* and half *aa* individuals (with no heterozygotes).

Heterozygosity can be shown, by rearrangement of the equation, to decrease at the following rate (the rearrangement involves substituting $H = 1 - f$ in the equation):

$$H' = \left(1 - \frac{1}{2N}\right)H \qquad ...(3)$$

That is, heterozygosity decreases at a rate of $1/(2N)$ per generation until it reaches zero. The population size N is again important in governing the influence of genetic drift. If N is small, the march to homozygosity is rapid. At the other extreme, we reencounter the Hardy-Weinberg result. If N is infinitely large, the degree of heterozygosity is stable, and there is no march to homozygosity.

Although this derivation might seem to apply for only a particular, hermaphroditic breeding system, the result can, in fact, be extended to all cases (a small correction is needed for the case of two sexes). The march to homozygosity in a small population proceeds because two copies of the same gene may combine in a single individual. In the hermaphrodite, this combination happens obviously with self-fertilization. If there are two sexes, a gene in the grandparental generation can appear as a homozygote, in two copies, in the grandchild generation. The process by which a gene in a single copy in one individual combines in two copies in an offspring is called *inbreeding*. Inbreeding can occur in any breeding system with a small population, and its likelihood increases as the population size decreases. With inbreeding, homozygosity is increased over the Hardy-Weinberg level generated by random mating because it is impossible for two copies of the same gene to combine to form a heterozygote. The important point is that, in a small population, homozygosity is increased relative to an infinite population. In an infinite population, two copies of the same gene never fertilize each other; in a small population, there is some chance they will cross-fertilize.

Polymorphism and Neutral Mutation

It might appear that the theory of neutral drift predicts that populations should be completely homozygous. However, new variation will be contributed by mutation, and the equilibria} level of polymorphism (or heterozygosity) will actually be a balance between its elimination by drift and its creation by mutation.

We can now work out what that equilibrium is. The *neutral* mutation rate = u per gene per generation. (As before, u is the rate at which selectively neutral mutations arise, not the total mutation rate.) To find out the equilibrial heterozygosity under drift and mutation, we modify equation (2) to account for mutation. If an individual was born a homozygote, and if neither gene has mutated, it remains a homozygote and all of its gametes will have the same gene. (We ignore the possibility that mutation produces a homozygote—for example, by a heterozygote *Aa* mutating to a homozygous *AA'*, we are assuming that mutations produce new genes.) For a homozygote to produce all of its gametes with the same gene, *neither* of its genes must have mutated. If either of them has mutated, the frequency of homozygotes will decrease. The chance that a gene has not mutated = $(1 - u)$ and the chance that neither of an individual's genes has mutated = $(1 - u)^2$.

Now we can simply modify the recurrence relation derived earlier. The frequency of homozygotes will be as before, but multiplied by the probability that they have not mutated to heterozygotes:

$$f' = \left[\frac{1}{2N} + \left(1 - \frac{1}{2N}\right)f\right](1-u)^2 \qquad \ldots(4)$$

Homozygosity (f) will not increase to 1, but will converge to an equilibrial value The equilibrium lies between the increase in homozygosity by drift, and it' decrease by mutation. We can find the equilibrium value of f from $f^* = f = f'$. In this equation, f^* indicates a value of f that is stable in successive generation ($f^* = f$). Substituting $f^* = f' = f$ in the equation gives (after a minor manipulation):

$$f^* = \frac{(1-u)^2}{2N-(2N-1)(1-u)^2} \qquad ...(5)$$

The equation simplifies if we ignore terms in u^2, which will be relatively unimportant because the neutral mutation rate is low. Then

$$f^* = \frac{1}{4Nu+1} \qquad ...(6)$$

The equilibrial heterozygosity ($H^* = 1 - f^*$) is:

$$H^* = \frac{4Nu}{4Nu+1} \qquad ...(7)$$

This important result gives the degree of heterozygosity that should exist for a balance between the drift to homozygosity and new neutral mutation. The expected heterozygosity depends on the neutral mutation rate and the population size. Because the march toward homozygosity is more rapid if the population size is smaller, it makes sense that the expected heterozygosity is lower if *N* is smaller. Heterozygosity is also lower if the mutation rate is lower, as we should expect. In sum, the population will be less genetically variable for neutral alleles when population sizes are smaller and the mutation rate lower.

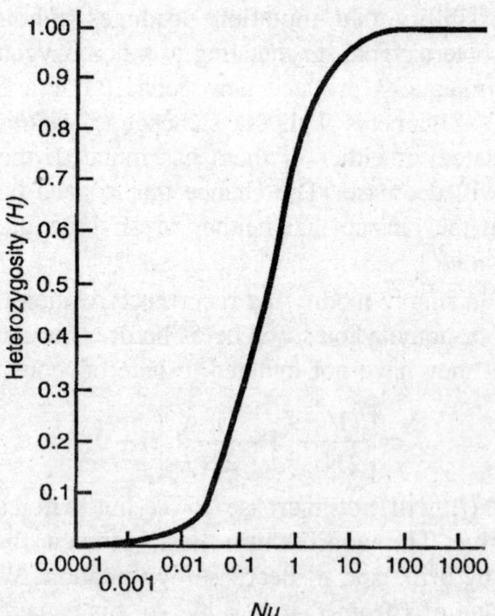

Fig. 10.6. The theoretical relation between the degree of heterozygosity and the parameter Nu (the product of the population size and neutral mutation rate).

Population Size

What is "population size"? We have seen that *N* determines the effect of genetic drift on gene frequencies. But what exactly is *N*? In an ecological sense, *N* can be measured by counting items, such as the number of adults in a locality. However, for the theory of population genetics with small populations, the estimate obtained by ecological counting is only a crude approximation to the "population size," *N*, implied by the equations. What matters is the chance that two copies of a gene will be sampled as the next generation is produced, which is affected by the breeding structure of the population. A population of size *N* contains 2*N* genes. The correct interpretation of *N* for the theoretical equations is that *N* has been correctly measured when the chance of drawing two copies of the same gene is $(1/2N)^2$.

If we draw two genes from a population at a locality, we may be more likely for various reasons to get two copies of the same gene than would be implied by the naive ecological measure of population size. Population geneticists there- fore often write N_e (for "effective" population size) in the equations, rather than *N*. In practice, effective population sizes are usually lower than ecologically observed population sizes. The relation between N_e, the effective population size implied by the equations, and the observed population size *N* can be complex. A number of factors are known to influence effective population size.

1. *Sex ratio*. If one sex is rarer, the population size of the rarer sex will dominate the changes in gene frequencies. It is much more likely that identical genes will be drawn from the rarer sex, because fewer individuals are contributing genes to the next generation. Sewall Wright proved in 1932 that in this case

$$N_e = \frac{4N_m \bullet N_f}{N_m + N_f} \qquad \ldots(8)$$

   where N_m = number of males, N_f = number of females in the population.

2. *Population fluctuations*. If population size fluctuates, homozygosity will increase more rapidly while the population goes through a "bottleneck" of small size. N_e is disproportionately influenced by *N* during the bottleneck, and a formula can be derived for N_e in terms of the harmonic mean of *N*.
3. *Small breeding groups*. If most breeding takes place within small groups, then the effective population size will be much smaller than if there is population-wide random mating (called "panmixis").

The degree of homozygosity will exceed that predicted from the size of the breeding groups drawn from the total population. Breeding groups can be small either because of geographic subdivision of the population or because of non-random mating. If certain types of individuals consistently choose a particular kind of mate, then the mating will be conducted through only a part of the whole population even without geographic subdivision. In either case, the chance that copies of the same gene will be "drawn" together is increased relative to the case with a large population. Several ways of creating population subdivisions are possible to derive expressions for N_e. Effective population size is mainly influenced by the degree of subdivision and the migration rate between the subgroups: the more migration occurs, the more N_e tends to N.

4. *Variable fertility*. If the number of successful gametes varies between individuals (as it often does among males when sexual selection operates), the more fertile individuals will accelerate the march toward homozygosity. Again, the chance that copies of the same gene will combine in the same individual in the production of the next generation is increased and the effective population size is decreased relative to the total number of adults, Wright showed that if k is the average number of gametes produced by a member of the population and σ_k^2 is the variance of k, then

$$N_e = \frac{4N - 2}{\sigma_k^2 + 2} \qquad \text{...(9)}$$

For $N_e < N$, the variance of k must be greater than random. If k varies randomly, as a Poisson process, $\sigma_k^2 = k = 2$ and $N_e = N$.

These are all quite technical points. The N_e found in the equations for neutral evolution is an exactly defined quantity, but it is difficult to measure in practice. It is usually less than the observed number of adults, N. $N_e = N$ when the population mates randomly, is constant in size, has an equal sex ratio, and experiences approximately Poisson variance in fertility. Natural deviations from these conditions produce $N_e < N$. How much smaller N_e is than N is difficult to measure, though it is possible to make estimates by the formulas we have previously used. Other things being equal, species with more subdivided and inbred population structures have lower N_e than more panmictic species.

11

Human Genome

In evolution, novel genes have arisen either by whole genome duplication or by regionally localized duplication events. Both mechanisms give rise to *paralogous* genes, genes that occur within the same species and which have a common ancestor. Thus, members of multigene families and superfamilies are paralogous and are distinct from *orthologous* genes which are found in different species and have diverged from their common ancestor over evolutionary time. In practice, the consequences of whole genome duplication and regionally localized gene duplication are often very hard to distinguish.

Ohno (1970) first proposed that the vertebrate genome had evolved to its present size through an ancient tetraploidization event. Evidence supporting this hypothesis now comes from a variety of different sources. Comparative nuclear DNA measurements in higher organisms are compatible with the idea of successive genome doublings. So are cytogenetic data, since the human karyotype can be divided into similar pairs of chromosomes on the basis of structure and banding patterns viz. chromosomes 1 and 2, 4 and 5, 7 and 8, 11 and 12, 14 and 15, 16 and 17, 19 and 20, 21 and 22. The postulated ancient tetraploidization event also appears to be reflected in the genetic information content of these chromosome pairs. This is exemplified by the distribution of the insulin and RAS oncogene family members in the human genome: the insulin (INS), insulin-like growth factor 2 (IGF2) and Harvey RAS (HRAS) genes are located on chromosome 11p whilst the insulin-like growth factor 1 (IGF1) and Kirsten RAS (KRAS2) genes are located on chromosome 12.

That the genome duplications occurred prior to the adaptive radiation of the vertebrates is evidenced by the similar gene number exhibited by both fish and mammals, the survival of syntenic linkage groups over quite long periods of evolutionary time and the tetralogy exhibited by many vertebrate loci. This tetralogy is evident from the observation that the basic set of ~15,000 genes found in all primitive metazoans varies only moderately from *Caenorhabditis elegans* to *Drosophila* to *Ciona intestinalis*, a tunicate but this gene complement is approximately four-fold smaller than the total number of genes in vertebrate genomes. At the level of the individual gene, there are numerous instances where an invertebrate (*Drosophila*) gene has up to four related genes (paralogues) in vertebrates, implying two rounds of genome duplication. These paralogues represent quadruplicated loci on different chromosomes that are more similar to each other than they are to members of other tetralogous groups. In humans, these so-called 'tetralogues' include the HOX genes (HOXA, 7p14-p15; HOXB, 17q21-q22; HOXC, 12q12-q13; HOXD, 2q31), the epidermal growth factor receptor genes (EGFR, 7p13; ERBB2, 17q11.2-q12; ERBB3, 12q13; ERBB4, 2q34), the Jak family of tyrosine kinases (jAK1, 1p31.3-32.3; jAK2, 9p24; jAK3, 19p12-p13; TYK2, 19p13.2), the MADS box enhancing factors (MEF2A, 15q26; MEF2B, 19p12; MEF2C, 5q14; MEF2D, 1q12-q23), the Src family of nonreceptor tyrosine kinases (SRC, 20q11.2; YES1, 18p11.22-p11.31; FGR, 1p36.1-p36.2; FYN, 6q21) and the syndecans (SDC1, 2p; SDC2, 8q22-q23; SDC3, 1p32; SDC4, 20q12-q13). Many further examples of triplicated loci occur in the human genome. In these cases, it may be that one paralogue has been lost or alternatively, still remains to be characterized.

The results of studies of the genes encoding the homeobox proteins, insulin-like growth factor genes and high mobility group proteins in the primitive chordates *Amphioxus* and *Ciona*, and a jawless vertebrates *Lampetra fluviatilis* (lamprey), are consistent with the occurrence of one genome duplication in the common ancestor of all jawed and jawless vertebrates after the lineage leading to *Amphioxus* had diverted and a second genome duplication occurring in the common ancestor of jawed vertebrates after their divergence from the jawless vertebrates. Studies of HOX gene number suggest that the duplication events are likely to have occurred before the radiation of the teleosts.

Gene number does not however automatically distinguish between tandem duplications and polyploidization events. Postlethwait et al. (1998) mapped 144 zebrafish (*Danio rerio*) genes; comparison of the resulting map with their mammalian counterparts led to the identifications of orthologous chromosome segments for at least three chromosome paralogy groups in zebrafish and mammals. This finding is consistent with the hypothesis that these segments were duplicated prior to the divergence of zebrafish and mammals. The presence of more than two copies of each paralogous chromosomal segment, is suggestive of at least two rounds of duplication which would have occurred after the divergence of the cephalochordates and cranial chordates, but before the divergence of the ray-finned and lobe-finned fishes, which is thought to have occurred about 420 Myrs ago.

Table 11.1. Possible paralogies between parts of human chromosomes 4 and 5

4	5
FGFR3	FGFR4
	HTR1A
	ADRB2
ADRA2C	ADRA1B
DRD5	DRD1
QDPR	DHFR
GABRA2	GABRA1
GABRB1	
STATH	SPARC
KIT	PDGFRB
PDGFRA	CSF1R
AREG	C7
EGF	C9
AGA	HEXB
FGF5	
FGF2	FGF1
IF	F12
F11	GZMA
KLK3	
	CSF2
IL2	IL3
	IL4
	IL5
	IL9
	CSF1
MLR	GRL
ANX3	
ANX5	ANX6

Several extensive regions of paralogy have been identified in the human genome which have been claimed to result from ancient tetraploidization events. The 13 groups of paralogous genes found on chromosomes 4 and 5 provide one example. Lundin (1993) identified several other possible examples of paralogous pairs or groups of genes on different human chromosomes; (i) parts of chromosomes 2, 7, 12, 14, and 17, (ii) parts of chromosomes 8, 10, and 16, and (iii) parts of chromosomes, 1, 11, 12, 15, and 19. Although the extensive paralogy noted between chromosomes 11 and 12 is explicable by a model of chromosome duplication resulting from tetraploidization, there are some discrepancies in the location of genes on these chromosomes. These can however be accounted for by the occurrence of a pericentric inversion on chromosome 12.

Paralogy may be explained by mechanisms other than tetraploidization. Indeed, some paralogous gene loci are explicable by regional duplication. The relative importance of regional duplication/translocation as compared to tetraploidization is unclear and ambiguity even extends to individual cases. Thus, the relative locations of the tyrosine hydroxylase (TH; 11p15.5), tryptophan hydroxylase (TPH; 11p14.3-p15.1) and phenylalanine hydroxylase (PAH; 12q22-q24) genes have been explained in terms of both mechanisms.

The two highly related regions on the proximal and distal long arms of human chromosome 21 (21q22.1 and 21q11.2) appear to have arisen as a result of an intrachromosomal duplication of >200 kb. This duplication is thought to have arisen between 15 and 30 Myrs ago after the separation of the orangutan from the other great apes. By contrast, the origin of the paralogous 2-20 Mb segments on human chromosomes, 1, 6, and 9 is unclear. Regardless of the mechanism, at least two intra-chromosomal duplications must have occurred resulting in the triplication of a series of genes, for example the retinoid X receptor genes and RXRG (1q22-q23) RXRB (6p21.3) and RXRA (9q34.31), the pre-B cell leukemia transcription factor genes PBX1 (1q23), PBX2 (6p21.3), and PBX3 (9q34), and the tenascin genes TNR (1q25-q31), TNXA (6p21.3), and HXB (9q32-q34). Interestingly, *Alu*- and LINE-dense clusters flank the boundaries of the 6p21.3 segment, a finding which may be significant in view of the recombinogenic potential of these sequence elements. In this context, it may be significant that a sequence related to the pseudoautosomal boundary of the human sex chromosomes has also been noted at the centromeric boundary of the 6p21.3 segment.

Table 11.2. Possible paralogies between parts of human chromosomes 1, 11, 12, 15 and 19

1	11	12	15	19
RYR2		CACNA1C		RYR1
		BCAT1		BCAT2
EN01		EN02		
GDH			SORD	
EBVS1	EBVM1			
				CEAL1
				CEA
PTPRF	NCAM1			MAG
				NCA
				PSG1-13
TNFR2		TNFR1		
SLC2A1		SLC2A3		
SLC2A5				
GOT2L1		GOT2L3		
GOT2L2				
GNAI3		GNA12L		
GNAT2				
GNB1		GNB3		
	LMO1	LMO3		
	LMO2			
MYCL1		MYF5		LYL1
MYOG	MYOD1	MYF6		TCF3
CHRM3	CHRM1,4		CHRM5	
	FGF3	FGF6		
	FGF4			
		A2M		
		PZP		
	ESA4	ELA1		C3
	F2	C1S		KLK1
		C1R		KLK2
			LIPC	LIPE
		TP11	MPI	GPI
LDHAL2	LDHA, LDHC	LDHB		
NRAS	HRAS	KRAS2		
RAP1A		RAP1B		RRAS
RAB3B				RAB3A

1	11	12	15	19
RAB4				
	CALCA	IAPP		
	CALCB			
	PTH	PTHLH		
FGR				
LCK	SEA		FES	
ABL2				
JUN				JUNB
				JUND
INSRR		ERBB3	IGF1R	INSR
NTRK1			LTK	TYK2
C8A, C8B		LRP1	THBS1	LDLR
GSTM1	GSTP1	MGST1		
C1QA, C1QB				
COL11A1		COL2A1		
COL8A2				
H1F2		H1F4		
	PGR	HMR		
		RARG	CRABP1	
		VDR		
HKR3	WT1	GLI		HKR1
				HKR2
PEPC		PEPB	ANPEP	PEPD
KCNC4		KCNA1,2,5		KCNA7
		KCNC2		KCNC3
ACADM		ACADS	IVD	
	TH, TPH	PAH		
NGFB	INS			
	IGF2	IGF1		
	INSL2			
	TRV2	TRV3		
	FRV1	FRV3		
PLA2G2A		PLA2G1B		
PLA2G2C				
TSHB	FSHB			LHB
				CGB
	ACP2		ACP5	
PKLR			PKM2	
ATP1A1				ATP1A3
ATP1A2				

1	11	12	15	19
ATP1B1				
ATP1AL2		ATP2A2		
ATP2B2		ATP2B1		
TRAP2		TRA1	TRAP1	
CD48	THY1			A1BG
	CD3D			
	CD3E			
CD3Z	CD3G			
CD1AE	FCER1B			FCER2
FCER1A				
FCER1G				
FCGR2A, 2B				
2C, 3A, 3B		CD4		
PIGR				
	APOA4			APOC1
APOA2	APOA1			APOC2
	APOC3			APOE
	FUT4			FUT1
	SIAT4C			FUT2
			MANA1	MANB
			CKMT1	CKM
	GANAB		GANC	
CAPN2	CAPN1			
			CAPN3	CAPN4
MUC1	MUC2,			
	MUC5B			
	MUC5AC			
AT3	C1NH			
AGT				
PFKM		PFKX		
FDPSL1			CHR39B	
ACTA1			ACTC	
POU2F1				POU2F2
TNNI1				TNNT1
SNRPE			SNPRN	SNRPA
				SNRP70
				AMH
TGFB2				TGFB1
CTSE	CTSD		CTSH	
REN	PGA3-5			

Table 11.3. Locations of human gene loci indicating large regional duplications of chromosome 1

TRN	tRNA, asparagine	1p36
TRN	tRNA, asparagine-like	1q12-q22
TRE	tRNA, glutamic acid	1p36
TREL1	tRNA, glutamic acid-like 1	1q21-q22
RNU1	Small nuclear U1 RNA	1p36
RNU1P1-4	Small nuclear U1 RNA pseudogenes 1-4	1q12-q22
FGR	Feline sarcoma oncogene	1p36
LCK	Lymphocyte protein tyrosine kinase	1p32-p35
ABLL	Abelson murine lukemia oncogene-like	1q24-q25
C8A, C8B	Complement component, 8, α/β chains	1p22-p36
C1QA, C1QB	Complement component 1q, α/β chains	1p
C4BPA, C4BPB	Complement component 4 binding protein	1q32
CR1, CR2	Complement component receptor 1/2	1q32
AK2	Adenylate kinase 2	1p34
GUK1	Guanylate kinase 1	1q32-q42
GOT2L1	Glutamic-oxaloacetic transaminase 2-like 1	1p32-p33
GOT2L2	Glutamic-oxaloacetic transaminase 2-like 2	1q25-q31
RAB3B	Member RAS oncogene family	1p31-p32
NRAS	Neuroblastoma RAS oncogene	1p13
RAP1A	Member RAS oncogene family	1p12-p13
RAB4	Member RAS oncogene family	1q42-q43
FTHL1	Ferritin, heavy polypeptide-like 1	1p22-p31
FTHL2	Ferritin, heavy polypeptide-like 2	1q32-q42
CD58	Lymphocyte function-associated antigen	1p13
DAF	Decay accelerating factor for complement	1q32
ATP1A1	ATPase, Na^+ K^+, α1 polypeptide	1p13
ATP1A2	ATPase, Na^+ K^+, α2 polypeptide	1q21-q23
ATP1B1	ATPase, Na^+ K^+, β polypeptide	1q22-q25

Consequences of Genome Duplications for Gene Evolution

In principle, the genetic redundancy created by a genome duplication would have allowed evolutionary experimentation, in that while one gene copy continued to function as before, the other was freed to acquire mutations, irrespective of whether they were adaptive or inactivating. If the newly duplicated gene acquired mutations that modified either the expression pattern of the encoded gene or the function of the encoded protein in an advantageous way, the novel allele could have become fixed in the population. Ohno (1970) expressed this idea rather elegantly: 'An escape from the ruthless pressure of natural selection is provided by the mechanism of gene duplication. By duplication, a redundant copy of a locus is created. Natural selection often ignores such a redundant copy, and, while being ignored, it accumulates formerly forbidden mutations and is reborn as a new gene locus with a hitherto non-existent function. Thus, gene duplication emerges as the major force of evolution.'

In this context, evidence for positive selection has come from the observation of accelerated evolution in some genes subsequent to gene duplication. In such studies, positive selection is implicated in the process of evolutionary change by the observation of a higher frequency of non-synonymous over synonymous substitutions. Changes in the expression patterns of the duplicated genes are sometimes also apparent as in the neuronal and muscle expressed genes of the nicotinic acetylcholine receptors family. In principle, gene conversion may either promote diversification of proteins encoded by duplicated genes or promote homogenization in which case the molecular evolutionary record is automatically erased. In practice, however, at least for the HLA and immunoglobulin genes, gene conversion is notable more by its absence: inserted new genes tend to be created by a 'birth-and-death' process of duplication and deletion.

A more likely scenario, however, is that the duplicated gene rapidly acquires inactivating mutations and becomes a pseudogene. Indeed, assuming that a gene duplication is not selectively disadvantageous, the duplicated genes can survive in the genome for quite long periods. Arguably the best available model system to assess whether duplicated genes will be retained or inactivated over evolutionary time is yeast (*Saccharomyces cerevisiae*), the organism with the best characterized genome duplication. Sequencing data from the 12 Mb yeast genome are consistent with the occurrence of a whole genome duplication that occurred ~100 Myrs ago, after the divergence of *S. cerevisiae* from *Kluyveromyces*. Most duplicated genes were subsequently deleted; only 13% of yeast proteins are now represented as homologous pairs encoded by homologous genes. Of these homologous gene pairs, only a few possess functions that have clearly diverged under the influence of selection viz. the mitochondrial and peroxisomal isozymes of citrate synthase (CIT1 and CIT2), RAS1 and RAS2, the transcription factors ACE2 and SW15, the phosphatidylinositol kinase TOR1 and TOR2, and the myosins MYO3 and MYO5. Conclusions drawn from yeast may not however be applicable to vertebrates and in this phylum, many more genes may have survived the aftermath of duplication to acquire new functions. Thus, Nadeau and Sankoff (1997) analyzed the frequency of distribution of family size for gene families present in humans and mice and which arose putatively by genome duplication early in vertebrate evolution. They concluded that duplicated genes were as likely to have survived and acquired a novel function as to have been lost through the acquisition of inactivating mutations. In agreement with these findings, studies of fish and *Xenopus* have both suggested that about 50% of newly duplicated genes are retained after tetraploidization.

Mammalian Genome Evolution

The best estimates of divergence times for the various mammalian orders and the other major vertebrate lineages have come from the use of extant gene sequences to calibrate a 'molecular clock' of vertebrate evolution. Gene-specific evolutionary rates often vary quite dramatically, but the use of multiple genes to derive mean divergence times should yield more accurate and reliable estimates. Kumar and Hedges (1998) therefore employed 658 genes from 207 vertebrates specie to derive a molecular timescale for vertebrate evolution. The divergence times corresponded well to previous estimates based upon the fossil record. Thus the calculated divergence time for the jawless fish (Agnatha) was 564 Myrs ago in the Precambrian era. Interestingly, the molecular data indicated that at least five major lineages of placental mammals [Edentata (armadillos, anteaters and sloths), Hystricognathi (porcupines and guinea pigs), Sciurognathi (squirrels), Paenungulata (hyraxes) and Ferungulata (carnivores)] could have arisen in the early to middle Cretacious between 130 and 90 Myrs ago. This represent an important revision of previous estimates of the timing of the adaptive radiation of the mammals. Since this now appears to have predated the Cretacious/Tertiary extinction of the dinosaurs 65 Myrs ago, the adaptive radiation of the mammals could not have been simply a consequence of the filling of niches vacated by the departing super-lizards. Other factors such as climatic changes and the continental breakup must also have played a role. This question notwithstanding, the adaptive radiation of the mammals has been very successful, resulting in the emergence of >4600 living species that occupy a very diverse range of habitats and environments.

It has been suggested that the rate of mammalian speciation may have been influenced by the rate of karyotypic change. However, mammalian genomes still contain significant regions of genetic linkage

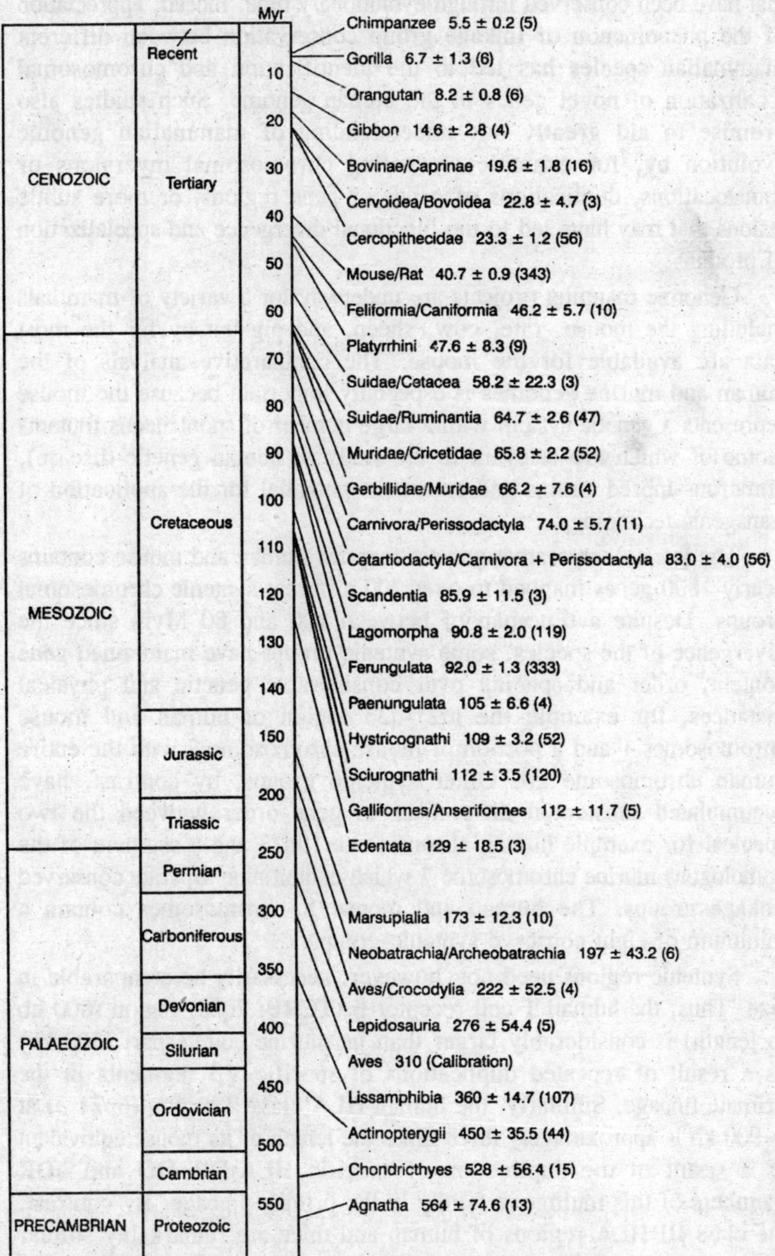

Fig. 11.1. A molecular timescale for vertebrate evolution.

that have been conserved through evolutionary time. Indeed, appreciation of the phenomenon of linkage group conservation between different mammalian species has led to the identification and chromosomal localization of novel genes in the human genome. Such studies also promise to aid greatly our understanding of mammalian genome evolution by, for example, revealing chromosomal inversions or translocations, duplications of genes or gene regions, or more subtle lesions that may have led to the functional divergence and specialization of proteins.

Genomic mapping projects are underway for a variety of mammals including the mouse, rate, cow, sheep, and pig but by far the most data are available for the mouse. The comparative analysis of the human and murine genomes is especially important because the mouse represents a genetic system with a large number of spontaneous mutants (some of which are relevant to the study of human genetic disease), numerous inbred strains and enormous potential for the application of transgenic technology.

The latest comparative genetic map for human and mouse contains nearly 1800 genes mapped to over 200 different syntenic chromosomal groups. Despite a timespan of between 100 and 80 Myrs since the divergence of the species, some syntenic groups have maintained gene content, order and spacing over considerable genetic and physical distances, for example the lq21-q23 region of human and mouse chromosomes 1 and a portion of mouse chromosome 2 with the entire human chromosome 20. Other syntenic groups, by contrast, have accumulated substantial differences in gene order between the two species, for example human chromosome 19q13 and a segment of the homologous murine chromosome 7 which exhibit nine separate conserved linkage groups. The human and mouse X chromosomes contain a minimum of eight conserve syntenic groups.

Syntenic regions need not, however, necessarily be comparable in size. Thus, the human T-cell receptor β (TCRB; 7q35) region (800 kb in length) is considerably larger than its murine counterpart (500 kb) as a result of repeated duplications of specific Vβ segments in the primate lineage. Similarly, the human HLA class II region (6p21.3) at ~900 kb is approximately three times the length of its mouse equivalent as a result of the duplication of specific HLA-DP,-DQ and -DR members of this multigene family in the primate lineage. By contrast, the class III HLA regions of human and mice are remarkably similar in structure.

Many syntenic blocks extend across human centromeres. Thus, gene order in the pericentric regions of human chromosomes 1, 2, 4, 6, 11 and 19 is conserved in the homologous regions of murine chromosomes 3, 6, 5, 1, 2, and 8 respectively. The majority of rearrangements appears to be due to inversions within chromosomes rather than rearrangements between chromosomes but more detailed mapping data will be required to obtain the map resolution necessary for a definitive assessment. One *caveat* which should be borne in mind is that the linkage map may have been broken up to a large extent in rodents than in primates as compared to the ancestral mammalian genome. One way to study this type of rearrangement is by the comparative analysis of the chromosomal locations of the individual gene loci involved. For example, Maresco et al. (1998) determined the locations of the high-affinity immunoglobulin receptor genes (FCGR1A, 1q21; FCGR1B, 1p12; FCGR1C, 1q21) in the rhesus monkeys (*Macaca mulatta*), baboon (*Papio papio*) and chimpanzee (*Pan troglodytes*) thereby providing evidence for the occurrence of two pericentric inversions during the evolution of human chromosome 1.

Conservation of synteny may extent beyond the mammals. For example, 11/18 genes from the chicken Z chromosome have orthologues on human chromosome 9pter-q22, albeit in a different order. What are the reasons for this degree of conservation? One reason may have been functional, for example in order to allow coordinate regulation of the genes involved or in order to avoid possible meiotic disturbance consequent to a major chromosomal rearrangement. In the case of the HLA system, synteny may have indirectly promoted the generation of diversity by optimizing the potential for gene conversion. Alternatively, it is possible that insufficient time has passed for ancestral linkages to have been broken up completely. Synteny may be conserved in one phylogenetic group but not in another. Thus, the surfeit genes are tightly clustered in human (SURF1, SURF2, SURF4, SURF5; 9q34.1) and chicken but not in invertebrates where the Surf genes are unlinked.

Assuming a human-rodent divergence time of 80 Myrs, Collins and Jukes (1994) estimated the rate of silent substitution to be 2.9×10^{-9} site^{-1} year^{-1}. However, this is very much an average figure arrived at by comparison of the 4-fold degenerate sites in the protein-coding sequences of 337 human/rodent gene comparisons. Sequence conservation does vary quite considerably between proteins: indeed a study of 1196 orthologous mouse and human protein sequences revealed sequence conservation of between 36% and 100% with an average of 85%.

Large scale genomic DNA sequence comparisons between human and mouse are still difficult owing to the paucity of orthologous pairs of sequences > 20 kb in length. However, Koop (1995) noted three distinct patterns of sequence divergence in the noncoding DNA sequence of human and rodent genomes:

(i) A high level of sequence similarity in gene regions contrasting with divergent non-coding regions, for example β-globin (HBB; 11p 15.5) and γ-crystallin (CRYGA; 2q33-q35) genes.

(ii) A conserved pattern of sequence similarity noncoding regions, for example T-cell receptor-α (TCRA; 14q11.2) and -δ (TCRD; 14q11.2) genes and α- and β-myosin heavy chain (MYH6, MYH7; 14q11.2-q13) genes. At least some of this conservation may be attributed to regions which bind T-cell nuclear proteins which may play a role in the control of gene transcription.

(iii) A mixed pattern of sequence similarity, for example the immunoglobulin heavy chain J-Cμ-Cδ gene region (14q32.33): the J-Cμ portion exhibits ~64% sequence homology between human and mouse while the Cδ region shows little if any sequence conservation.

This 'mosaic model' of genome evolution may reflect differing rates of mutation or differential repair efficiencies between different regions of the genome. As the sequences of further syntenic regions become available [e.g. the Bruton's tyrosine kinase (BTK; Xq21.33-q22) gene region], this question can be addressed.

PRIMATE EVOLUTION

Adaptation and Adaptive Radiation

The order of primates originated in the Palaeocene some 50-60 Myrs ago and contains the monkeys, apes and humans. Originally adapted for arborial life, extant primates include low canopy runners (guenons), high canopy acrobats (spider monkeys) and the brachiating great apes whilst some have become exclusively terrestrial (baboons, mandrills, and humans). These habits are responsible for the adaptation of the skeleto-muscular system to allow jumping, swimming and grasping. Primates exhibit a wide variety of characteristics which have equipped them to exploit their various niches optimally. The Anthropoidea, with their relatively large brain size, possess binocular vision and high visual acuity but a relatively poor olfactory sense. Their high intelligence allows rapid and measured reaction to external stimuli. The extension of the time periods for gestation, developmental

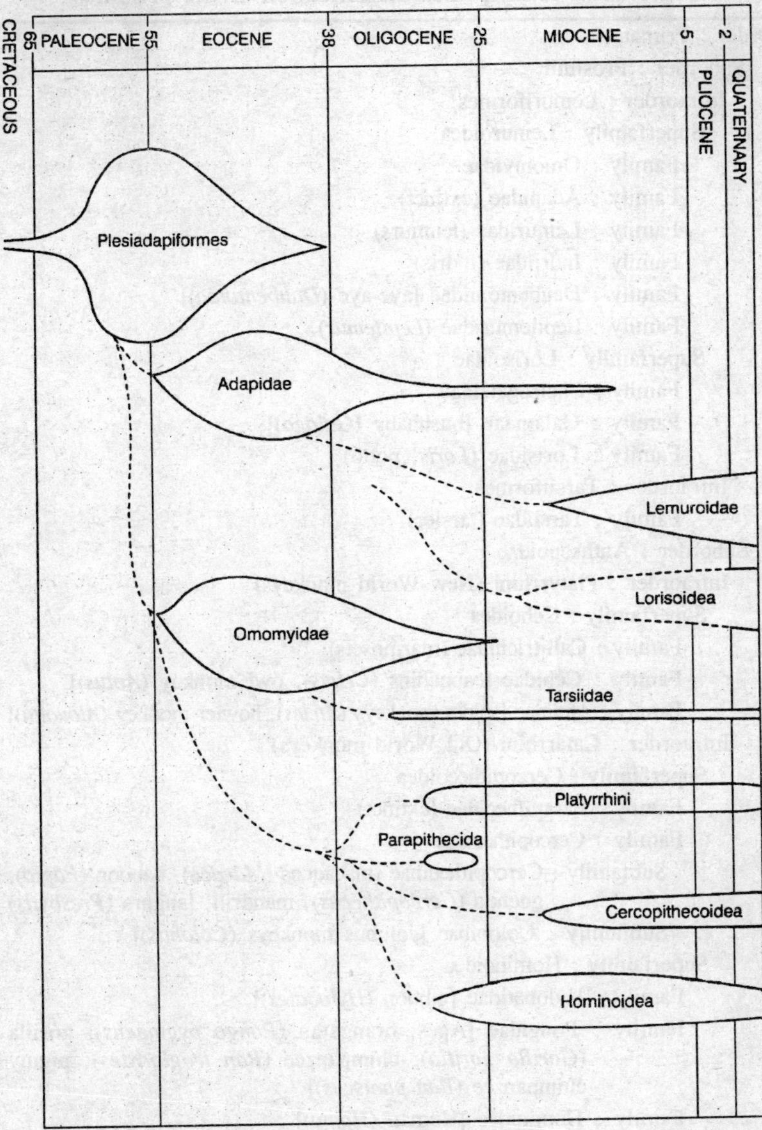

Fig. 11.2. Longevity of primate groups.

growth and parental care are associated with increased powers of learning whilst communication by use of elaborate vocal signals has allowed the development of complex patterns of social behaviour. The menstrual cycle with ovulation has been accompanied by the development

Table 11.4. A simplified classification of the primates

Order : Primates
- Suborder : Prosimii
  - Infraorder : Lemuriformes
    - Superfamily : Lemuroidea
      - Family : Omomyidae
      - Family : Adapidae (extinct)
      - Family : Lemuridae (leumurs)
      - Family : Indriidae (indris)
      - Family : Daubentoniidae [aye-aye (*Daubentonia*)]
      - Family : Lepilemuridae (*Lepilemur*)
    - Superfamily : Lorisoidae
      - Family : Cheirogaleidae
      - Family : Galagidae [bushbaby (*Galago*)]
      - Family : Lorisidae (*Loris*, potto)
  - Infraorder : Tarsiiformes
    - Family : Tarsiidae (tarsier)
- Suborder : Anthropoidae
  - Infraorder : Platyrrhini (New World monkeys)
    - Superfamily : Ceboidea
      - Family : Callitrichidae (marmosets)
      - Family : Cebidae [capuchins (*Cebus*), owl monkey (*Aotus*)]
      - Family : Atelidae [spider monkeys (*Ateles*), howler monkey (*Alouatta*)]
  - Infraorder : Catarrhini (Old World monkeys)
    - Superfamily : Cercopithecoidea
      - Family : Parapithecidae (extinct)
      - Family : Cercopithecidae
        - Subfamily : Cercopithecinae [macaques (*Macaca*), baboon (*Papio*), guenon (*Cercopithecus*), mandrill, langurs (*Presbytis*)
        - Subfamily : Colobinae [colobus monkeys (*Colobus*)]
    - Superfamily : Hominoidea
      - Family : Hylobatidae [gibbon (*Hylobates*)]
      - Family : Pongidae [Apes; orangutan (*Pongo pygmaeus*), gorilla (*Gorilla gorilla*), chimpanzee (*Pan troglodytes*), pigmy chimpanzee (*Pan paniscus*)]
      - Family : Hominidae [Human (*Homo*)]

of sexual behaviour and signaling in the female. Finally, the omnivorous diet of primates may have been responsible for the development of their characteristics manual dexterity which has allowed them both to explore and manipulate their environment.

The order Primates contains about 200 living species which are divided into two suborders: the Prosimii and the Anthropoidea. The Prosimii, which originated in the Palaeocene have retained the insectivoran characteristics of long face, lateral eyes and small brain. The Anthropoidea comprise the Old World monkeys, the New World monkeys and the great apes. The New World or platyrrhine (flat nosed) monkeys are thought to have been isolated since the Eocene. The Old World or catarrhine monkeys share a common ancestor in the late Eocene and do not differ markedly in either habitats or organizations from the New World monkeys. The great apes include the gibbon and orangutan from East Asia and the chimpanzee and gorilla from Africa.

Primate Phylogeny

The phylogeny of the hominoid primates was initially investigated by means of DNA-DNA hybridization. These studies employed single copy nuclear DNA to calculate the temperature ($T_{50}H$) in degrees Celsius at which 50% of all single copy DNA sequences were in the hybrid form and 50% had dissociated. The delta $T_{50}H$ between chimpanzee and human is 1.6. Assuming a relationship of delta $T_{50}H$ = 1% base mismatches, this translates into $\sim 3.2 \times 10^7$ mismatches between the chimpanzee and human genomes. Sibley and Ahlquist (1987)

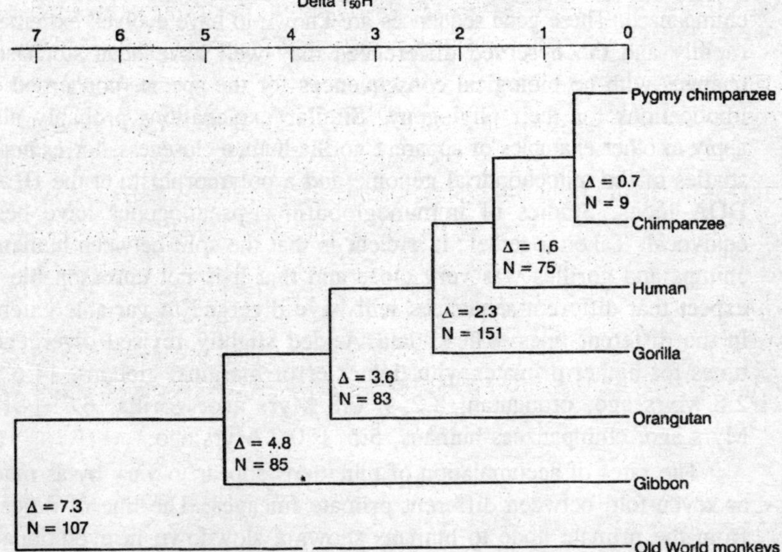

Fig. 11.3. Phylogeny of the hominoid primates as determined by average linkage clustering of delta $T_{50}H$ values derived from DNA-DNA hybridization.

estimated times of divergence for higher primates as: Old World monkeys, 25-34 Myrs ago; gibbons, 16.4-23 Myrs ago; orangutan, 12.2-17 Myrs ago; gorillas, 7.7-11 Myrs ago, chimpanzees-humans, 5.5-7.7 Myrs ago. It is now recognized that the chimpanzee is actually represented by two distinct species, the common chimpanzee (*Pan troglodytes*) and the pigmy chimpanzee (*Pan paniscus*) which diverged from each other ~2.3 Myrs ago.

The studies of Sibley and Ahlquist (1984, 1987) received support from Caccone and Powell (1989). However, the validity of conclusions drawn from DNA-DNA hybridization data has been challenged and the interpretation of these studies is still somewhat contentious. Sibley and Ahlquist's scheme is nevertheless in broad agreement with the fossil record, comparative morphology, immunological studies and chromosome phylogeny as well as being compatible with studies of individual DNA sequence. Thus, a similar picture has emerged from maximum parsimony analysis of DNA sequences derived from primate β-globin gene regions, the c-*myc* oncogene, the ribosomal DNA genes, the α-1,3-galactosyltransferase gene, the cytochrome P450 CYP21 gene and mitochondrial DNA as well as from transversion rates in nuclear and mitochondrial DNA.

However, divergence data from primate protamine P1 and α-fetoprotein gene sequences place gorilla closer to human than to chimpanzee. These gene sequences are known to have evolved extremely rapidly and the observed differences may well have been stochastic changes with no biological consequences for the species concerned or implications for their phylogeny. Similar explanations probably also apply to other examples of apparent gorilla-human closeness, for example studies of the mitochondrial genome and a polymorphism at the HLA-DQA locus. Studies of immunoglobulin ε pseudogenes have been equivocal. Taken together, it is clear is that the split between humans, chimps and gorillas was very close and that it is not unreasonable to expect that different sequences will have diverged to variable extents in the different lines. These data yielded slightly revised divergence times for higher primates with tighter error margins: gibbons, 14.6 ± 2.8 Myrs ago; orangutan, 8.2 ± 0.8 Myrs ago; gorilla, 6.7 ± 1.3 Myrs ago, chimpanzees-humans, 5.5 ± 0.2 Myrs ago.

The rates of accumulation of mutations appear to vary by as much as seven-fold between different primate lineages. The line of descent from the primate node to humans shows a slowdown in evolutionary rates from 7.7×10^{-9} fixed changes site^{-1} year^{-1} for the first 15 Myrs

(55-40 Myrs ago) to 1.3×10^{-9} for the next 15 Myrs (40-25 Myrs ago) 1.0×10^{-9} for the last 25 Myrs. The average evolutionary rate for the hominoids (1.1×10^{-9}) is lower than the rates for macaque, a catarrhine (1.9×10^{-9}) and for spider monkey, a platyrrhine (1.8×10^{-9}). By comparison, the line of descent from primate node to *Tarsius* shows an evolutionary rate of 3.4×10^{-9} fixed changes site^{-1} year^{-1} which is approximately half the stem-simian rate. The *hominoid slowdown* is at its greatest in human although anatomically, humans are quite divergent. Clearly, changes in certain key genes must have assumed a critical importance. As if perhaps to emphasize this point, some gene sequences appear to buck the trend; thus, the evolutionary rate of the noncoding region of the immunoglobulin-α gene is greater in hominoids than in Old World monkeys.

Chromosome Evolution in Primates

Old World primates

The evolution and probable phylogeny of primate chromosomes has been extensively reviewed by both Rumpler and Dutrillaux (1990) and Clemente et al. (1990). The interested reader is referred to these reviews for detailed accounts. Some chromosomes appear to have been relatively protected from change during primate evolution, for example human chromosomes 19 and X. By contrast, other chromosomes have been prone to significant reorganization, for example human chromosomes 1, 3, and 7.

The most frequent types of chromosomal change detected in primate evolution are inversions, changes in the amount and localization of heterochromatin, fusion and fissions, and changes in the location of centromeres due to activation/inactivation. Reciprocal translocations, deletions and insertions are much less frequent. Human chromosome 18 differs from the homologous chromosomes in the great apes by a pericentric inversion and it is thought that one inversion breakpoint may have been located at or within the centromere. Pericentric inversion breakpoints have also been identified on the chimpanzee equivalents of human chromosomes 4 (4p14, 4q21), 9 (9q22) and 12 (12p12 and 12q15) and these appear to coincide with the locations of either fragile sites or tumor-associated break-points. Pericentric inversions may have played an important role in establishing reproductive isolation and speciation during the evolution of the higher primates.

There are at least a dozen blocks of X-Y sequence homology outwith the pseudoautosomal region of humans but these blocks occur in a very different order and orientation on these chromosomes. This

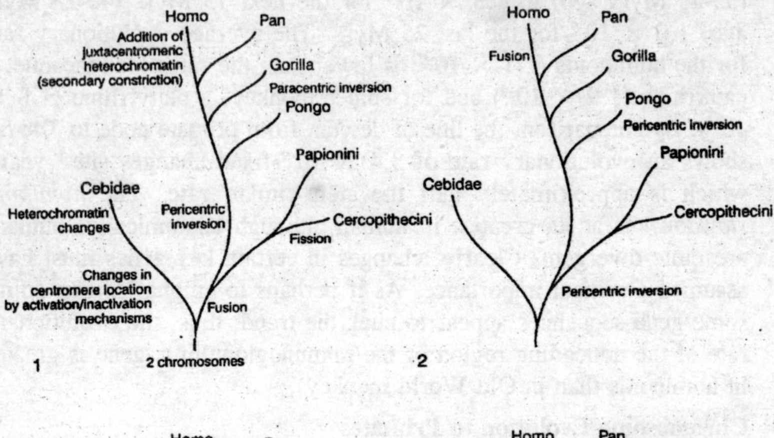

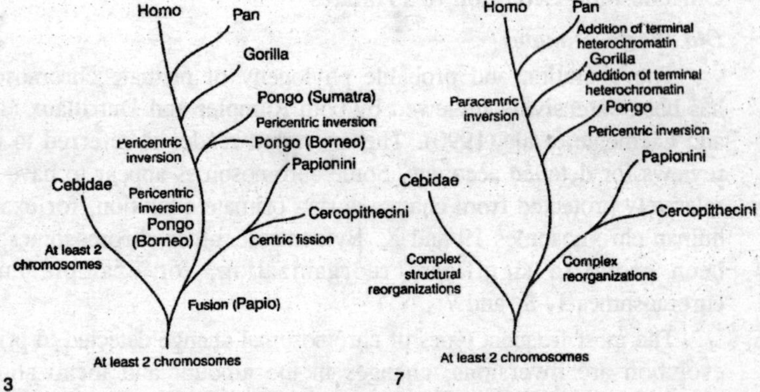

Fig. 11.4. Chromosome evolution in primates.

may be accounted for in terms of the occurrence of a number of different inversions, transpositions and other rearrangements during primate evolution. One example of a human-specific inversion is that involving the short arm of the Y chromosome. Sine humans from different racial groups all possess this Yp inversion, the rearrangement must have occurred prior to the divergence of the human racial groups.

Studies of chromosome banding patterns and hybridization homologies between ape and human chromosomes have provided evidence for human chromosome 2 having arisen form the fusion of two ancestral simian chromosomes showed that this probably occurred by telomere-telomere fusion at 2q13 rather than by translocation after chromosome breakage. This fusion has subsequently been confirmed by chromosome painting and since it accounts for the reduction in chromosome number from 24 pairs in the great apes (chimpanzee, orangutan, and gorilla) to 23 pairs in humans, it must have been a relatively recent event. Fusion must have been accompanied or followed by inactivation or removal of one of the ancestral centromeres. Consistent with this postulate, IJdo et al. (1991) found evidence by hybridization for the residual presence of an ancestral centromere at 2q21. Clearly, this reduction in chromosome number would have been a critical event during the speciation process; if it was not in itself responsible for bringing about reproductive isolation, it would certainly have helped to maintain it.

Another major chromosomal rearrangement to have occurred in the great apes is to be found in the gorilla. Using human chromosome-specific libraries as probes for *in situ* hybridization, Stanyon et al (1992) described a reciprocal translocation in the gorilla lineage which is not present in the chimpanzee. Thus, chromosomes 4 and 19 of the gorillas were derived from a reciprocal translocation between the ancestral chromosomes homologous to human chromosomes 5 and 17. Wienberg et al. (1990) used chromosomal *in situ* suppression hybridization to demonstrate that the centromere of human chromosome 17, the long arm of the chromosome and a small part of the short arm

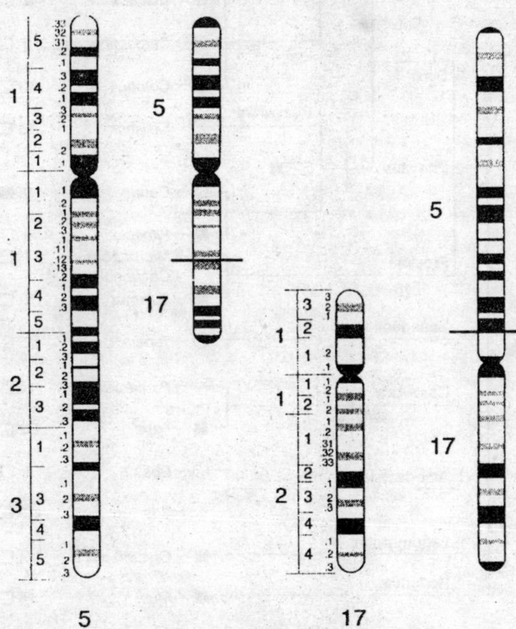

Fig. 11.5. Idiograms of human chromosome 5, gorilla chromosome 19, human chromosome 17, and gorilla chromosome 4.

all contribute to gorilla chromosome 4 whilst most of the short arm of chromosome 17 contributes to gorilla chromosome 19.

Probably the best understood human chromosome in terms of its evolution is chromosome 21. The equivalent of human chromosome 21 (HSA21) formed a large and unique chromosome together with chromosome 3 (HSA3) in the eutherian ancestor. This chromosome was conserved without significant alterations only in lemurs, the civet and the pig. It underwent inversions in the tree shrew and the cow. Various translocations involving the portion corresponding to HSA3 occurred in the brown lemur, cat, rabbit and mouse. In the primates, two independent fissions occurred. In New World monkeys, a small segment of HSA3 remained attached to SHA21 and this chromosome

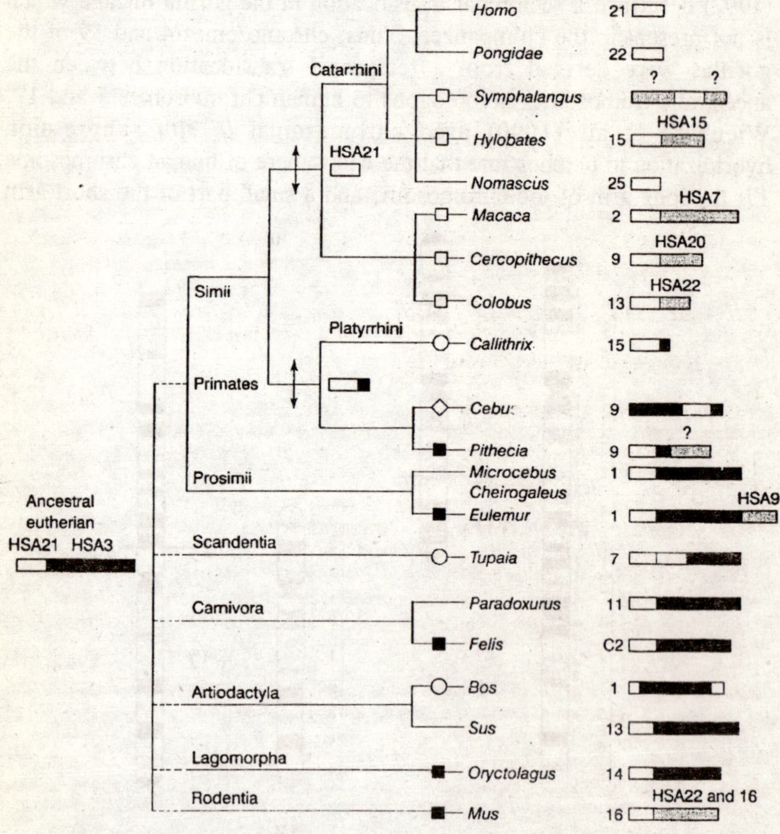

Fig. 11.6. Evolution of the equivalent of human chromosome 21 (HSA21) in eutherian mammals.

then underwent further rearrangements; an inversion in the marmoset, the addition of heterochromatin in the capuchin monkey and a translocation in the saki monkey. HSA21 was formed in the common ancestor of Old World monkeys and underwent translocations with various equivalents of human chromosomes in all the Cercopithecidae. HSA21 was conserved without visible alteration in the black gibbon and the great apes.

One technique which is proving extremely useful in primate cytological studies in cross-species chromosome painting (CSCP; also known as comparative painting or ZOO-FISH). CSCP involves the hybridization of a chromosome-specific paint from one species (usually human) onto metaphase spreads of another species. This approach allows the rapid construction of chromosome maps from the primate species in question which can reveal cytogenetic homologies with the human karyotype. By these means, a high degree of synteny has been found between human and baboon chromosomes. By contrast, numerous translocations are apparent in the gibbon (*Hylobates lar*, 2n = 44) genome as compared to human and the great apes; the 22 human autosomes have to be divided into 51 elements in order to recombine them into the 21 gibbon autosomes. Similarly, in the Concolor gibbon (*Hylobates concolor*, 2n = 52), the 22 human autosomes have to be divided into 63-67 segments in order to recombine them into the 25 gibbon autosomes.

Despite the degree of genetic similarity between the great apes, the chimpanzee genome is approximately 10% larger than that of the human of gorilla. Nevertheless, the structure of the human and chimpanzee genomes is highly conserved at both chromosomal and sub-chromosomal levels. Even at lower resolution, the extent of evolutionary conservation is readily apparent. Human microsatellite DNA sequences are sufficiently well conserved in chimpanzees that human PCR primers can be used to amplify $(CA)_n$ repeats in the chimpanzee. Differences in microsatellite allele length between humans and other primates have however been noted. Crouau-Roy et al. (1996) studied microsatellites within a 30 cM region of human chromosome 4p and found that all informative loci which are linked in human were also linked in the chimpanzee, indicating that evolutionary conservation extends to the locus level. In general, heterozygosity was found to be greater in chimpanzees, a reflection perhaps of the greater genetic diversity in chimpanzee populations. Some loci, however, appeared to be less heterozygous than in human, a phenomenon that appears to be caused by interruptions of the repeat elements at these loci.

Human chromosomes are not always identical. Indeed many exhibit heteromorphism, especially in the centromeric and satellite regions of the acrocentric chromosomes. Chromosomes 1, 9, 13, 14, 15, 16, 19, 21, 22, and Y are the most heteromorphic whilst chromosomes 2-8 and X are the least heteromorphic. Inter-chromosomal variation can be substantial; two homologues of chromosome 21 having been noted to vary in healthy individuals by as much as 21 Mb or 40% of the length of the chromosome. Family studies have shown that such heteromorphisms are not artefactual and can be inherited in mendelian fashion. It would appear that this variation can be largely ascribed to variation in the size of repeat sequence arrays and probably results from unequal crossing over between different classes of repetitive element. Bivariate flow karyotyping has been used to study the relative DNA content of homologous chromosome pairs in individuals from different racial groups. Significant variation in DNA content, ranging from 10 to 40%, was found for chromosomes 1, 13, 14, 15, 16, 19, 21, 22, and Y. However, the spectrum of variation observed in the different racial groups was very similar.

New World primates

Like gibbons, many New World primates possess highly rearranged genomes. All new World primates have a translocation between chromosomes 8 and 18. The Cebidae possess translocations between chromosomes 10/16 and 2/16 whilst the Atelidae are characterized by translocations between 3/15 and 4/15. Significant synteny is nevertheless apparent between human and the capuchin monkeys, *Cebus*, between human and the Colobus monkey, between human and the howler monkey, between human and the black-handed spider monkey, *Ateles geoffroyi* and between human and the marmoset *Callithrix*.

Evolution of the Human Sex Chromosomes and the Pseudoautosomal Regions

The human sex chromosomes are heteromorphic; the X chromosome contains ~160 Mb DNA and perhaps 3000 genes whereas the Y chromosome contains only 60 Mb DNA and probably only a handful of genes. The Y chromosome is largely composed of constitutive heterochromatin harboring different families of repetitive DNA. Despite the size difference between the sex chromosomes, they are nevertheless able to pair successfully during meiosis. Recombination, however, is largely confined to the two *pseudoautosomal regions* (PARs); a major PAR (2.6 Mb in size) at the tips of the short arms of X and Y chromosomes which is the site of an obligate crossover during male meiosis, and a minor PAR (320 kb) at the tips of the long arms of the X and Y chromosomes. Regular X-Y recombinational exchanges have served to maintain homology between the chromosomes in the PAR regions.

Gene mapping studies have shown that part of the eutherian (placental) mammalian X chromosome ('conserved region'; XCR) is shared by the X chromosomes of marsupials and monotremes. Since a series of genes on the short arm of the human X are clustered in two autosomal groups in marsupials and monotremes, these loci define a region (XRA) that has been recently added to the X chromosome in eutherian mammals. Thus, the X chromosome of the common mammalian ancestor was smaller than that found in extant eutherians and at least two autosomal regions have since been added to it. The location of this X-autosome fusion corresponds to the border between the XCR and the XRA and lies at Xp11.23.

The human X and Y chromosomes also exhibit substantial homology outwith the pseudoautosomal region. Although this homology is consistent with these chromosomes having once constituted a homomorphic pair, the observed homologies also reflect intra-chromosomal duplication events followed by inter-chromosomal translocations. For example, the

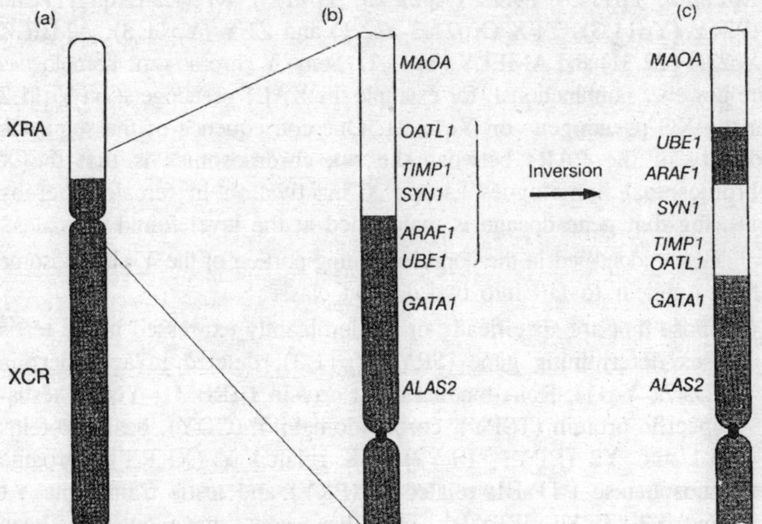

Fig. 11.7. (a) Evolutionary origins of the human X chromosome. XCR, conserved region of the X. XRA, recently added region. (b) Details of original gene order at the fusion point. (c) Possible inversion event that placed the XCR genes, UBE1 and ARAF1, between the XRA genes, SYN1 and MAOA.

minor PAR is thought to have originated during human evolution by the translocation of 320 kb of X chromosomal sequence to the Y chromosome, an event which may have been mediated by recombination between LINE elements.

The sex chromosomes however vary dramatically in terms of their evolutionary conservation. The Y chromosome has undergone significant changes during mammalian evolution. By contrast, the X chromosome exhibits conservation of synteny between human and mouse. This is considered to be a sequence of X inactivation because X-autosome translocations would have tended to be disadvantageous by virtue of their interference with the dosage compensation mechanism. Despite this, gene order on the X chromosome has changed significantly between mouse and human with numerous inversions altering the relative position of genes.

The human Y chromosome contains a number of active genes. Most notably it hosts the rapidly evolving sex determining gene, SRY (Yp11.3) which occurs in the sex chromosome-specific region, about 5 kb from its boundary with the major PAR. In human, a few X-linked genes have functional homologues on the Y chromosome (e.g. CSF2RA (Xp22.32, Yp11.3), MIC2 (Xp22.23, Yp11.3), RPS4X (Xq31.1) and RPS4Y (Yp11.3), ZFX (Xp21.3-p22.1) and ZFY (Yp11.3), AMELX (Xp22.1-p22.31) and AMELY (Yp11.2). Some Y chromosome homologues are however nonfunctional, for example the KAL1 pseudogene on Yq11.2 or the XG pseudogene on Yq11.21. One consequence of the sequence identity of the PARs between the sex chromosomes is that the X chromosomal homologues escape X inactivation in females thereby ensuring that gene dosage is maintained at the level found in males.

Genes identified in the nonrecombining portion of the Y chromosome (NRY) appear to fall into two distinct classes.

(i) Those that are specifically or predominantly expressed in the testis [sex determining gene (SRY; Yp11.3), deleted in azoospermia (DAZ; Yq11), RNA-binding motif protein 1 (RBM1; Yq11), testis-specific protein (TSPY), chromodomain Y (CDY), basic proteins Y1 and Y2 (BPY1, BPY2), XK related Y (XKRY), tyrosine phosphatase PTP-BL related Y (PRY) and testis transcripts Y1 and Y2 (TTY1, TTY2)]. That these genes have not only been retained on the Y chromosome, but in two cases have also been amplified, may have been part of an evolutionary strategy to optimize male reproductive fitness.

(ii) Those that are widely or ubiquitously expressed and which have closely related counterparts on the X chromosome [dead box Y (DBY), thymosin β4 (TB4Y), translation initiation factor 1A (EIF1AY), ubiquitous TPR motif Y (UTY) and *Drosophila* fat facets-related (DFFRY; Yp11.2), AMELY (Yp11.2), RPS4Y (Yp11.3), zinc finger protein Y (ZFY; Yp11.3) and SMCY]. Conservation of specific X–Y gene pairs may have been associated with a requirement to maintain comparable expression levels for certain housekeeping genes between males and females. Consistent with the predictions of this postulate, the X chromosome homologues of these Y-borne genes escape X inactivation.

The mammalian sex chromosomes are thought to be descended from a homologous pair of autosomes. This process could have been initiated with the evolutionary appearance of the testis-determining SRY gene on the nascent Y chromosome, probably by duplication, translocation and subsequent divergence of the X-linked SOX3 (Xq26-q27) gene. Suppression of recombination with its homologous chromosome (the nascent X) then led to the gradual degeneration of the Y chromosome owing to its inability to segregate genes carrying deleterious alleles. Evidence for this degenerative process comes from several sources. The rate of nucleotide substitution in Y chromosome genes appears to be ~2-fold higher than the rate exhibited by X chromosomal genes although the frequency of DNA sequence polymorphism may be lower in the sex-specific region of the Y chromosome than in the PARs. The Y chromosome also exhibits a high frequency of retroviral insertion in humans, chimpanzees and orangutans. Finally, there is emerging evidence for gene loss from the Y chromosome during mammalian evolution. In mouse and human, the ubiquitin-activating enzyme (UBE1) gene is located on the X chromosome (Xp11.23-p11.3). A copy of the *Ube1* gene is also located on the Y chromosome in the mouse, ring-tailed lemur, squirrel monkey (*Saimiri sciureus*) and marmoset (*Callithrix jacchus*) but not in the Old World monkeys, chimpanzee or human, indicating loss of the Y-linked gene >35 Myrs ago during the evolution of the primates. Similarly, the Y-linked copy of the EIF2S3 gene (Xp22.1-p22.2) encoding the eukaryotic translation initiation factor EIF-2γ was lost 35-60 Myrs ago in a common ancestor of the simian primates. The gradual degeneration of the Y-chromosome implies that the retention of functional gene copies on this chromosome for a significant period of evolutionary time could have conferred some selective advantage.

It should however be noted that the Y chromosome can also acquire genetic material from other chromosomes. For example, the multicopy DAZL1 gene (Yq11.23; deleted in azoospermia) was transposed to the Y chromosome from an autosome during primate evolution as was the multicopy RNA-binding motif (RBM1; Yq11) gene. It may be that transfer to a male-specific location provided protection against inactivation or loss. Another example of the duplicational transposition of a gene to the Y chromosome is that of AMELX (Xp22.1-p22.31) and its Y-chromosome counterpart AMELY (Yp11.2); the latter gene, which appears to be fully functional, is present on the Y-chromosomes of bovids and primates but not rodents thereby dating the transpositional event to at least 40 Myrs ago.

In humans, the XG blood group gene (Xp22.32) spans the major PAR on the X chromosome—the first three exons are pseudoautosomal whereas the remaining seven are X chromosome-specific. In humans and the great apes, an *Alu* sequence is located at the boundary between the major PAR and the Y chromosome-specific DNA but this sequence is not present in Old World monkeys. The *Alu* sequence was therefore inserted into the pre-existing boundary after the divergence of the great apes from the Old World monkeys. Although it did not create the boundary, the *Alu* sequence does serve to demarcate it.

Ellis et al. (1994) proposed a model for the formation of the boundary of the major PAR. They hypothesized a pericentric inversion of the Y chromosome with one breakpoint in the ancestral XG gene and the other breakpoint 5 kb distal to the ancestral SRY gene. In a refinement of this postulate, Fukagawa et al. (1996) suggested that the inversion occurred by illegitimate recombination between two PAR boundary sequences, one in the ancestral XG gene and the other near the ancestral SRY gene.

The PAR has undergone quite rapid change during mammalian evolution involving both gene duplication and translocation events in the region (e.g. STS, MIC2, XG, CSF2RA, IL3RA, ARSD, ARSE) and resulting in the movement of the PAR boundary to create X-unique regions. The evolution of the PAR and the divergence of the mammalian X and Y chromosomes may be viewed in terms of the 'addition-attrition' hypothesis. This states that the incorporation of autosomal sequences into the PAR of either the X or Y chromosome initially served to generate homologous regions which could pair at meiosis. Recombination with an homologous partner could then result in PAR enlargement. Alternatively, the steadily accumulating mutations on the Y chromosome would have served to decrease the level of homology to the X chromosome thereby reducing PAR size. Fukagawa et al. (1996) proposed a further twist to this argument in that once divergence had reached a certain level, recombination frequency would have decreased thereby further increasing the rate of divergence.

Evidence in favour of the addition-attrition theory comes from the dynamic nature of the major PAR region during mammalian evolution. Thus, the STS gene which is X-linked in humans and the great apes (Xp22.32) is autosomal in prosimians as the ANT3 gene which is pseudoautosomal (Yp11.3) in humans. Similarly, the human pseudoautosomal genes CSF2RA (Xp22.32/Yp11.3), SHOX (Xp22/Yp11.3) and IL3RA (Xp22.3/Yp11.3), are autosomal in the mouse. Further evidence for the process of attrition may come from the finding that whereas the *Fxy* gene spans the PAR on the murine X chromosome, its human counterpart (FXY; Xp22.3) lies proximal to the human PAR.

The 'X-driven' hypothesis of Graves (1995; 1998) essentially proposes that the rapid evolutionary spreading of X inactivation preceded the decay of Y chromosomal genes and even drove its initial steps. This hypothesis predicts that inactivated X-linked genes with functionally comparable Y-linked homologues should exist as evolutionary intermediates, but as yet no such gene has been found in any mammalian species. Indeed, a considerable number of human X-linked genes escape X-inactivation but have no detectable Y-borne counterpart.

An alternative 'Y-driven' pathway of X-Y gene evolution has been proposed by Jegalian and Page (1998). Briefly, these authors suggested that many extant genes represent intermediates on a general pathway by which X-Y genes or gene clusters evolved form autosomal genes. Autosomal genes would have entered the pathway either by virtue of their presence on the emergent sex chromosomes or via translocation of an autosomal gene. This would have been followed by suppression of X-Y recombination. These steps occurred either at the chromosomal or sub-chromosomal level and gave rise to functionally equivalent X-linked (but not inactivated) genes and Y-linked genes. Subsequently, three different processes (Y gene decay, upregulation of X-linked gene expression, and X-inactivation) interacted resulting in an inactivated X-linked gene accompanied by the loss of the Y gene. Expression of the X-linked gene then increased as an adaptation to the reduced or restricted expression of its Y-linked counterpart. This compensated for the loss of Y gene function and restored optimal expression levels in males. X-inactivation, on the other hand, may be viewed as a counter-

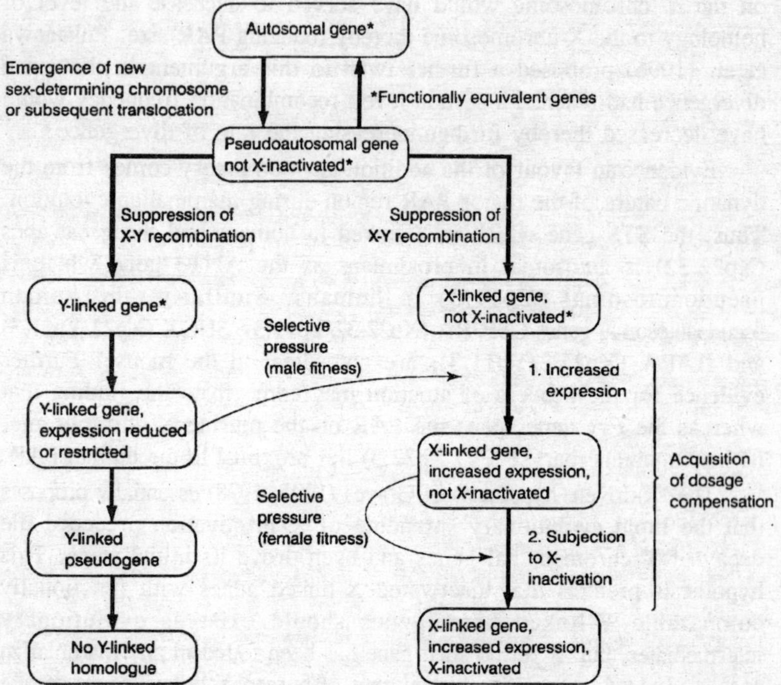

Fig. 11.8. A proposed Y-driven pathway for X-Y gene evolution in mammals.

response which restored optimal expression levels in females. This explanation could in principle account for most X-linked and X-Y homologous genes in extant mammals, many of which exist at intermediate steps in the pathway. Jeglian and Page (1998) pointed out that only one gene cannot be accommodated in their pathway schema: the human pseudoautosomal gene, SUBL1, which is X-inactivated and transcriptionally silenced on the Y chromosome.

Evolution of the Mitochondrial Genome

The 16 569 bp of the human mitochondrial genome encodes 13 polypeptides, all subunits of the enzyme complexes of the pathway of oxidative phosphorylation, and a total of 22 tRNAs. The mitochondrial genome is characterized by its high proportion of coding DNA, the paucity of repetitive DNA sequence, the absence of introns within its genes and its own genetic code distinct from that of the nuclear genome. Mitochondrial genes experience a mutation rate that has been estimated to be up to 17 times higher than the corresponding rate for nuclear genes. This is thought to be due to the fact that the mitochondrial genome undergoes many more rounds of replication but may also be a consequence of the relative error proneness of the mitochondrial DNA polymerase γ and increased exposure to the potentially mutagenic products of oxidative metabolism. Whatever the explanation, one consequence of the higher mutation rate has been that the mitochondrially encoded subunits of the oxidative phosphorylation enzyme complexes have evolved at a much higher rate than their nuclear encoded counterparts.

The origin of mitochondria is thought to have been through endocytosis by an anaerobic eukaryote of an aerobic eubacterium possessing on oxidative phosphorylation system. During the evolutionary transformation of endosymbiont to organelle, there has been significant transfer of DNA sequences from the mitochondrial to the nuclear genome. Interestingly, the mitochondrial genome of the slime mould *Dictyostelium* has retained a gene encoding a NADH: ubiquinone oxidoreductase subunit which was transferred to the nuclear genome in the common ancestor of other eukaryotes. Some mitochondrial DNA sequences present in the human genome as pseudogenes have been incorporated only relatively recently in primate evolution. It has been suggested that coevolution may have occurred between nuclear and mitochondrial genes, for example, cytochrome c oxidase subunit IV (COX4; 16q22-qter and mt genome).

Mitochondrial DNA variation has been very informative for the study of the evolution of human population. The interested reader is referred to Stoneking (1996) for a review of the topic.

Evolution of Human Populations

Discussion of the fossil record of early hominids is outwith the remit of this volume and the interested reader is referred to Wood (1992, 1996) for readable preces. Similarly, the origin of modern humans and the movement of human populations have been well reviewed by a number of authors including Cavalli-Sforza et al. (1994), Cavalli-Sforza (1998), von Haeseler et al. (1995), Jones, Martin, and Pilbeam (1992) and Lewin (1998).

There are currently two different views of the origin of modern humans. The first, which is not inconsistent with the fossil record, proposes that different populations ('races') of *Homo-sapiens* evolved independently from their ancestor *Homo erectus* in different parts of the Old World ('multiregional model'). Migration of *H. erectus* from Africa to the rest of the Old World may have occurred ~ 1 Myrs ago. The alternative hypothesis is that *H. sapiens* arose once in Africa and that the species may have been subjected at some stage to a severe population bottleneck. With both models, the geographical separation of populations has led to the emergence of morphological differences although continued gene flow between populations has served to ensure that human gene diversity remains graded rather than discrete.

One of the most dramatic demonstrations so far of the utility of molecular genetics to the study of human gene evolution has been the analysis of mitochondrial DNA (mtDNA) from a fossil Neanderthal specimen. Neanderthals and humans are considered to have shared a common ancestor between 550,000 and 690,000 years ago. Since the Neanderthal mtDNA sequence was found to be quite distinct from those of modern humans, it would appear as if Neanderthals became extinct without contributing mtDNA to human populations, a finding consistent with the Out of Africa hypothesis.

The question of the age of the human gene pool has been approached by studying variation associated either with the mitochondrial genome or with Y chromosome-derived DNA sequences (mtDNA exhibits a high rate of evolutionary change and, along with most of the Y chromosome, is transmitted without recombination). Studies of mtDNA have yielded dates of the order of 200,000 years ago for the origin of modern humans, broadly consistent with estimates derived from Y-chromosome-derived DNA sequence data. Use of intronic variation in the ZFX (Xp21.3-p22.2) gene yielded a figure of 306,000 years that is almost certainly discrepant since the 95% confidence interval was extremely broad. Using 30 microsatellite loci to construct a phylogenetic tree for 14 different human populations, Goldstein et al. (1995) estimated that the time since divergence of African and non-African populations was 156,000 years. The above estimates are broadly compatible with those derived from polymorphisms associated with the CD4 (12p12-pter) gene, microsatellite data and protein polymorphism data. It must be remembered, however, that these results simply reflect the time elapsed since the most recent common ancestor for the sample population rather than the most recent common ancestor of all humans. Small populations and/or population bottlenecks will have served to obscure the actual timing of the 'origin' of modern humans.

As to the place of origin of modern humans, Wainscoat et al. (1986) claimed that the relative frequencies of haplotypes of the β-globin (HBB; 11p15.5) gene is African and nonAfrican population provided evidence for a migration out of Africa by a fairly small population. Further evidence for a recent African origin for modern humans comes from the observation that African populations have a ~20% greater microsatellite sequence diversity as compared with Asian and European populations. Studies of mitochondrial DNA have also shown that there is greater genetic diversity between African populations than among Asian or European populations. Indeed, these authors showed that genetic variation among humans on all continents are subsets of the variation present in Africans. However, some polymorphism lineages do not show deep branches for African populations which has made the Out of Africa hypothesis somewhat contentious. The higher level of genetic diversity manifested by African populations may simply be a reflection of their greater population size over the last million years.

Y chromosome variants appear to be more highly clustered geographically than those of mtDNA. One explanation for this difference could be that male migration has been more limited than that of women.

About 84% of human genetic diversity exists as differences between individuals within populations but the remaining 16% can be used to distinguish between populations. By comparison with apes, the extent of the genetic variation exhibited by modern humans is relatively low. This lack of genetic diversity is likely to be reflection of long-term small population size [before the introduction of agriculture 10,000 years ago, the entire human population probably did not exceed 100,000 and is thought to have been around 10,000 for most of its history], the effects of past population bottlenecks and the explosive population growth particularly during the last 10,000 years.

In human history, demographic expansions have often occurred as a result of technological developments affecting food availability and transportation fuelled by the pursuit of military or economic objectives. In terms of establishing genetic differences between populations, linguistic barriers have also been important.

Action of Natural Selection in Human Populations

Evidence for the recent effects of natural selection on human populations comes indirectly from observed variation in allele frequencies with infectious disease often serving as the selecting agent. The classical example of pathogen-driven selection is the heterozygote advantage accruing to carries of the Glu6→Val sickle cell mutation in the β-globin (HBB) gene which confers resistance to *Falciparum* malaria. It has been suggested that the high frequency of certain disease in specific populations may be explained in similar ways.

Heterozygote advantage has been invoked to account for the spread of the common cystic fibrosis mutation (δF508) in the CFTR (7q31.3) gene in Caucasian populations. The basis for heterozygote advantage was proposed to be increased fitness of heterozygous carriers during cholera epidemics. However, it is difficult to see how such overdominant selection could have brought about the extremely high prevalence of one specific CFTR lesion (δF508) relative to the large number of alternative CFTR mutations known. It would also be difficult to explain the gradient in δF508 frequency across Europe unless there was also a gradient of selective pressure. Thus, the most parsimonious explanation is probably genetic drift.

A large body of data has accumulated on HLA polymorphism and its relationship to disease susceptibility, resistance and progression. Polymorphic alleles at both HLA class I and II loci have been shown to be under selection. One recent example is selection for specific HLA class II (HLA-DR and HLA-DQB; 6p21.3) alleles as a result of hepatitis B virus infection. More than 90% of the 135 known HLA-DRB1 alleles appear to have been generated since the divergence of human and chimpanzee such changes appear to have arisen both by point mutation and by gene conversion and are consistent with the existence of substantial selective pressure.

Pathogen-driven selection may also have been responsible for increasing the frequency of genetic variants at other loci, for example the CCR2 (3p21) and CCR5 (3p21) chemokine receptor genes and the stromal cell-derived (SDF1; 10q11.2) gene. With the advent of the HIV epidermic, these variants may become advantageous in that they appear able to restrict HIV-1 infection and decrease the progression of HIV-1 infections to AIDS. Genetic susceptibility to parasitic infections may be under oligogenic control (e.g. *Leishmania*, *Schistosoma mansoni*, *Mycobacterium tuberculosis* and *Plasmodium falciparum*) implying that a number of different variants at different loci may be subject to selection. Infectious and parasitic disease is currently estimated to kill up to 20 million people in the world every year with acute respiratory infection, tuberculosis, diarrhoea, malaria, measles, hepatitis B, whooping cough and tetanus responsible for ¾ of this toll. Clearly, it is likely that selection will continue to operate on existing genetic variation at a wide range of genetic loci that serve to determine both the host's susceptibility and resistance to the disease in question.

Polymorphism of the human ABO blood group system (ABO; 9p34) may owe its origins to balancing (overdominant) selection mediated by infectious agents. Intriguingly, the same substitutions that differentiate the A and B alleles in humans are also present in the great apes and Old World monkeys leading to speculation that these polymorphic antigens may have arisen early in primate evolution. However, intronic sequence data point instead to an origin for the human alleles about 4 Myrs ago which argue for a model of convergent evolution of the blood group antigens in the higher primates.

Whilst pathogen-driven selection may be an important factor in increasing the frequency of certain genetic variants at specific human loci, susceptibility to infectious disease is unlikely to be the sole means by which natural selection influences allele frequencies. One common genetic variant not thought to be associated with infectious disease has been found in factor VII. Plasma levels of factor VII vary significantly in the general population and are known to be influenced by a number of different environmental factors including sex, age, cholesterol, and triglyceride. An Arg/Gln polymorphism at residue 353 of factor VII (F7; 13q34) which occurs with a frequency of about 10% in various populations, is associated with a 20-25% reduction in the level of plasma factor VII activity as a result of the impaired secretion of Gln variant. This high frequency is suggestive of a balanced polymorphism and could indicate that the Gln variant confers some benefit, for example protection against thrombosis, myocardial infarction or arterial disease. In support of this postulate, Silveira et al. (1994) have shown that the Gln allele is associated with a reduction in the amount of activated factor VII (FVIIa) generated in response to fat intake; individuals with the Arg/Gln genotype were found to possess FVIIa levels 48% of that exhibited by individuals homozygous for the Arg allele. Interestingly, a decanucleotide insertion polymorphism at –323 in the F7 gene promoter has been shown to be associated with a 33% reduction in promoter activity *in vitro* and a lower level of plasma factor VII activity and antigen *in vivo*. The occurrence of two functionally significant polymorphisms in the same gene is consistent with the idea of a selective advantage accruing to individuals with reduced factor VII activity.

A similar hemostatic system polymorphism is factor V Leiden. This variant, which underlies the phenomenon of activated protein C resistance, results from the substitution of Arg506 by Gln in coagulation factor V (F5; 1q23). Factor Va serves as a cofactor in the activation of prothrombin by factor Xa and the factor V Leiden variant is relatively resistant to activated protein C-mediated inactivation. Between

1% and 7% of the Caucasian population possess the factor V Leiden mutation which may therefore be regarded as a fairly frequent polymorphism with phenotypic effect. Since the factor V Leiden mutation is also associated with a relative risk of ~6.0 for venous thrombosis, this is also a polymorphism with clinical effect. Why is this factor V variant so common? Its high frequency in the general population suggests that it confers, or has conferred, some selective advantage on its bearers. Dahlback (1994) speculated that a slight hypercoagulable state associated with possession of the factor V Leiden variant might have been advantageous in certain situations such as traumatic injury and childbirth. Consistent with this postulate, Lindqvist et al. (1998) have shown that carries of the factor V Leiden variant have a significantly reduced risk of bleeding during childbirth. It may be that other common polymorphic variants in hemostatic factor genes e.g. the G20210A transition in the 3' untranslated region of the prothrombin (F2; 11p11-q12) gene, are explicable by similar models.

Other examples of polymorphic variants which may have conferred a selective advantage on carriers are the 'insertion' (I) allele of the angiotensin-converting enzyme (DCP1; 17q23) gene which appears to be associated with improved human endurance and the Glu487/Lys polymorphism in the aldehyde dehydrogenase 2 (ALDH2; 12q24) gene which is associated with alcohol sensitivity and alcohol avoidance. Selection is likely to have also operated on a variety of other human characteristics including cognitive ability, both form, skin pigmentation and pharmacogenetic variation.

Sequencing the Genomes of Model Organisms and Humans

The characterization of the genomes of a number of different and disparate species should aid significantly our understanding of the human genome, its structure, function and evolution. Such species ('model organisms') include a bacterium (*Escherichia coli*), a yeast (*Saccharomyces cerevisiae*), a nematode (*Caenorhabditis elegans*), the fruitfly (*Drosophila melanogaster*), the pufferfish (*Fugu rubripes*), the mouse, and the rate. Sequence o the genomes of these model organism is essential for the discovery, description and characterization of all genes within those genomes and proteins that these genes encode. It will provide information not only on the chromosomal organization of genes and gene families but also on the elements.

The 4.64 Mb genome of *E. coli* has been sequenced and encodes 4288 protein coding genes. By comparison, the 12.1 Mb genome of *S. cerevisiae*, which is organized into 16 chromosomes, contains 5885 protein-coding genes, ~140 ribosomal RNA genes, 40 snRNA genes and 275 tRNA genes.

The entire 97 Mb genome of *C. elegans* has also been sequenced and represents the first fully characterized genome of a multicellular eukaryote (*C. elegans*). This sequence predicts a total of 19,099 protein-coding genes and at least several hundred further genes specifying noncoding RNAs. At least 36% of *C. elegans* proteins exhibit a match in humans whilst 74% of characterized human proteins exhibit a match with a *C. elegans* protein. Each *C. elegans* gene has an average of five introns and the exons constitute some 27% of the nematode genome.

Comparison of the complete gene/protein sets of yeast and nematode has revealed that for a substantial proportion of the two organisms' genes, one-to-one orthologous relationships are identifiable. This suggests that the functions of many gene products were already established in the common ancestor of fungi and the metazoa. By contrast, most of the *C. elegans* signaling and regulatory genes that are known or expected to be involved in multicellularity have no yeast orthologue even though they may contain domain sequences present that are in yeast.

The possession of the compete genome sequences of various model organisms is proving of enormous benefit in identifying the human homologues of genes that are shared between these organisms and humans. Thus, the expressed sequence tags (EST) database (dbEST) can be screened using model organism genes as 'probes'. An example of this approach (termed 'cyberscreening' or '*in silico* cloning') is provided by the cloning of five human orthologues of yeast genes encoding proteins of the mitochondrial respiratory chain complex.

Within the next 5 years, the 3200 Mb human genome sequence should also become available. This will permit integration of cytogenetic, genetic, physical and transcriptional maps of the genome, information on inter-individual polymorphic variation, and the genotype-phenotype relationship particularly in the context of complex traits. The availability of the human genome sequence will lead to the identification of novel genes encoding new proteins and the characterization of diseases gene which should provide new insights into mechanisms of disease. Comparative genome mapping will also provide important insights into the evolution of the mammalian genome, its chromosomal architecture and its genes and gene families.

12

Human Evolution

The major trends in human evolution include a restructuring of the pelvis and lower limbs permitting upright bipedalism, exploitation of the terrestrial environment, increasing brain size, tool use and manufacture, and the addition of meat protein to the diet. Bipedal locomotion is the major trait differentiating early humans from the common ape–human ancestor. The first evidence of hominid bipedality in the fossil record appears about 4.2 mya. The effective exploitation of the habitat by humans involved such behavioral adjustments as life in a social group, a sexual division of labour, and the development of a capacity for symbolic language. The increase in brain size was neither rapid nor consistent and was likely associated with increasingly complex social organizations and means of environmental manipulation.

Stone tools first appear in the archaeological record about 2.5 mya, approximately 2.5–3 million years later than the first identified human fossil remains. Primate visual acuity, manual dexterity, and inquisitiveness were preadaptations to human tool use and manufacture. Humans are the only primate to rely on tool technology for survival. Most nonhuman primate diets include heavy amounts of vegetable materials, unlike many human diets, which include large amounts of meat. It is not entirely clear when meat protein began to play a large role in the human diet, nor is it clear how much meat was obtained by hunting or scavenging.

Evolution of Bipedalism

Trunk erectness developed early among primates; All major primate groups include species that sit or sleep in an upright position. On occasion, many primates assume an upright posture. (Although birds, some lizards, and some dinosaurs assumed a bipedal posture, it is quite different from that exhibited by humans).Primate bipedalism usually takes the following forms: (1) Consistent bipedalism characterized by standing erect with straightened knees is practiced by humans (2) Bipedal running occurs in many nonhuman promates However, to compensate for the restrictions of the pelvis and hind limb musculature, they use a bent-knee gait. (3) Bipedal walking is less common among monkeys than among the great apes, whose gait, however, is a bent-knee gait. Only humans can stand bipedally erect for long periods of time. The distinguishing feature of human locomotion is that we are bipedal all the time as our normal mode of locomotion.

Because we are the only primate to have intensively taken up bipedalism, we are interested in how and why we did so. The anatomical changes necessary for bipedalism have been detailed in many places, and there is general agreement with Washburn's (1971) scheme presented here. There were major changes in the human lower limbs to accompany the shift to habitual bipedalism not only in general skeletal proportions but also in the form of the muscles and in general limb functioning. Structural changes in the lower limbs include an elongated femur (upper leg bone) and restructuring of the foot. Major changes in the lower limb skeleton include elongation of the bones (our lower limbs are longer than our upper limbs, opposite of the great ape configuration), reorientation of bones, and different positioning of the muscles on the bones.

The human foot has shifted from the nonhuman primate pattern of a grasping organ to a weight-bearing platform. Major anatomical changes in the foot region occurred early in human evolution. The structure of the human foot indicates that it evolved from the kind of foot typical of apes and atypical of quadrupedal monkeys. The essential points are that weight is borne on the first toe and that in walking and standing, the foot toes out rather than points in. Monkeys bear almost no weight on the first toe, and the axis of the weight-bearing stress is through the middle toes rather than between the first and second toes as in humans.

A number of important features distinguish the human from the ape pelvis. The pelvis is actually comprised of three distinct bones; the *ilium*, *ischium*, and *pubis*. The structural basis of bipedalism is anatomically complex and involves a reorganization of the pelvis. The ilium appears to have evolved to human form prior to the ischium. The essential problem for bipedal humans was the transfer of the

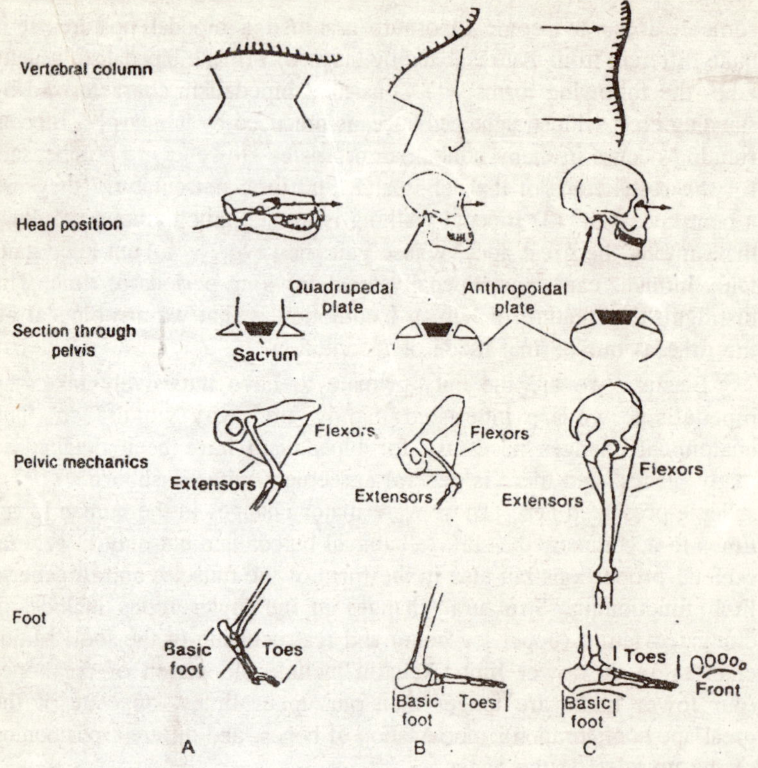

Fig. 12.1. Structural changes associated with bipedalism. A—Quadrupedal animal; B—Chimpanzee; C—Modern human.

landing and balancing functions from the forelimbs to the hind limbs. In quadrupedal monkeys or knuckle-walking chimpanzees, the principal locomotor force derives from muscles posterior to the femur. If the main drive is from the right leg, the animal lands on the left forelimb. When the human foot strikes the ground, it must first perform the landing–balancing functions of the ancestral forelimb, then give the push resulting in the step.

The major pelvic changes involved a shortening and broadening of the ilium. The ilium tilted backward, allowing the trunk to be held vertically, followed by rotation of the sacral vertebrae (vertebrae at the end of the spinal column) and compensated for by a curving of the spine (referred to as a *sigmoid* or *S-shaped* curve). The change in curvature developed with the onset of walking. This change allows an opening in the birth canal and the maintenance of an erect posture.

Table 12.1. Important skeletal Modifications Needed for Human Bipedalism

Skull

Foramen magnum moves forward with flexion of the cranial base. Position is common in primates generally and is related to upright sitting.

Pelvis

Major change is the shortening and broadening of the ilium.

Vertebral Column

Rotation of sacral vertebrae and sigmoid-shaped spine.

Lower Limbs and Feet

Feet change from grasping to weight-bearing platform.

Change in muscle size and structure, especially gluteal and hamstring muscles.

Elongation of lower limbs, most noticeable in genus *Homo*.

Muscular changes essential for the maintenance of balance and stabilization of the trunk accompanied skeletal reorganization. Although most human and ape thigh muscles do not differ with regard to the type of action produced,. they do differ in the effect of the muscle's action. Of special importance are the gluteus maximus, the largest muscle in the human buttocks, and the hamstring muscles (the semitendinous, semimembranous, and biceps femoris), which are important thigh extensors in humans and apes. Upon contraction, extensor muscles tend to straighten a bone around a joint. These muscles are also powerful flexors of the leg at the knee joint (when they contract, they allow bending at the knee, decreasing the angle between the thigh and calf) and are important rotators of the thigh. The muscles that move the leg forward are the flexors because they bend the leg at the hip; those that move the leg backward are the extensors because they extend the leg at the hip joint. In humans these muscles are more developed than in other prinatesw because humans alone depend on them for their locomotion. Also in humans the leg is proportionately heavier and larger.

The ape's ischium is relatively longer, and the femur relatively shorter, than in upright bipeds. This relationship influences the functioning of the hamstrings and gluteus maximus. With a long moment arm (stable element)—the ischium—and a shorter lever arm (movable elements) the femous the muscles produce power of action. In upright bipeds with the reverse proportions (short ischium and long femur), the hamstrings produce a speed rather than power action. Gluteus maximus in humans is more a speed-of-action muscle than it is among apes.

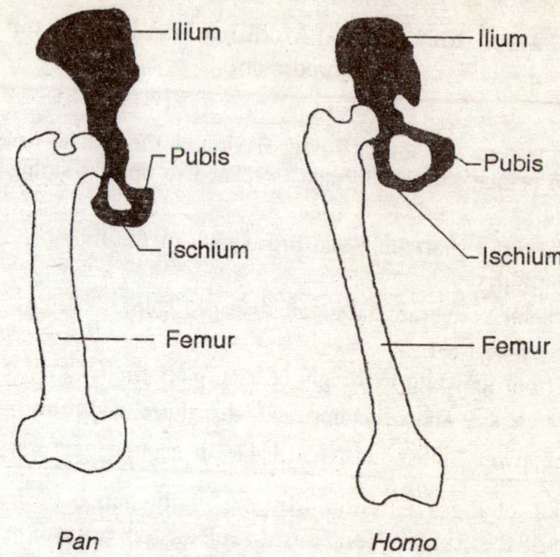

Fig. 12.2. Pelvis and femur of chimpanzee.

Power is apparently more valuable in animals that spend time climbing; speed and range of movement are of greater importance to a biped less dependent on tree life and more dependent on its abilities to cover long distances in the shortest possible time. The upright biped sacrificed power of action for endurance.

Ideas on the origin of human bipedalism, the first change that differentiated apes and humans, require the reconstruction of many unknowns. Darwin (1871) suggested that when primates came to the ground they could have become bipedal or quadrupedal. He suggested that the reason that humans alone among the terrestrial primates became bipeds is tied to how they use their hands.

But the hands and arms could hardly have become perfect enough to have manufactured weapons, or to have hurled stones and spears with a true aim, as long as they were habitually used for locomotion and for supporting the whole weight of the body, or,.... so long as they were especially fitted for climbing trees. Such rough treatment would also have blunted the sense of touch, on which their delicate use largely depends..... it would have been an advantage to man to become a biped; but for many actions it is indispensable that the arms and whole upper part of the body should be free; and he must for this end stand firmly on his feet. Shipman (1984) suggested that human bipedalism was an adaptation to the pattern of scavenging meat.

Although bipedal running is neither fast nor efficient when compared to the quadrupedal gaits, bipedal walking is more energetically efficient than quadrupedal walking.

Bipedalism increased the energetic efficiency of human travel, and this increased efficiency was an important factor in the origin of bipedalism. Bipedalism is an efficient means of covering large areas slowly, and Shipman argued that it is an appropriate adaptation for a scavenger who must cover large areas. Bipedalism elevates the head, thereby improving the ability to locate distant items. Combining bipedalism with agile climbing ability (of which there may be evidence in such early hominids as *Australopithecus afarensis*) further improves the opportunities to exploit the environment. Bipedalism frees the hands and makes them available for carrying items.

Table 12.2. Some Explanations for Bipedalism

Explanation	*Source*
Energetically more efficient	Rodman and McHenry, 1980
Reduces incidence of solar radiation	Wheeler, 1984
Facilitates tool use and making	Darwin, 1871
Infant dependency on mother	Lovejoy, 1981
Posture for small tree feeding	Tanner, 1981
Energetically efficient for meat scavenging	Hunt, 1994
Long-distance migration and scavenging	Sinclair et al., 1986

According to Shipman (1984;26–27), "Bipedalism is compatible with a scavenging strategy. I am tempted to argue that bipedalism evolved because it provided a substantial advantage to scavenging hominids. But I doubt hominids could scavenge effectively without tools and bipedalism predates the oldest known tools by more than a million years."

Leonard and Robertson (1995) have shown that at normal walking speeds, the energetic efficiency of human bipedalism relative to generalized quadrupedalism is greater than previously thought. The relative efficiency of bipedalism varies with speed of gait and gender. The energetic effects are most evident at slower speeds, consistent with the argument that bipedalism among hominids is an adaptation to long-distance walking. Energetic efficiency is likely part of a suite of adaptative traits that allowed hominids to expand into an open African habitat.

Sinclair and others (1986) suggested that bipedalism developed along with long-distance migration, and they agree with Shipman that scavenging was important to the evolution of bipedalism. They argued that early hominids scavenged migrating ungulate (hooved mammal) populations, the only population that existed in large enough numbers to provide sufficient food for scavengers. Because many migratory animals die from starvation, carcasses are available to scavengers, who would not necessarily have to contend with predators for a kill. A migratory scavenger has access to an abundant and constant food supply.

There was an unfilled niche in Africa for a mammalian scavenger that could follow migratory ungulates. However, if the scavenger was a bipedal human, it had to carry its young. Access to a rich and constant food supply encouraged bipedalism among humans. Those who developed the ability to walk long distances might have rapidly increased their numbers and displaced the less numerous sedentary quadrupeds dependent on plant gathering.

Sinclair and other' (1986) protohominid was a quadrupedal plant gatherer and occasional scavenger, much like the baboon. Two changes were needed for such a form to follow migrating ungulates. First, the young had to be carried. Among humans this purpose was accomplished by using the hands and arms, which were freed from locomotion to carry the youngster, and by assuming an upright posture. A prolonged upright stance for efficient carrying required adaptations of the pelvis. Second, they must be able to travel long distances efficiently. The primate foot with an opposable big toe changed to the modern human foot as a propulsive lever. This change must occur coincident with the hip changes.

The migration hypothesis suggests that habitual tool use developed from a need to quicken the butchering of carcasses and avoid competition with other, stronger mammal predators. The opportunity for hominid migration was enormous because savanna Africa was dominated by migration ecosystems.

Lovejoy (1981, 1984) suggested that perhaps bipedalism was necessary to a new and effective reproductive strategy—pair bonding in a kind of rudimentary family unit. The link between reproductive strategy and bipedalism lies in primate sexual behaviour, according the Lovejoy. Some female monkeys and apes have a well-defined and rather short estrous period, the time of ovulation. Males attracted to estrous females seek the best chance of mating by competing (not necessarily through physical altercations) for them. The degree and kind of such competition varies among species. Except among human males and a few other primates, if males provide infant care they usually do so with minimal skill and interest. The offspring's care is almost exclusively left to the mother, who must care for both herself and her infant. The mother usually cannot care for more than one infant at one time and normally does not become sexually receptive again until the infant can forage on its own.

Birthrates of monkeys, whose offspring mature rapidly, are not much reduced because of a mother's preoccupation with her infant. For apes and early humans, however, the problem is and was more critical. An ape mother may spend five years raising one infant; this long period of care lowers the birthrate because primate mothers rarely give birth to another infant while still nursing previous offspring. Because of the long birth intervals among apes, when compared to monkeys, Lovejoy argued that apes would be at an evolutionary disadvantage if they were in direct competition with monkeys.

According to Lovejoy, the solution to this dilemma is to space infants more closely, as monkeys do, while still providing each infant with quality care. That solution necessitates adopting a better strategy of offspring care, for example, providing the mother with a food supply so that she does not have to forage for her own and her infant's needs. If this occurs, the mother can care for two or more infants simultaneously. Because she moves less, her offspring may have a greater chance of survival.

One possible auxiliary food source is the male. If an adult male is to be incorporated into a food-sharing role, his attention must turn to the female and infants—or to one female and one infant. A monogamous male could indirectly help his infant survive by bringing food to its mother, giving her more time to be a parent and to protect the infant. To provide greater quantities of food, the male must be able to carry it; thus he would find bipedalism adaptive.

According to Lovejoy's scenario, the female benefited from male provisioning. For example, her ability to care for her offspring was increased. She could spend less time searching for food and more time directly caring for and protecting her young. The provisioning bipedal male was able to obtain food farther away from the area occupied by the female and infant; thus he increased their food supply without depleting local food resources. The female was thereby able to collect more foods locally, reducing possible dangers both she and her young might encounter from roaming over a larger area.

Because the bipedal male increased the female's food supply through his provisioning behaviour, she had more energy available for parenting. According to Lovejoy (1984:26), "Therein lies the ultimate advantage of provisioning." Females who benefited from provisioning were better able to manage overlapping offspring. Selection would have rapidly favoured females that chose mates whose interest in them continued after fertilization. Any behaviour of a potential mate that improved a female's reproductive rate would be favoured.

Lovejoy found anatomical evidence among early human ancestors favouring his position of a monogamous situation. Among extant polygynous primates, where male conflict determines reproductive success, the male's canine teeth are enlarged. Females of the same species, however, have relatively small canines. In *A. afarensis*, an early example of the human lineage, males also show relatively small canines. Lovejoy suggested that this is evidence for the presence of a monogamous unit.

Lovejoy's hypothesis concentrated on the acquisition of bipedalism by males as a means of supplying food for a female encumbered by a slowly maturing infant, but Tanner (1981) suggested a hypothesis that centers on the acquisition of bipedalism be females. Tanner noted that bipedalism accommodated the increased infant dependency on its mother. "With effective baby and child care, the young can be born more physically immature—which was certainly fortunate, for this was becoming necessary due to changes in the pelvis for bipedalism, Because infants and young must in any case be cared for already, increasing immaturity at birth would not by selected against".

The ability to walk great distances while carrying items was essential to early humans. Mobility over long distances and effective carrying for both females and males were made feasible by the development of bipedalism. There was much to be carried by the gathering females—their infants, digging implements, and the gathered foods. It was already suggested that because of these demands on the female some sort of carrying device was a likely first tool.

Tanner argued that with the development of bipedalism, learning to walk required more time to develop motor coordination prior to independence. Prolonged infant dependency on the mother became necessary. "This meant that the mother—or older sibling, mother's sibling, or mother's friend—had to carry a child that could no longer cling as effectively because the changed anatomy of bipedalism required loss of the foot's ability to grasp the mother. Even young who were already able to walk would have to be carried frequently and often, because they would tire". Tanner's hypothesis suggested a restructuring of social life in response to the rigors imposed by bipedalism.

McHenry (1982) questioned Lovejoy's scenario. For example, (1) evidence indicates that early humans were polygynous and not monogamous, as Lovejoy suggested, and the provisioning of immobile females and their offspring by males is unlikely; (2) Among monogamous primates, sexual dimorphism is nearly absent, but in early humans it is pronounced, as one might expect in a polygynous species. (3) All monogamous nonhuman primates are territorial, the adult male and female with their young living in an exclusive territorial social unit. Early hominids apparently lived in larger groups. (4) In no known monogamous nonhuman primate do we have a pattern whereby a male provides food for the female and her offspring. (5) Parental care of infants is not limited to monogamous primates, and a decrease in birth spacing may not be a sufficient cause for the origin of monogamy and parental care.

Wheeler (1994) suggests that the body of bipedal animals is raised off the ground away from heat radiating from the surface and that additional cooling might occur through contact with air movements several feet above ground. However, Chaplin et al. (1994) argue that the increased time in open habitats that thermoregulation supposedly afforded to early hominids because of bipedalism was relatively short and of little or no adaptive significance. They suggest that thermoregulatory considerations, including those implicated with brain growth, cannot be cited as a primary cause in the evolution of bipedalism.

Hunt (1994) proposes an explanation for bipedality that incorporates feeding ecology and for which there is some proof in the fossil record. Hunt's explanation is titled the small-tree postural feeding hypothesis. Hunt suggests that bipedalism evolved during climatic changes in the Miocene. Using extant chimpanzees as his models, Hunt notes that bipedalism was observed most commonly when chimpanzes fed erect on the small fruits of small, open-forest trees while standing bipedally erect. Animals fed bipedally by reaching to pick fruit while standing on the ground, or from within the tree, standing bipedally and being stabilized by grabbing an overhead branch. According to Hunt, the food-gathering function of chimpanzee bipedalison suggests that hominid bipedalism may have evolved in conjunction with arm-hanging as a feeding adaptation that allowed fruit harvesting in open-forest or woodland trees.

Hunt's suggestion has support in the anatomy of *A. afarensis*, in which anatomical features of the hand, shoulder, and torso have been related to climbing. While there are differences of opinion, Hunt suggests that *A afarensis's* pelvis and hind limb indicate a less than optimal bipedal adaptation compared to modern humans. If such locomotor insufficiency existed, according to Hunt, it supports the hypothesis that bipedalism evolved more as a terrestrial feeding posture than as a locomotor posture.

Several problems in locating the origins of bipedalism can be tied to the fact that investigators tend to look for one cause. As in most cases of evolutionary change, a number of interrelated pressures probably result in the onset of a new adaptation.

Effective Exploitation of the Terrestrial Habitat

Behavioural adjustments were needed to cope with the terrestrial habitat. Major adjustments were probably reflected in the social group. Based on comparative evidence from extant foragers, nonhuman primates, and the archaeological record, early humans may have lived in groups with an average of twenty-five members per group, and there was a sexual division of labour and sexual role differentiation.

Tanner's (1981) picture of early hominid communities assumed a plant gathering economy in which mothers were the regular gatherers. Tanner suggested that the early hominid social group probably contained a mother and one to several young. This genealogical unit could have comprised three generations, an old mother, adult daughter (s), and her infant or juvenile offspring. This unit is typical of many nonhuman primates. Because generations were short and these creatures relatively short-lived, units of more than three generations would be rare. This mother-centered (matrifocal) genealogical unit was the most stable group; individuals within the unit would probably travel and gather together. Several units, especially those composed of sisters, may have met frequently and camped together with some regularity around well-known water sources.

Among most nonhuman primates each animal is a separate subsistence unit. Once infants are weaned, they must depend on their own skills in food acquisition for survival. Among humans, however, weaned young still depend on adults for food. Because of this long-term dependence, much of the day for both male and female humans is spent in activities that provide food for the young: "Because of the long-term dependence of children, a division of labour evolved in which the adventurous, wandering male became the hunter and the female developed the less mobile role of gatherer and mother".

Females have the major responsibility for the early care, feeding, and rearing of the young. Therefore, any sexual division of labour, by increasing reproductive fitness (that is, the infant's survival), is beneficial. The human infant is born relatively immature and unable to fend for itself. The human mother lacks the body hair common to nonhuman primates that the infant can grasp, allowing the mother the usual range of movement. The human mother must carry the infant. Given the long period of immaturity among human infants, once foraging developed as a way of life, a sexual division of labour must have occurred.

The division of labour allowed a flexible system of joint dependence on plant and animal foods, which were provided by both sexes. The sexual division of labour was an efficient coping strategy quite different from that characterizing nonhuman primates. The human hunting-gathering pattern provided great flexibility in coping, for it allowed adjustments to daily, seasonal, and cyclical variations in food supply and geographical and habitat variations.

There is no archaeological evidence supporting the contention that early humans exhibited a sexual division of labour (although there is evidence of anatomical sexual dimorphism). The adaptive advantages provided by this system, however, suggest the likelihood that such a division existed. A sexual division of labour may help explain why early humans could successfully compete and establish themselves in their new habitat.

A concomitant of the human sexual division of labour was increasing parental care, which can be explained by referring to what evolutionary biologists call K and *r selection*. The *r* stands for the intrinsic rate of population increase and is a measure of how polific a population can be. The K symbolizes the saturated carrying capacity of the environment. K and r selection may have a bearing on understanding the evolution of the human family. Stable, predictable environments with a great deal of complicated structure tend to select for larger adult size, long life-spans, and periodic reproduction with reduced numbers of young. If food supplies are scarce, specialized, and otherwise difficult to extract, selection favours prolonged immaturity to provide more time for learning. Predation pressures also call for increased parental protection. All these factors combine to favour increased parental care for survival, which probably enhanced the sexual division of labour. It is likely that early hominids were under K selection; they exhibited adult sexual dimorphism, prolonged immaturity, heavy parental investment, and a reduced number of young at birth.

All these factors are essentially primate traits, and all are elements of K selection. K selection factors are self-reinforcing. Longer life, as part of a function of year-round resources, selects for longer developmental periods and larger body sizes, both useful for effectively exploiting the environment. Prolonged nurturing brings pressure for reduced numbers of young and greater birth spacing to allow parental care to aid in learning the complicated environment, in this case also to learn tool use and manufacture. Because offspring live to an age when they compete with their parents for food, there is increasing pressure to reduce birthrates.

The acquisition of tool use among humans, with its concomitant muscular and neural requirements, helped initiate other adaptations for effective environmental manipulation and exploration. Tool use and terrestriality probably placed a premium on the development of an effective signaling system. Language can be seen as a means of environmental exploitation. Language may be a partial response to continuing pressures to communicate effectively about increasing complexities of life—for example, how to make tools, where to find food, and passing social traditions from a mother to her infants.

Although much has been written about the evolution of human language, we do not know when language first evolved. Furthermore, few of the anatomical modifications necessary for human speech fossilize because many of the speech structures are either muscle or cartilage.

Many have tried to link an increasing brain size with the onset of language, which was supposedly tied to coordinating hunting behaviour, and Poirier has tried to link the onset of language to food sharing. Calvin (1983) argued that it was the sequential skills and coordination developed in throwing that subsequently made language possible. Calvin suggested that learning to throw with one hand produced the very first lateralization of the brain, the first concentration of a brain function in one hemisphere. Language skills eventually become lateralized, and both throwing and language are most commonly concentrated in the brain's left hemisphere.

Calvin further suggested that the lateralization of throwing behaviour was the spurt for rapid brain enlargement. He saw a link between the muscle sequencing and visual perception that allowed one to hit a rapidly moving distant target by throwing a rock with one hand and an increasingly sophisticated tool complex. "The same rapid-movement sequences in the left brain could have gone from throwing to talking to chipping rocks into scrapers and knives, incidentally providing the sparks that made possible fire on demand."

The communication system and brain structure of some nonhuman primates indicate that vocalizations carry information about the sender's sex, group membership, and social relationship. Such vocalizations can also refer to external objects or events. Some species show hemispheric asymmetries in auditory perception as well as in vocalization. As a result, Stekelis (1985:157) has commented that "these data suggest that the vocal-auditory machinery of the earliest hominids was far more ready to take on 'primordial' speech function than has been previously supposed." If true, then selection may have occurred rather early in hominid evolution for an increase in these capacities in conjunction with tool use and manufacture, food-getting activities, problem solving, and other activities.

Laitman (1984) attempted to explain the anatomical modifications involved in the evolution of human language by noting the shift in position of the larynx (voice box) in the neck. The position of the larynx determines how an animal breathes, swallows, and vocalizes. In almost all mammals, at all stages of development, the larynx is situated high in the neck, lying roughly opposite the first to third cervical (neck) vertebrae. In this position, the larynx allows a direct passage of air between the nose and lungs. While an animal is breathing, liquids can still be swallowed. A nonhuman primate, for example, can simultaneously breathe and swallow liquids. However, the larynx's position severely limits the range of sound production.

In contrast, the position of the larynx shifts downward around the second year of life in modern humans. From birth to approximately two years of age, the human larynx is high in the neck, much like any other mammal. The human newborn can breathe, swallow; and make sounds much like monkeys and apes. Around two years of age, however, the larynx begins to descend, altering forever the way the child breathes, swallows, and vocalizes. Once descent occurs, the larynx is in a position unlike that of any other mammal. In human adults, the position of the larynx can appear from the fourth to almost the seventh cervical vertebrae.

Although human adults cannot breathe and drink simultaneously, the descent of the larynx has produced a greatly enlarged pharyngeal chamber (the pharynx is the internal passage from the nose to the throat) above the vocal folds. Sounds produced in the larynx of human adults can be modified to a greater degree than is possible for new borns and any other nonhuman mammal. The expanded pharynx seems to be the key to our articulate speech.

Laitman reconstructed the laryngeal positions among early hominids. He suggested that the early hominids, the australo-pithecincs, had a condition approaching that of monkeys or apes. Consequently, they probably had a very restricted vocal repertoire compared to modern humans. The high position of their larynges would have made it impossible for them to produce some of the universal vowel sounds found in human speech, and it is unlikely that they could speak the way we do today. Laitman suggested that the modern laryngeal condition allowing articulate speech appeared first among archaic *H. sapiens* some 300,000 to 400,000 years ago.

Until recently, there was no good fossil evidence that could be used to solve the dilemma of which early humans had speech. However, the discovery of a well preserved hyoid bone from Kebara Cave, Israel, dating to about 62,000 years ago suggests to some researchers that Neandertals may have had the capacity for speech. The hyoid is a small, U-shaped bone that lies between the chin and larynx. It anchors the muscles that move the tongue and the larynx. The fossilized hyoid is almost identical to that of the modern human hyoid in its size and shape. At least for some researchers, the discovery of this fossil bone negates any arguments against the Neandertals having speech.

Arensburg et al. (1990) conclude that the Kebara hyoid is quite similar to that of modern humans. The relation of the hyoid to the mandible and cervical vertebrae probably did not differ from that of modern humans. These findings suggest that Middle Paleolithic peoples did not differ much from modern humans in terms of the anatomy of the vocal tract. However, Lieberman (1992) disputes Arensburg et al.'s contentions. The same linear measurements that Arensburg et al. used on the Kebara hyoid apparently would identify a pig's hyoid as that of a modern human.

However, Laitman et al. (1990) stated that although the fossilized hyoid may hold valuable clues for interpreting vocal tract evolution, the absence of both other fossil hyoids for comparison and an understanding of what metric analysis of an individual bone actually signifies severely limits the value of the Kebara hyoid.

Kay et al., (1998) suggested that Neandertals may have had the nerve complex needed to control the subtle and varied movement of the tongue needed for speech. Kay et al. said that the hypoglossal canal, the bony canal that carries a nerve connection between the brain and the tongue, is about the same size in Neandertals and early humans as it is in modern humans. The nerve that goes through the hypoglossal canal controls most voluntary muscles in the tongue. Without such control, the tongue is unable to perform the motions necessary for speech. The presence of the canal and its size allows one to reasonably suggest that the Neandertals may have been capable of speech; however, it does not prove that they had speech.

Increasing Brain Size And Complexity

There has been a trend toward increased brain size and complexity during human evolution; however, the trend was neither consistent nor steady. The size increase was slight during the approximately 3 million years of the evolutionary history of the *Australopithecus* and early *Homo* lineages but rather rapid during the middle Pleistocene in *H. erectus* and later *H. sapiens*. Increased brain size and complexity were probably related to tool use and manufacture, increasing environmental challenges, and more complex social groups, among other factors.

Increased brain size and complexity may be related to the infant's slower maturation rate, which required extended parental investment. Extending the period of maternal care allowed more time for infant socialization and placed premium on learning abilities. Learning may have been enhanced by increasing brain sizes and complexity. Size increases could have led to earlier births requiring longer learning periods.

Table 12.3. Cranial Capacity Comparisons among some Primates

From	*Cranial Capacity Range (cc)*	*Average Cranial Capacity (cc)*
Great Apes		
Chimpanzee	282–500	383
Gorilla	340–752	505
Hominidae		
Australopithecus afarensis	380–500	413
A. africanus	435–530	441
A. robustus	—	530
A. boisei	506–530	513
Homo habilis	590–752	640
H. erectus	750–1250	937
H. s. sapiens	1,0002,000	1,330

Culture has as a major component learned behaviour that is transmitted across generations. One prerequisite for culture is adequate memory storage facilitating relatively complex learning. A threefold increase in brain size occurred from early hominids to modern humans. This increase likely enhanced the capacity for information storage and the ability to learn. Although early hominids had relatively small brains, even the smaller-brained forms had relatively larger cranial capacities than contemporaneous apes. The functional significance of the larger cranial capacities of later humans is that they probably reflect increased abilities for technology and long-term memory.

Anthropologists have long speculated on the amount of feedback between increased brain size and complexity and tool use and manufacture. Although some aspects of this interaction have been questioned there is little doubt about its importance. Tool use and manufacture are best understood as related to hemispheric specialization, especially brain lateralization (specialization of the right and left hemispheres for different functions). We cannot say which appeared first, hemispheric specialization or manual skills. However, increasing verbal ability, skilled tool use and fabrication, and later hemispheric specialization probably evolved together.

Martin (in Lewin, 1982) suggested that the upper limit of brain development is determined while the infant is still in the womb. After birth, the brain follows a set trajectory, which in nonhuman primates typically involves doubling of the brain's weight; in humans the brain weight quadruples. Differences in gestation times apparently have consequences for potential brain growth.

Dietary patterns and the nature of the resources exploited for food seem to affect brain size. Among bats, for example, fruit eaters have larger brains in relation to their body size than do insectivores. The reason typically offered is that as fruit is more widely dispersed and difficult to find than insects, fruit eaters need to be smarter to gather their food.

Milton (1981) hypothesized that the element of predictability associated with the spatial and temporal distribution patterns of plant foods in tropical forests stimulated primate intelligence. What appears to set most primates apart from other relatively long-lived and large-brained animals is their ability to store and retrieve a great amount of independently acquired information about their environment. Milton (1981) noted, "The extreme diversity of plant foods in tropical forests and the manner in which they are distributed in space and time have been a major selective force in the development of advanced cerebral complexity in certain higher primates".

Given the dynamic nature of tropical forests, it seems maladaptive for a tropical forest dweller to genetically code a large amount of dietary information. What appears to be required is a great deal of behavioral flexibility, allowing a response to continually changing forest conditions. Increasing mental complexity, which places a strong emphasis on learning and retention, might have been selected for.

Milton noted that among later members of the human line, for example, the genus *Homo*, a behavioral shift favouring an increased ability to exploit mobile big game, so that it could become a dependable addition to the diet, may have required some major changes in certain areas of the brain. Additionally, mobile prey typically evolve at the same or a faster rate than their predators: "Lacking powerful jaws and claws characteristic of hunting carnivores and already *predisposed* to solve their dietary problems primarily through behavioral rather than morphological or physiological adaptations, these hunting hominids may have depended heavily on mental acuity to outwit and capture prey".

The reliance upon intelligence as a primate way of life has been stressed by researchers. Jolly (1985) noted, for example, "if there is an essence to being a primate, it is the progressive evolution of intelligence as a way of life." She may have been the first to link sociality to the evolving primate brain. As has been noted, primates are social animals whose daily existence depends on their ability to socialize with their peers. Primates are socially sophisticated animals who recognize and differentiate between their kin, their friends, and their mates. This social acumen "may be the fundamental selective force behind the evolution of primate intelligence" (Small, 1990). However, Small also noted that "any proposal that the need for these social machinations must have favoured large and complex primate brains is equivocal, because no one has yet demonstrated a direct relation between social acumen and individual reproductive success". This caveat not withstanding, it must be remembered that increasing encephalization characterizes the order Primates.

Holloway (1982, 1983) questioned the commonly held assumption that the brain was one of the last structures to change in human evolution. He suggested that the brain was reorganized early in human evolution and that only the enlargement of the brain occurred late. The development of bipedalism was probably integrated with neuropsychological restructuring. Tobias ((1982) noted that cerebral enlargement was not striking for the first 100,000 generations of human existence, but cerebral changes other than enlargement occurred.

Given Holloway's position, analyses examining brain size alone are likely to provide misleading and possibly erroneous conclusions regarding the dynamics of human evolution. Natural selection has operated on the evolution of the human brain, and the selection pressure has been in the realm of social behaviour. Holloway suggested that selection pressure on brain organization and size continued throughout most of human evolution, or at least until the Neandertals appeared. The brain was not the terminal organ to evolve in the overall mosaic of human evolution, according to Holloway, unless one talks only about absolute size and ignores relative brain size, cerebral cortical organization, and cerebral asymmetry.

Absolute and relative increases in brain size were probably interspersed with evolutionary episodes of brain reorganization and other changes. Holloway's understanding of brain evolution changes our understanding of its mode and importance in human evolution.

Falk, like Holloway, spent years studying primate endocranial casts (or endocasts); however, her views differ from those of Holloway. Falk (1984) noted that endocasts belonging to members of an early hominid lineage, the South African australopithecines, resemble ape brains in both size and external form. This resemblance does not mean, however, that the australopithecine and the ape had the same thought processes or mental capacities. In contrast to apes, australopithecines were bipedal and may have had tools and other cultural attributes. Some evolutionary changes occurred in the early human brain that were not evident in the gross organization of the brain's outermost portion.

Australopithecine brains were still very apelike; the first humanlike endocast belonged to a more advanced species, *Homo habilis*, from northern Kenya, which dates to approximately 2 mya. This endocast exhibits a humanlike frontal lobe, including what appears to be Broca's area, an area on the cerebral cortex named after Paul Broca. Broca's area is crucial for human speech and Broca's area appears in the australopithecines. On the parietal lobe of *H. habilis* is also an expansion of Wernicke's area, an association area of the brain necessary for comprehension of spoken language. Tobias (1987, 1994) suggested that the developed presence of Broca's and Wernicke's areas of the brain implies that *H. habilis* was the first hominid to have the neural capacity for language. There is, however, no agreement on the issue of whether or not *H. habilis* had language.

Car Wernicke was a German neurologist who studied brain damage associated with the loss of comprehension in aphasia. The site he studied, now called Wernicke's area, is located on the dominant hemisphere in the temporal lobe. A bundle of nerve fibers sends signals from Wernicke's to Broca's area, allowing the vocal repetition of heard and memorized sounds. Broca's area is located toward the front of the dominant hemisphere. Broca's area sends the code for the succession of speech sounds to the motor cortex of the brain, which controls the muscles of the face, jaw, tongue, pharynx, and larynx, thus helping to set the speech apparatus in motion.

Falk suggested that until early humans acquired language, "early ancestors may not have been very human" (1984). Falk did not accept Holloway's view that the brains of the South African australopithecines were reorganized along human lines some 2.5 to 3 mya. She suggested that their brains were still apelike, judging from their external appearance, and that this apelike condition persisted for a long time. Given the scanty nature of australo-pithecine cranial materials, no resolution to this argument is in sight.

With the continuing increases in brain size and complexity and with continuing and perhaps heightened awareness of the surroundings that such increases allowed, all distinctive features of a protocultural stage were being elevated to a new level of sociopsyhological integration. This new level would eventually affect every aspect of the earlier modes of protocultural adaptation and lead, according to Hallowell (1961), to "more inclusive, complex, and diversified sociocultural systems".

Not too long ago, the leading theory of hominid brain evolution was tied to a combination of bipedalism, forelimbs freed from locomotion and used for other purposes (such as tool use and manufacture), hunting and meat consumption, and brain enlargement. Washburn (1960) best exemplified this viewpoint. This theory has lost favour, largely because of the discovery of 3.5- 3.7-million-year-old footprints indicating the presence of hominid bipedalism long before brain size began its dramatic increase.

Many other theory attempt to explain hominid brain reorganization and enlargement. Some include parts of Washburn's earlier explanation. One of the more interesting recent theories is called the "radiator" theories. Human brains have relatively large cooling needs, and the evolving brain had to make crucial adjustments for heat dissipation. Because humans have more stamina for exercise than do many other mammals, they must be able to dissipate metabolic heat for longer durations. Furthermore, early humans occupied high-light and high-heat regions in the African savannas. Fialkowski (1986) suggested that increased brain size during hominid evolution provided structural redundancy that prevented cognitive impairment during heat stress. Fialkowski argued that increased brain size was a by-product of heat dissipation. Falk (1990) added that the increased brain size and danger of heat stress led to important changes in blood flow to and from the brain.

Wheeler (1984) drew a connection between upright body posture and heat stress. He argued that an upright hominid presented to the sun only about 40 percent of the body area that it would present if it were a quadruped. Thus, the bipedal hominid would reduce its heat load, and bipedalism would have been a factor allowing early hominids to occupy a noonday scavenging niche that was out of harm's way and avoided competition from other scavengers who rested during the heat of the day.

Combining elements from these different ideas, there seems to be a connection between a scavenging daytime savanna niche, bipedalism, increasing brain and body size, heat dissipation, and change in blood flow patterns to and from the brain.

Extensive Manipulation of Natural Objects and the Development of Motor skills to Facilitate Toolmaking

Extensive manipulatory behaviour was facilitated by hands freed of locomotor functions, stereoscopic vision, increasing brain size, and more effective hand–eye coordination. The selective pressure for extensive environmental manipulation probably grew out of increasing tool use, which was, in turn related to increasing problems of survival. Manipulation of the environment was an outgrowth of primate inquisitiveness.

The primate grasping hand permits detachment of objects from the environment. Knowledge about our environment stemmed from, among other things, just naturally picking up objects and examining them. This appears to be a natural outgrowth of primate curiosity, which is often expressed in play behaviour.

Grooming, a primate behavioural pattern, uses the hands to pick through another animal's hair and requires sophisticated hand–eye coordination. Most nonhuman primates who engage in grooming can

Table 12.4. Approximate Time and Characteristics of stone Tool Industries

5.0-2.5 mya
No known stone technology. Suspected use and discard of sticks and stones with little or no modification.
2.5-2.0 mya
Oldowan tools in Africa. All-purpose tools. Associated with *Homo habilis* fossils.
1.8 mya-200,000 ya
Developed Oldowan traditions in Africa, with Acheulian tools first recorded at 1.4 mya. Perhaps use of bamboo tools in Asia. Large bifacially flaked tools associated with *H. erectus* (or *H. ergaster*) in Africa.
150,000-40,000 ya
Often referred to as the Middle Stone Age. Several regional tool traditions, such as the Mousterian, associated with many Neandertal sites.
35,000-Present
Often referred to as the Late Stone Age. Specialized and composite tools. Many regional traditions are named. Associated with modern *H. sapiens*.

oppose the thumb to the forefinger. Such opposability is required in holding and handling objects. These seemingly incauspicious beginnings involving play and grooming behaviour were preparation for picking up and examining objects. In time, such actions might have led to tool use and manufacture. Although other arboreal animals have some of the same adaptations to the arboreal niche as nonhuman primates, rarely do any approach the primate's fine manipulative abilities and hand-eye coordination.

Let us return to the proposition that the ability to detach objects from the environment using the hands is a major advance that was most important for the development of primate perceptual skills. We owe recognition of different kinds of objects to our visual acuity and tactile examination. Given the upright sitting position of nonhuman primates, vision became "a supervision, a guide and control of fine manipulations". The relation between the evolution of keen vision—which was possibly an adaptation to locomoting through the trees and grasping at swift insects—and fine manipulation is two-dimensional. Vision itself became more refined, and intellectual absorption and mental utilization were more complete and lasting as the skilled movements became more complex and efficient.

Primates extract objects from the environment with their hands; they smell and taste them, visually and tactilely examine them, and then perhaps replace them; "In this way the higher primates have come to see the environment not as a continuum of events in a world of pattern but as an encounter with objects that proved to make up these events and this pattern".

Susman (1994) may have provided anatomical proof for early hominid tool making abilities. By noting that the human first metacarpal (thumb) bone has a broad head in relation to its length, Susman may have provided a way to determine which early ancestors had hands that functioned like our own. Human thumbs are much stronger than the thumbs of other primates, so the thumb, like some of the wrist bones, can resist increased pressures. Compared to the thumbs of other primates, the human thumb has three additional muscles that add strength and refined motor control. The broad head in relation to the length of the metacarpal bone in the human hand has no peers among the other primates. When such a configuration is matched in the hominid fossil record, Susman concludes that the bone must have come from a hand capable of generating the force needed for human tool use and making.

Susman investigated the metacarpals of *Australopithecus afarensis*, *A. robustus*, *Homo habilis*, and *H. erectus*. He concluded that only A. atarensis was anatomically unable to made and use tools. However, A. *robustus* had a thumb that could make and use tools. The problem here is that the metacarpal Susman attributes to A-robustus is attributed by others to *H. erectus*, widely acknowledged to be a consistent tool user and maker.

It is widely recognized that some of the nonhuman primates make and use tools of both stone and vegetation. These primates, however, do not have the stout human thumb. The stout metacarpal allows humans a type of manual dexterity thought necessary for stone use and manufacture.

The Increased Acquistion of Meat Protein

Although most nonhuman primates are vegetarians, we have already noted that some kill and eat other vertebrates. The time when meat first assumed importance in the human diet is not known. Consistent meat eating required behavioural adjustments and changes in dentition and jaw musculature. Once meat became basic to the diet, means would have developed whereby it could be most efficiently obtained and butchered. Methods for carrying meat, rules for sharing (perhaps one of the first sets of societal rules), and even a linguistic system naming those who were to receive shares would have appeared.

Sometimes analyses of tooth surfaces can tell us the role of certain foods in an animal's diet. Using an electron microscope one can examine the minute scratches left on tooth enamel and dentine as the animal chews and crushes food. Tooth surfaces of an early hominid *Australopithecus* (=*Paranthropus*) *robustus* for example, suggest that it was basically vegetarian. More precisely, the tooth wear patterns resemble those of chimpanzees. Because the chimpanzee ingests a large number of dietary items, the statement that the teeth of *A. robustus* resemble those of a chimpanzee under an electron microscope does not conclusively define its diet. It does, however, exclude grass eating, bone crushing, and root eating, all of which would have left distinctive scratches and pits on the tooth enamel.

A shift in enamel wear patterns on the teeth of a later human species, *Homo erectus*, suggests that a large amount of grit was incorporated into the diet. Perhaps this grit came from the discovery that roots and tubers were a good food source. The grit may also have been adhering to meat that was lying on the ground, but the teeth show no signs of bone crushing.

Too much emphasis may have been placed on hunting as an important factor in human evolution. With most interest focused on the hunting half of the hunter-gatherer food-getting complex, gathering has been underemphasized. Many researchers have ignored the probable evolutionary importance of gathering vegetable foods or have dismissed its results as "casually collected foods." Coon (1971) referred to the primacy of hunting and contended that it had more impact on social structure than did gathering. However, Lee (1968), among others, criticized this view. Food sources other than those offered by meat were important, and it is very unlikely that any human group ever relied primarily on meat.

When scenarios of human evolution refer to ingestion of meat protein, they usually refer to meat that has been hunted. The idea that scavenging might have represented a complete ecological adaptation has only recently come to the fore front. Several researchers, such as R. Potts (1984) and P. Shipman (1984), who worked with animal bones found at the 1.8-million-year-old Bed I site at Olduvai Gorge (also spelled Oldupai Gorge), Tanzania, have indicated that scavenging was probably quite an important component of early human life.

Based on an analysis of more than 2,500 antelope bones from Bed I, Olduvai Gorge, Shipman (1984) suggested that "instead of hunting for prey and leaving the remains behind for carnivores to scavenge, perhaps hominids were scavenging from carinvores". Scavenging has different costs and benefits than hunting. For example, the scavenger can be relativcly sure that the prey is dead, for predators have already performed the task of chasing and killing the prey. However, "while scavenging may be cheap, it's risky". Both predators and scavengers face the danger of possibly fighting over a carcass. The major energetic costs to scavengers come from the fact that they must survey much larger areas than predators to find food. Predators tend to be specialized for speed, scavengers for endurance. Scavengers also need an efficient means of locating carcasses. Shipman suggested that bipedalism was an adaptation to this need for endurance and for locating carcasses.

There are few full-time mammalian scavengers. The bulk of the scavenger's diet comes either from other food sources or from hunting small game. Such game as rats or hares are consumed on the spot, eliminating the problem of having to defend a carcass against larger carnivores. Because small carnivores such as jackals and striped hyenas often cannot defend a carcass, much of their diet consists of fruits and insects.

There is dental evidence that early humans ate both meat and fruit. Shipman noted, "The evidence of cut marks, tooth wear, and bipedalism, together with our knowledge of scavenger adaptations,...is consistent with the hypothesis that two million years ago hominids were scavengers rather than accomplished hunters" (1984:27). Hunting as a way of life may not have appeared until 1.5 mya with *H. erectus*.

Blumenschine (1987) provided an interesting discussion of the potential niche of an early hominid scavenging strategy, although he, like others, is not convinced that early hominids did in fact scavenge. Hominid scavenging success would have been influenced by the size of the carcass scavenged in relation to the animals scavenging from the carcass and the degree of competition for the carcass.

If early hominids did scavenge meat, they could have secured carcasses during the dry season from within or on the margins of riparian woodlands, that is, woodlands on the banks of rivers or streams. Most carcasses that would have been encountered there would be from medium-sized adults that were killed, partially eaten, and abandoned by fields during the dry season. Cavallo (1990) suggested that early hominids may have scavenged from leopard bills catched in tress. If, as has been suggested, some early hominids were more parboreal than their ancestors, scavenging carcasses from trees would not have presented a problem: "By scavenging from the leopard's temporarily abandoned larder, early hominids could have obtained the fleshy and marrow-rich bones of small-to-medium-sized prey animals in relative safety". Cavallo also suggested that the sharp, broken limb bones from partially eaten prey could have been used by hominids to peel back the hide to expose the flesh of a carcass. This activity might have provided early hominids with the initial impetus to manufacture and utilize tools to prepare and extract meat from a carcass.

The "man the hunter" hypothesis argues that the development of hunting had a profound effect on human evolution. This hypothesis suggests that males are both the cooperative and competitive sex, cooperating with one another in pursuit of big game and competing (this competition is often mistakenly viewed as aggressive physical confrontation) for access to female mates. Male bonding is viewed as having an old evolutionary history, and females are portrayed as infant-producers, whose reproductive functions demand neither cooperation nor competition with members of their sex.

This long-standing interpretation of human evolution has been vigorously challenged by those arguing that food gathering, traditionally a female task, was easily as important as hunting in the evolution of human behaviour. Food gathering exerted strong selective pressures on intelligence and technological skills, but proponents of this view are divided about the importance of competition and cooperation in the evolution of human behaviour. Female reproductive success is assumed to be less dependent than that of males on competition for mates, and it is often argued that there has been little selection for competition or aggression in females. This stress on noncompetitive females ignores the important element of competition among them for access of the "most desirable" male mates.

HOMO ERECTUS

Homo erectus fossils were first recovered in Java in 1891, but skeletal and cultural remains have since been recovered in Africa, Asia, and Europe. The oldest H. erectus fossils are from East Africa and possibly Java and date to about 1.8 mya. Some argue that the African fossils are more modern than their Asian counterparts and that they warrant a separate species named H. ergaster. Supposedly the more modern H. ergaster and not H. erectus is ancestral to H. sapiens.

Homo erectus is distinguished from earlier members of the genus Homo by a number of features including larger cranial capacities, which range from 750 to 1,250 cubic centimeters. The flattened skull vault is distinctive, and behind large brow ridges is a marked postorbital constriction. There is a sagittal ridge and thick skull bones, and the face and jaws are prognathic.

Beginning with H. erectus, the story of human evolution is increasingly associated with cultural elaborations and geographic spread into new environments. In Africa its tool tradition is called the Acheulian. H. erectus's apparent migration out of Africa was aided by its cultural adaptations. It seems to have hunted big game, and one of H. erectus's most important cultural innovations was the use of fire.

The first fossils eventually called *Homo erectus* were found in 1891 in Java, Indonesia, by Eugene Dubois, a Dutch physician. Dubois joined the medical division of the Dutch Colonial Service and traveled to Indonesia hopeful of recovering the remains of the link between apes and humans—the so-called missing link. Several years before Dubois's journey, the German biologist Ernst Haeckel postulated a

Fig. 12.3. Distribution of Homo erectus by country in which fossils or artifacts have been found (black). The European sites have transitional forms that some may refer to archaic Homo sapiens or Homo heidelbergensis.

theoretical human ancestral line based on fragmentary information. The only well-known fossil remains were comparatively recent bones discovered years earlier in the Neander Valley of Germany. Haeckel suggested that the human evolutionary line began among some extinct Miocene apes and reached *Homo sapiens* by way of an imagined speechless group of "ape-men" that he called *Pithecanthropi*.

Soon after his arrival in Indonesia, Dubois reported the discovery of a skull, femur, and several bone fragments at Trinil, a locale on the Solo River. Dubois eventually applied the name Haeckel suggested, "Pithecanthropus," to the bones that he recovered. Dubois first called his find "Anthropopithecus erectus" (upright manlike ape), the genus name indicating that he thought the form was closely related to chimpanzees. He changed the name to "Pithecanthropus" (upright apelike man) when he realized he had underestimated the brain size, which was larger than that of a modern ape. The name *Pithecanthropus* indicates that Dubois thought the fossil to be a link between humans and apes. Because the *Pithecanthropus* femur resembled that of modern *Homo*, Dubois concluded that his "ape-man" walked erect in the same manner as *Homo sapiens*. He christened his from *Pithecanthropus erectus*. The original generic name was subsequently dropped, and the form renamed *Homo erectus* (subspecies designation *erectus* to distinguish it from other *H. erectus* finds). For years many scientists doubted Dubois's contention that he had recovered an ancestral human. Most were hesitant to accept the association of an archaic skull with a relatively modern looking femur. By the time most scientists became convinced of Dubois's claims, he had rejected his original interpretation and died unconvinced of the immense value of his discovery.

Dubois's role with *Pithecanthropus* has been badly misunderstood and misrepresented. Readers interested in this story should consult Theunissen's (1989) excellent biography or Gould's (1990) shorter explanation of Dubois's role in the understanding and acceptance of *P. erectus* as our ancestor.

Quite some time has passed since *H. erectus* was first described by Dubois, but the story is still not done. The record is incomplete, and recently there have been unprecedented finds with fascinating dates from heretofore unsuspected regions in the world. There are continuing problems with the chronology in Asia because in China, one of the most important centers for *H. erectus*, datable strata are rare. The nature of evolution in *H. erectrus* and the question of the number of species are intriguing. The origins of *H. erectus* are unclear. It must have descended from an earlier *Homo*, probably *H. habilis*. But *H. habilis* is still poorly known, and the species may incorrectly include fossils that belong elsewhere. If *H. erectus* is the true ancestor to *H. sapiens*, where did the transition occur and when? There seem to be a number of intermediate fossils, sharing traits with both the *H. erectus* and *H. sapiens* species. This group may include *H. heidelbergensis* and *H. antecessor*.

The number of species in *Homo* prior to *H. sapiens* is debated but continues to increase. Added rather recently were *H. rudolfensis* and *H. ergaster*. The most recent addition is the Spanish material from Atapuerca that is being called *H. antecessor* by some researchers. Even the number of genera is proliferating with *Homo* in Africa, *Pithecanthropus* and *Meganthropus* being revived in Indonesia, and some Chinese colleagues continue to use *Sinanthropus*. Others use different generic names for the North Africa forms. We really must ask if the names are taxonomically relevant.

Why *Homo* appeared is debatable, but many pin its appearance on disputable climatic changes, Some find evidence for cooling and drying and a shrinking of forests, constituting an environmental thrust. Not everyone agrees that this change happened, over what areas it may have happened, and whet its effect was. There is general agreement that *H. erectus* (or H. ergaster) was adapted to the more arid, open grasslands and less stable environments that spread across tropical Africa. Its long and linear body build in Africa signals its adaptation to tropical and subtropical regions. By at least 1 mya it was adapted to the arid highlands of Ethiopia and the cold regions of China and Siberia. Because of the range of habitats, modern human diversity may have had its start with *H. erectus*.

H. erectus is differentiated from the earlier *H. habilis* or *H. rudolfensis* by continuing changes in the face and increases in cranial capacity. The tool assemblage, the Acheulian handaxe culture, is primarily African based. In Asia it may have relied heavily on bamboo. *H. erectus* seems to have been the first hominid to inhabit caves and to exploit fire. Intelligence seems to have reacted to climatic stress, although fire was being exploited in less extreme climates in Africa as well. Fire was exploited for warmth, light, cooking, and perhaps art and magic.

Perhaps manipulation of fire where to find it and how to bup it alive, led to the first social stratification among hominids. Fire exploitation may be a root of magic, ritual, and religion. Some

Table 12.5. Cranial Traits of Homo erectus

Long and low skull whose greatest breadth is situated low down on the skull. Skulls generally referred to as platycephalic.

Midline of skull often exhibits a sagittal ridge or elevation.

A relatively flat frontal bone characterized by prominent brow ridges continuous across the brow (supraorbital torus). Postorbital constriction is prominent.

Relatively small mastoid process.

Cranial bones almost twice as thick as those in modern humans.

Mean value of adult cranial capacity about 833 cc.

Very broad nasal bones.

Dental and Jaw Traits

Relatively robust mandibles, absence of a chin, broad ascending ramus (indicating large masseter muscles).

Molar that generally have large pulp cavities (taurodontism); upper central incisors that are typically shovel-shaped; enamel of molars in Asian forms generally wrinkled.

researchers recognize more than the species *H. erectus*—for example, *H. ergaster* in Africa and *H. heidelbergensis* and *H. antecessor* in Europe. Some retain the Chinese fossils in the originally suggested genus, *Sinanthropus*, and others retain the original Indonesian genus *Pithecanthropus*. We refer to all of the materials as *H. erectus*.

General Morphology

The genus and species designation *H. erectus* is based on skeletal remains of samples from Africa, Asia, and Europe. Although there is a moderate degree of individual and geographical variation, these remains comprise a group whose morphological characters are sufficiently consistent and distinctive to justify their inclusion in the genus *Homo*. That being the case, what morphological features distinguish them from *H. sapiens* and *H. habilis*?

H. erectus has a cranial capacity varying between 750 and 1,250 cc. The lower values are found among earlier representatives and the higher values among later representatives of the species. The mean cranial capacity of twelve Asian skulls in 929 cc. Considering the *H. erectus* sample as a whole, an increase in cranial capacity over the preceding australopithecines may be partly the result of increased body size (Wolpoff, 1980).

Leigh (1991) has shown a significant cranial capacity increase through time in both *H. erectus* and early *H. sapiens* (where there is a relatively marked increase in a short period of time). Among *H. erectus*, the increase is especially apparent in some Chinese and Indonesian specimens, although other *H. erectus* samples do not show the increase.

It has been argued that with increased body size *H. erectus* also had an increased cranial capacity. However, Alan Walker (1993) estimates that WT 15000 (material from Lake Turkana, Kenya) had a cranial capacity of 909 cc, equivalent to that of earlier members of *Homo*. Perhaps the relative brain size often thought to characterize *Homo* is only found in *Homo* after the Middle Pleistocene.

There is marked flattening, *platycephaly*, of the *H. erectus* skull vault. A sagittal ridge runs across the skull's midline, and the skull bones are quite thick. The greatest width of the skull is low on the skull vault; in modern humans it is higher on the skull vault because of decreasing bone thickness, reduced mass of the chewing muscles, and increased cranial capacities. Posterior to the large brow ridges above the eye orbits, the skull is marked by *postorbital constriction*. The nasal bones are broad and flat, and the face more prognathic than in modern *Homo*.

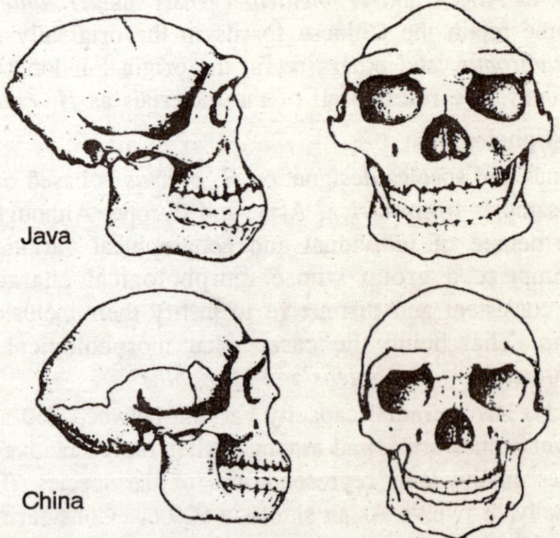

Fig. 12.4. Comparison of Homo erectus skulls from Java and China. Note the large brow ridges, the sagittal ridge and the cranial flattening on the side of the ridge, the constriction of the skull behind the brow ridges, and the amount of prognathism.

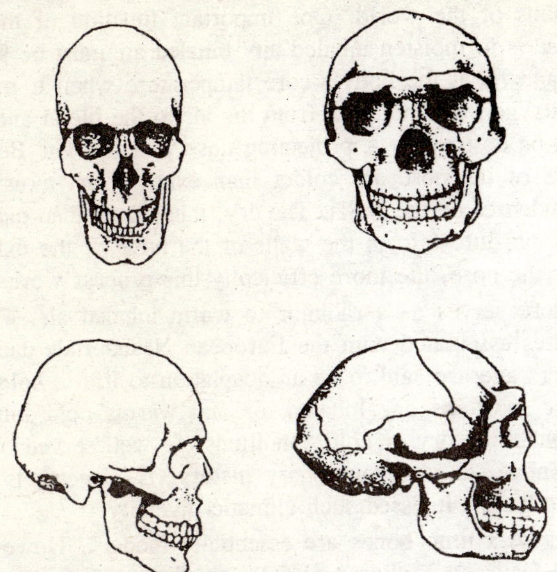

Fig. 12.5. A modern skull (left) and a Homo erectus skull (right).

The heavily constructed *H. erectus* mandible lacks a chin. The teeth are larger than those found in *H. sapiens*. The canines are sometimes slightly projecting, with a small diastema in the upper dentition; however, the first lower premolar is bicuspid with subequal cusps. The molars have well-differentiated cusps complicated by wrinkling of the enamel surfaces. This wrinkled enamel pattern found in many extant Asian populations has been used by some scientists to show a direct evolutionary link between modern Asian populations and Asian *H. erectus*. The second upper molar may be larger than the first, and the length of the third lower molar may exceed the second. Premolar and molar tooth size reduction is one of the most dramatic changes that occurred between the australopithecines and *H. erectus*, and an equally great reduction occurred over the *H. erectus* time span. The mandible also reduced in size.

H. erectus is perhaps the first hominid to have displayed a projecting nose (Waters, 1990). Although there are no fossils of noses because they are comprised of soft cartilage, the nose's profile can be determined from the shape of the bones around the nasal opening. With a projecting nose, *H. erectus* may have found it easier to conserve water in a hot and dry climate. This may have been one factor that allowed *H. erectus* to move across the hot African savannas to colonize other regions of the world. One important function of the modern human nose is to moisten inhaled air. Inhaled air must be 90 percent saturated and be at the body's core temperature when it reaches the lungs, or oxygen will not pass from the air to the blood and the lung tissue will be destroyed. A projecting nose wets the air. Because the inner walls of the nose are colder than exhaled air, moisture from this air condenses on the walls. The dry, inhaled air then picks up the condensed moisture left on the walls of the nose by the exhaled air. The larger the nose, the more efficiently this process works.

The nose serves as a radiator to warm inhaled air. The larger nasal cavities associated with the European Neandertals dating about 55,000 years ago are said to be an adaptation to life in cold regions. If the nose moistens dry inhaled air and warms cold inhaled air, hominid inhabiting dry or cold conditions are well served by a large projecting nose. In its evolutionary history, *H. erectus* is the first hominids to have witnessed such climatic diversity.

H. erectus's limb bones are essentially modern. However, Day (1971) and Day and Molleson (1973), on the basis of material from Olduvai Gorge, suggested that there are some differences distinguishing the postcranial anatomies of *H. erectus* and *H. sapiens*. As will be noted later, a similar conclusion is drawn from preliminary study of *H. erectus* material (WT 15000) from Lake Turkana. Whatever the differences between the postcranial anatomies of *H. erectus* and *H. sapiens*, both species were bipedal.

The *H. erectus* morphology can be summarized as follows:

1. The cranial capacity is about the size of a modern one-year-old child.
2. The rib cage is like that of modern humans.
3. The hips are more narrow than those of other hominids, giving it greater speed. Its longer legs allowed it to cover more distance. The robust bones of the leg indicate a demanding life.
4. *H. erectus* was generally a tropic dweller. The finding that *H. erectus* was physiologically adapted to warm and humid climates means that a humanlike physiology with a dependence on sweating as a major cooling mechanism was established by 1.5 mya. African *H. erectus* are at least 158 centimeters tall and weigh no less than 51 kilograms. WT 15000 is estimated to be 185 cm tall and weighed 68 kg, a well-nourished and healthy creature. African *H. erectus*, with a mean stature of 170 cm, would be in the tallest

17 percent of modern populations, even if we make comparisons only with males.

5. The lower thoracic vertebra, T7, had a small opening through which the spinal cord passed. Nerves in the T7 area allow fine control of the rib cage muscles used in exhaling. Walker (1993) suggests that WT 15000's spinal cord could not carry enough nerve tissue for him to control his breathing as well as modern humans. Linking words into long sentences would have been impossible.

The overall similarity between early *H. erectus* and modern human is clear. The limb proportions are like those of modern humans who inhabit extremely dry, hot places. The facial skeleton is quite modern, although the face is certainly more prognathic. The upper limb skeleton is quite like modern humans. The major exceptions to modern morphology are in the size and shape of the brain-case, size of the thoracic neural canal, narrowness of the pelvis, long femoral necks, and other limb traits,

Although a number of traits distinguish H. erectus from *H. sapiens*, there is a general consensus that the former or some closely allied species stands in an ancestral relationship to the latter, based on the following observation: (1) The morphological traits of *H. erectus* conform very well with the theoretical postulates for an intermediate stage in the evolution of later species of *Homo*; (2) the existence of *H. erectus* in the early part of the Pleistocene, antedating well-authenticated fossil remains of *H. sapiens*, provides it with an antiquity conforming well with its supposed evolutionary position; and (3) some *H. erectus* materials appear to illustrate a graded series of morphological changes leading from *H. erectus* to *H. sapiens*. In fact, there is disagreement about whether some of the fossils belong with the species *H. erectus* or *H. habilis*.

Questions have been raised concerning the exact role *H. erectus* played in subsequent evolution and the validity of maintaining the African and east Asian samples in the same species. Some African fossils such as KNM-ER 3733 and 3883 are particularly different from the classic east Asian *H. erectus*. For example, the African skulls have thinner skull bones. The African fossils generally have larger brains than Asian forms. These larger brains are housed in flat and long but elevated braincases that are expanded higher up in the parietal region. Because they are older than east Asian specimens, these specimens could represent an African population that was ancestral to Asian *H. erectus*. However, when the African skulls differ from the east Asian *H. erectus*, they tend to be more like *H. sapiens*. This observation raises the possibility that the African specimens are directly ancestral to *H. sapiens* and that the east Asian *H. erectus* represents a specialized evolutionary side branch that became extinct. Louis Leakey was perhaps the first researcher to call attention to morphological differences between African and Asian early *Homo*. Although Leakey's contention for different taxonomic status for the African and Asian forms originally did not gain wide support, there is currently considerable favour for placing the African materials into the species *H. ergaster*, different from the Asian *H. erectus*.

Table 12.6. Some Critical Morphological Features of Homo erectus

Cranial capacity: Sange of 750–1,250 cc. The lowest values are for the earliest forms.

Skull: platycephaly, sagittal ridge, thick skull bones; greatest width of skull is low down on the skull.

Facial skeleton: Supraorbital torus and postorbital constriction, heavy mandible lacking a chin, prognathism, projecting nose.

Dentition: large teeth, wrinkled dental enamel quite apparent in Chinese fossils.

Body size: increased body size (weight and stature), mean stature of 170 cm for African fossils.

Limb skeleton: modern upper limbs, narrow hips, long and robust leg bones.

This table assumes that all the fossil are *H. erectus*, whereas some of the traits listed for the African species are referred to as *H. ergaster* by other researchers.

Review of Sites

Africa

North Africa. North African *H. erectus* material was recovered in Morocco, Algeria, and Tunisia. The time span of the North African sample is estimated to be from the late Middle Pleistocene back to the older but uncertain age of the Ternifine (Algeria) remains. The relative ages of this material, from the oldest at Ternifine to the most recent at Sale, is as follows: Ternifine, Sidi Abderrahman, Rabat, and Sale (the latter three are from Morocco). The Ternifine remains may date to 730,000 to 600,000 years ago. Sidi Abderrahman may be 500,000 years old. Middle remains. The exception is the Sale partial cranial vault and lower face, which dates to about 160,000 years ago. Pleistocene materials from Algerian and Moroccan coastal sites is mostly mandibular.

These material are variously referred to as late *H. erectus* or early *H. sapiens* because they combine traits from both species. Primitive *H. sapiens* traits are especially noticeable in the most recent remains from Sale and Rabat.

Lake Turkana. The oldest *Homo erectus* remains come from Lake Turkana, Kenya. A pelvic fragment, KNM-ER 3228, and two incomplete femurs (upper leg bones) labeled KNM-ER 1472 and 1481 date to about 1.8 mya. These femurs show a strong resemblance to later *H. erectus* from Olduvai.

Homo erectus remains (cataloged as WT 15000) from Lake Turkana found in 1984 are from a site known as Nariokotome III. The remains of a young male are dated to approximately 1.6 mya. Its chronological age is based on tooth wear and tooth eruption patterns. WT 15000 has a dental age of 11, but the skeletal age corresponded to a modern 13–13.5-year-old, and his stature corresponded to a modern 15-year-old. The remains were found in what was once a marshy habitat. The skeletal remains are the most complete known for *H. erectus*. Missing are the left arm and hand, the right arm from below the elbow, and most of both feet. Postmorem damage occurred, and some parts of the bones are crushed. There are no signs of carnivore or scavenger damage.

One of the most interesting features of WT 15000 is his estimated adult height. He had already grown to 5 feet 4 inches tall, the height previously postulated for adult *H. erectus*. When fully grown he might have reached a height of 6 feet, much taller than previously envisioned for any of our early ancestors. In fact, WT 15000 was taller than most modern people. WT 15000 weighed about 48 kg at the time of death and might have weighed about 68 kg fully grown. The height and weight estimates reinforce the view of early African *H. erectus* as tall, relatively thin individuals, physically similar to some modern human populations living under the same general climatic conditions. *Australopithecus* was apparently shorter and relatively heavier, and this difference may reflect a shift in both the ecology and behaviour of *H. erectus*.

Body size is a fundamental species adaptation that affects the development of the species' evolutionary strategies. More recent studies on body size in African and Asian *H. erectus* suggest that they were larger than originally thought (Gauld, 1996). The average estimated height of six African *H. erectus* skeletons dated between 1.7 and 0.7 mya is 170 cm, tall even by modern standards (Ruff, 1993). *H. erectus* from Zhoukoudian are shorter than their African brethren.

H. erectus was likely dwelling in relatively open, at least semiarid places, for this is where their physique is most adaptive. Australopithecines could have inhabited either closed and wet or open and dry habitats. Given their smaller size and small stride length, the australopithecines likely lived in relatively closed, wet environments—perhaps something like a riverine forest.

WT 15000's skull, assembled from about seventy pieces, is small, and his cranial capacity may be 909 cc (Lewin, 1984). The overall shape of the upper region of the thigh differs from that in extant or extinct humans. The vertebrae show some interesting differences from those of modern humans. The Turkana youngster has a narrow and flared pelvis and a combination of a large femoral head and long femoral neck. These features of the upper leg are not found in other human femurs, and the functional significance in terms of the locomotor pattern of WT 15000 is unknown.

Birth canal diameters of early *H. erectus* were significantly smaller than those in *H. sapiens*, and the passage of a modern-sized full-term fetus would have been impossible. *H. erectus* infants must have exhibited considerable postpartum development and long-term dependence upon their primary caretaker, probably the mother.

Excavations at Lake Turkana in 1978 revealed several deep, four-toed impressions of hippopotamus feet and shallow depression resembling a human footprint (Behrensmeyer & Laporte, 1981). Further excavations revealed two more complete humanlike footprints in line with the first, Excavations in 1979 revealed another two complete human tracks along with tracks of many other vertebrates. The footprints were apparently formed when the mud surface in the area was submerged under quiet water. Tracks of a small wading bird indicate that the water was shallow. The human prints date to about 1.5 mya and were made by an individual 5 to 5.6 feet tall and weighing about 120 pounds.

An almost complete *Homo erectus* skull, KNM-ER 373 , from Lake Turkana is dated to 1.5 mya. The skull looks almost identical to *H. erectus* skulls found in China that date to only 360,000 years ago.

Slightly older than KNM-ER 3733 is a very similar skull belonging to KNM-ER 3833. Its brow ridges, nuchal region, mastoids,* and cranial base are somewhat thicker than similar areas in KNM-ER 3733. Wolpoff (1980) suggested the possibility that KNM-ER 3733 is a female and KNM-ER 3833 is a male.

The KNM-ER 1805 skull from Lake Turkana dates to approximately 1.5 mya. The face is broken away from the skull, the brow ridge is missing, and there is an associated mandible. Although the specimen shows closer similarity to the preceding australopithecines than to KNM-ER 3733, Wolpoff suggested that it belongs with *H. erectus* and that it may represent a transitional stage.

Olduvai Gorge. In Bed II at Olduvai Gorge, in deposits near the margin of an ancient lake, L. Leakey (1961) recovered a human skullcap, which was labeled OH 9. Some 100 yards away lay numerous handaxes and abundant remains of large mammals, the long bones of which were apparently split for marrow. The material is dated to 1.2 mya. The skullcap resembles Far East *Homo erectus*. It is thick; the frontal region is flattened; and there are a prominent nuchal crest, small mastoids, and large, flaring brow ridges. There is no face.

OH 9 shows similarities to the smaller but more complex KNM-ER 3733 material. The two finds resemble each other in enough ways that Rightmire (1979) suggested the possibility of the Turkana population being ancestral to that at Olduvai.

Other artifactual and skeletal remains associated with *H. erectus* come from Bed IV at Olduvai and date between 700,000 and 400,000 years ago. A left femoral shaft and hipbone associated with an Acheulian industry (the stone tool industry of *H. erectus*) in Bed IV represented the first well defined artifactual assemblage of *H. erectus* at Olduvai. A sample of the occupational debris from Bed IV yielded 494 artifacts, 347 unmodified cobbles, and 257 largely fragmentary faunal skeletal remains. Handaxes from Bed IV resemble those of the Early Acheulian series in Middle Bed II. Handaxes, the characteristic tool of the Acheulian tradition, are pear-shaped implements with sharp edges all around.

*The mastoid process is a bony protuberance found on each side of the skull behind the ear. It carries the insertion of the sternomastoid (or sternocleidomastoid) muscles, which function to allow humans to rotate the head sideways. With the enlarging of the mastoids the muscles obtain leverage to pull the head forward and make it possible for the sternomastoid muscles to raise the head after it has tipped back. Femoral and pelvic remains attributed to *H. erectus* and referred to as OH 28 were found in situ during the 1970 Olduvai excavations.

The femur has most of the shaft but lacks the head and greater trochanter (a bony eminence on the upper part of the femur to which muscles that extend and flex the leg are attached). Femoral and pelvic fragments from the left side of the body are probably from the same individual. There is every indication of habitual bipedalism.

Ethiopia. H. erectus inhabited Ethiopia, as indicated at the sites of Omo, Konso, and the Plain of Gadeb. At the Konso site *H. erectus* and a robust australopithecine species have been shown to overlap. The *H. erectus* tool tradition, the Acheulian, has been recovered from the Plain of Gadeb on the southeast plateau in Ethiopia (Clark, 1987). Clark and Kurashina (1979) suggested that humans appeared on the Ethiopian high plateau (the altitude was 2,300–2,400 meters) perhaps about I mya, *Artifacts* (cultural remains) from this locale include tools of an advanced Oldowan assemblage, the *Developed Oldowan Industry*, and the Acheulian tradition. Although many elements of the artifact assemblage are provisionally classified as Developed Oldowan, some tools show a refinement more usual in the Acheulian tradition. If the Developed Oldowan assignation does hold, this is the first appearance of such assemblages outside Olduvai Gorge. The Gadeb sites show an alternating sequence of Acheulian and Developed Oldowan dating between 1.5 and 0.7 mya. These must have been contemporaneous traditions. The sites are generally found close to or in stream channels. Although remains of grazing animals are found at same sites, judging from the preponderance of the remains, the preferred food animal seems to have been the hippopotamus.

The Bodo material found in 1976 and hominid materials found in 1981 and 1990 have been dated to approximately 0.6 mya (J. D. Clark et al., 1994). The Ethiopian sites yielding this material also seem to show a transition from the Oldowan chopper and flake assemblages, present in lower stratigraphic units, to Acheulian bifaces, consistently prevalent and widespread in directly overlying deposits (or a relationship between these artifacts). These older dates from the Middle Awash region of Ethiopia and from Olorgesailie in Kenya combine with the older occurrences of Acheulian bifaces at other Ethiopian sites to emphasize the long duration of the Acheulian. Together with the older *H. erectus* from Java, the Ethiopian finds emphasize that the Old World origins and migration, as well as the culture, of *H. erectus* were complex.

Clark and Kurashina suggested that there was probably contact between groups inhabiting this region of the Ethiopian highlands and those to the south in the Rift Valley. They also suggested the possibility that these early humans were experimenting with fire.

Tanzania. The Ndutu skull discovered in Tanzania in 1973 is a fairly well-preserved specimen dated about 600,000 to 400,000 years ago. Ndutu was roughly contemporaneous with material from Bed IV, Olduvai, and Zhoukoudian, China, which it resembles. The Ndutu skull vault is thick, with prominent brow ridges, which are damaged, but with a rather vertical forehead. The back of the skull (the occipital region) is also thick. On the other hand, the skull shows less robusticity of the muscle markings on the occipital. There is no sagittal ridge across the midline. Ndutu appears to be morphologically intermediate between *H. erectus* and *H. sapiens*. Rightmire (1983) stressed Ndutu's affinities to archaic *H. sapiens*.

South Africa. At Swartkrans, South Africa, Broom and Robinson (1949, 1950) found three fragments of a lower and upper jaw. The best of the three pieces is a mandibular fragment (SK 15). There is also an almost complete skull (SK 847). When the material was first discovered, some argued that it was contemporaneous with members of the robust australopithecine lineage. W. L. G. Clark (1967) suggest that it belonged to a female *A. robustus*. The argument continues, but most suggest that Swartkrans provides evidence for the contemporaneity of *H. erectus* and *A. robustus*.

Olorgesailie Occupation Remains. Olorgesailie, located in the Rift Valley about an hour's drive from Nairobi, Kenya, is an extensive living site that may contain the relics of a *H. erectus* group. Because there are no human skeletal remains, any designation concerning the human species inhabiting the area must be made on cultural remains and on associated dating. Neither yardstick is satisfactory.

Geological and paleobotanical evidence from Pleistocene sediments indicates that for prolonged periods there was a stable freshwater lake in the area. Former climatic conditions were wetter than today. The site dates between 700,000 and 400,000 to 900,000 years ago.

The occupation areas have stone and bone accumulations that may have coincided with natural boundaries, such as the bank of a sandy brook or the limit of shade provided by a larger tree. The inhabitants might have built shelters of hedges that disappeared over time. Smaller occupation areas may have been used by four or five adults in a group; larger areas may have accommodated more than twenty adults.

Based on the accumulation of a ton or more of stone artifacts and *manuports* (materials carried into the site, but which do not necessarily show use), it seems the site was continuously occupied for two or three months. Archaeological assemblages at Olorgesailie are later than the lower Acheulian assemblages from Olduvai Bed II, Within one oval-shaped area, measuring 19 by 13 meters, excavators found 2,200 pounds of stone artifacts, rubble, and unmodified cobbles.

The abundance of bone of an extinct nonhuman primate (*Theropithecus*) seemed at first to attest to continued, extremely successful hunting of a single troop. One analysis of the primate remains stressed that butchering occurred but dismissed the contention that the animals were somehow killed as a troop by humans. Binford and Todd (1982) suggested that the primates were initially preyed on or at least scavenged by a relatively small predator-scavenger. Later, the bones were subjected to considerable attritional breakage by nonbiological agents or simply by trampling. Humans may have had little to do with this bone accumulation.

If the other faunal remains are the result of human activity, their abundance suggests that the Olorgesailie inhabitants were hunters and scavengers. Large mammalian remains predominate; rodent, bird, and reptilian bones are relatively scarce. However, the lighter bones of smaller rodents, birds, and reptiles are less likely to fossilize than the heavier and larger bones of larger mammals. Most bone is splintered; either the inhabitants pulverized the bone or scavengers moved in after they left camp.

Potts found two areas where hominids at Olorgesailie left major concentrations of fossil bones and tools. At one site a nearly complete elephant skeleton intermingled with stone tools was recovered. Potts thought he had uncovered a butchery site.

Potts also suggested that the Olorgesailie hominids had home bases. Bones scarred solely by stone tool marks have been recovered from Olorgesailie. These are unlike bones found at Bed I, Olduvai, that show both stone tool marks and carnivore tooth marks, indicating, according to Potts, that the Olduvai hominids shared carcasses with large predators and scavengers and were unable to keep them from their bases

The spread of *H. erectus* out of Africa probably involved moving through the Dardanelles, which were often dry during glaciations, facilitating movement from southwest Asia to Europe. The importance of the east Mediaterranean route through southwest Asia is indicated at the site of Ubeidiya in the Jordon Valley, Israel, which contains evidence that people emigrated from Africa I mya or earlier.

The strong volcanic activity in the ritt Valley region about 1.6 mya was perhaps one factor leading (forcing) *H. erectus* from Africa and into other parts of the world.

Large numbers of Pre-Acheulian stone tools were found in 1977 from sixteen sites near Ash Shuwahitiyah in northern Arabia. These tools are typologically similar to those from Lower Pleistocene sites in East Africa—especially Oldowan, Karari, and Developed Oldowan materials.

Another site, Najran, located on the eastern side of the Asir Mountains and the Rub's al. Khali desert also has tools. Artifacts at both sites are similar in type and percentage, raw material, and technology. The high percentage of choppers at Najran resembles Oldowan sites. However, the Najran sample is too small for positive attribution.

The site near Saffaqah has the largest collection of artifacts from a single Saudi Arabian site. It was found in 1979 and is located southeast of al-Dawadmi. The site is located in a valley and contains large areas of handaxes, cleavers, picks, spheroids, and other materials. It also has choppers and core axes. There are seven Middle Pleistocene artifact clusters—those referred to butchering and meat slicing, to bone splitting, to hide scraping, to plant gathering and processing to stone tool making, to wood making, and to bone working.

Saffaqah contained almost 8,000 stone tools. The site was abandoned when the large lake in the area disappeared. It had good plant and animal communities to exploit, and there was considerable stone for toolmaking. The period of human occupation stretched from the Middle Pleistocene through the Mousterian and into the Neolithic.

The Saudi sites were occupied by migrations out of Africa. The land route is the longest route and was down the Nile river from the Ethiopian highlands and east across the Sinai into Western Asia. A shorter route was through the Rift Valley and across the Afar depression to the narrow strait of Bab al Mandeb separating Asia and Africa. The strait is about 28 kilometers wide with a small island situated off the Yemen coast. If the migrants could use rafts to reach the island, they could penetrate southwestern Asia from that point.

Stone tools perhaps dating to 1 mya were recovered in Qatar (Abdul Nayeem, 1998).

If the new *H. erectus* dates in Indonesia are correct, *H. erectus* left Africa much earlier than assumed. Such an early migration is possible. Indonesia was once connected to the rest of Asia by lower sea levels that provided a land route from Africa. Java is 10,000 to 15,000 miles from Kenya, depending on the route traveled. At one mile of migration per year, *H. erectus* could have moved from Kenya to Java in 15,000 years.

Asia

Java. Human fossil remains have been recovered from Java, which among other islands of western and northern Indonesia received invasions of Pleistocene animals from both India and China. The Djetis faunal beds, the earliest Javanese human-bearing beds, contain fossil species thought to have come from the tropical region of south China. The overlying Trinil faunal beds, containing fauna largely of south Chinese origin, yielded a host of human fossil remains, including Dubois's original "Pithecanthropus erectus." The dating of the Trinil faunal beds is approximately 750,000 to 500,000 years ago.

The Trinil faunal beds contains fossils of widely varying age. Doubt exists about the relationship of the Trinil fossil skullcap of *H. erectus* and the femur attributed to the same individual. The site yielding the femur is about 15 meters away from the skullcap. The femur's antiquity can only be inferred from association and from its supposedly archaic features. Such indirect evidence, however, has not stilled criticism about the claimed association of the femur and skullcap. Modern chemical analyses can neither confirm nor deny the femur's antiquity—an enterprise not aided by Dubois, who boiled the specimen in glue in the mistaken notion that doing so would aid its preservation.

Dubois apparently mixed materials from two different levels, and the skullcap and femur were not associated. It is ironic that the femur indicating the modernity of *H. erectus's* bipedalism probably derives from *H. sapiens*. Later *H. erectus* finds do, however, support claims for bipedalism.

The Trinil faunal beds may contain examples of the earliest Javan tools, the Patjitanian (or Pacitanian) tools. These are advanced over earlier Anyathian tools found in Malaysia, which are not associated with human remains. Both industries are considered to be in the Acheulian tradition. Jacob and others (1978) announced the discovery of two stone tools from the Sambungmachan deposits that may range from 900,000 to 700,000 years old.

Tantalizing evidence suggests a far older appearance of *H. erectus* in Asia than previously thought. There are claims for old appearance of *H. erectus* in China at Gongwangling, for example. Recent redating of two important cranial samples from Java, the Modjokerto and the sangiran materials, further suggests a long time span. The modjoker to skull material (Modjokerto 1 or Perning 1) found in 1936 belongs to an approximately 5-year-old individual. It was thought to date to about 1 mya based on the age of the sediment layer with which it is associated.

Two fragments (S27 and S31), the Sangiran materials named *Meganthropus*, were found in the 1970s. This material consists of a crushed face and partial cranium from two individuals. Their robusticity led some to originally suggest that they were closely related to the African robust australopithecines. They were thought to date to 900,000 to 700,000 years ago.

Using argon-argon dating, Swisher et al. (1994) suggested that the Modjokerto material dates to 1.81 mya and the Sangiran to 1.6 mya. The argon-argon dating technique is similar to the potassium-argon technique, but it can be used on a single crystal that yields more accurate isotope ratios than do the larger samples, and, since the argon-argon technique compares two argon isotopes, not potassium, the rarity of potassium in Java is not a problem.

These dates are approximately the same as the earliest appearance of *H. erectus* (or *H. ergaster*) in Africa and suggest that either (1) *H. erectus* left Africa far earlier than imagined (if *H. erectus* evolved first in Africa and them migrated to Asia, the new Asian dates demand a much older *H. erectus* yet to be found in Africa) or (2) perhaps *H. erectus* evolved independently in Asia and in Africa. If this is true, then Asian must also contain *H. habilis*, currently the suspected ancestor of *H. erectus*, or even possibly *Australopithecus* as some suggest is the case in China.

The new dates may answer why no Acheulian tools are found in Asia. Acheulian tools are widely found with African *H. erectus* and should, therefore, be expected to appear in Asia. They do not. Redating of the Javan hominids suggests that if the Asian forms are related to African *H. erectus*, the latter may have left Africa before the invention of the Acheulian tool. If so, this explains its absence in Asia.

Redating the Java materials calls into question the exact evolutionary relationship of the Asian and African fossils normally called *H. erectus*. Noting morphological variation between the Asian and African remains, some suggest that the African forms should be placed in a species called *H. ergaster*. If this viewpoint holds, the Asian forms are put onto a side branch of human evolution. The possibility of an evolutionary line linking Chinese *H. erectus* and *H. sapiens* populations refutes this possibility.

One of the most unexpected discoveries is the suggestion that *H. erectus* lived on Java between 53,000 and 27,000 years ago. These dates indicate that *H. erectus* may have survived on Java at least 250,000 years longer than on the Asian mainland and perhaps 1 million years longer than in Africa. If the dating holds, this is the first time that *H. erectus* and *H. sapiens* have been shown to coexist. If this coexistence occurred, *H. erectus* likely became trapped on Java after having spread there over land from the rest of Asia. H. sapiens probably arrived after 40,000 years ago by boat. *H. sapiens's* arrival in Java perhaps led to the demise of *H. erectus*.

The new dates come from water buffalo teeth, not the hominid remains. The animal teeth have high levels of uranium that can compromise ESR dating. Furthermore, the bovid teeth may not be the same age as the human skulls, because flooding or erosion may have washed older skulls from their original resting places.

These findings challenge the multiregional evolutionary model, which proposes that *H. erectus* in Asia was among the ancestors of modern humans. Swisher et al.'s (1996) work questions whether the Solo skulls from Java are the ancestors of modern Australian aborigines as predicted by the multiregional model. If the new dates are correct, *H. erectus* persisted in Southeast Asia into the latest Pleistocene overlapping *H. sapiens*.

Morewood et al. (1998) suggest that *Homo erectus* in Indonesia actually used boat travel—perhaps bamboo rafts. This is a startling claim given most current thought on the abilities of *H. erectus*, and, if the claim is true, it will necessitate a serious rethinking of *H. erectus's* cognitive and technical skills. Prior to this announcement the earliest widely accepted evidence of major water crossing was the colonization of Australia by anatomically modern *H. sapiens* perhaps 60,000–40,000 years ago.

According to Morewood et al., *H. erectus* reached the island of Flores sometime between 900,000 and 800,000 years ago via a water crossing of at least 19 Kilometers from mainland South Asia. It will take considerably more evidence, more widely accepted dates, and time and argument before boat travel is an accepted means of travel for *H. erectus*.

China. A review of the Chinese hominids appears in Wu and Poirier (1995). *H. erectus* in China is found in the cold climates of the north at Zhoukoudian and in the warm and wet climes of southern China. This species had a wide geographical range and had cultural adaptations (e.g., fire, clothing, housing) allowing it to adapt to environmental diversity.

The skeletal remains of Chinese. *H. erectus* are mostly teeth and skull fragments. Limb bones are quite rare. Taphonomic analyses are needed at many sites, but some are certainly carnivore accumulations.

There are at least fifteen *H. erectus* sites in China. The time span is mostly the Middle Pleistocene, but some Lower Pleistocene sites are possible. Nine sites yield archaeological materials, sin sites yield skull and facial materials, twelve sites yield dental and jaw remains, and two sites yield postcranials.

The earliest representatives of human fossil remains from china belong to *H. erectus* and perhaps date to 1.7 mya. Although this date has been confirmed by three different laboratories, a date of 700,000 to 500,000 years ago also appears in the literature. This material from the Yuanmou basin in Yunnan Province consists of teeth and limb fragments. The incisors are large and may belong to a young male. The environment at Yuanmou was a grass-forest and the average temperature was about 10 degrees Celsius, lower than the present-day average. According to G. Zhou (1987, unpub. ms.), the material should be referred to as *H. erectus yuanmouensis*.

Until rather recently most Chinese representatives of *H. erectus* came from Zhoukoudian (once spelled Choukoutien), which lies about 25 miles southwest of Beijing (Peking). The remains are often referred to as "dragon-bone" (fossilized bone) collectors whose finds ended up in Chinese pharmacies. In the early 1900s, the German paleontologist K. A. Haberer found a human tooth in a Beijing drug-store and traced it back to Zhoukoudian. For twenty-four years various scientists worked in or watched the site, which had once been a large cave. In 1921 they were led to Locality 1, where they immediately found bits of quartz recognized as a tool. More quartz tools were found in subsequent excavations. The presence of tools at Zhoukoudian indicated human occupation.

In 1923 and 1926 human teeth were sifted from the Zhoukoudian debris. In 1927 the Swedish paleontologist B. Böhlin produced one more tooth. This was subsequently shown to Davidson Black (1931), who allocated the tooth to a new hominid genus and species that he named "Sinanthropus pekinensis." Zhoukoudian was excavated continuously from 1927 to 1937, and the remains were described by

Franz Weidenreich (1936, 1938, 1941, 1943). Work resumed in the 1950s, and a new mandible was recovered and described in 1959.

Human remains from Locality 1 included at least 14 skullcaps without faces, 12 mandibles, 150 teeth, and several postcranial remains. Much of the material was fragmentary; however, enough remained to allow a rather complete reconstruction. Early in World War II, at the beginning of the Japanese occupation of China, all the skulis were lost, and no one seems to know what happined to them shapiro, 1971, 1974). All that remained was a fine set of casts, a long tooth, and a very complete set of monographs by Weidenreich based on the original materials.

Deposits at Zhoukoudian were laid down over an extensive time period (Huang, 1960). The *H. erectus* remains seem to date back to approximately 360,000 years ago according to the K/Ar dating method. However, estimates range from 600,000 to 414,000 years ago. This postdates Djetis faunal beds and is perhaps contemporaneous with the later Trinil faunal beds from Java.

A number of animal bones were associated with the hominid material. It was a time when *Crocuta crocuta*, the living spotted hyena, was crowding its older cousin out of its habitats in Africa, Europe, and China. Deer bones were the most numerous faunal remains, but there were numerous elephant, rhinoceros, and giant beaver bones. High and low in the fill were layers of ash and burnt bone.

Scattered about were a few human trunk and limb bones and skull fragments A good many piece of the face and jaw were present, but these were not affixed to the braincase. Some suggested that the Zhoukoudian inhabitants were cannibals. If so, cannibalism was probably ritualistic; that is, human protein was not an important part of the diet.

Numerous pollen grains were collected at Zhoukoudian; 28 percent of the arboreal species were of beech and 11 percent of grasses. Seeds of the hack-berry, a relative of the wild cherry, were common. Pollen analysis indicates that Zhoukoudian lay near the border zone that separated the northern coniferous belt, or boreal forest, from the temperate steppe. The hills about Zhoukoudian were covered in pine and spruce, the slopes bearing trees that grow today in the fat north of Maine.

Floral and faunal remains suggest a northern temperate zone. Mammalian remains associated with a tropical or polar climate are rare. Evidence from different cave layers suggests climatic shifts during the long history of Zhoukoudian. There is, however, disagreement among Chinese researchers concerning the climate during *H. erectus* times at Zhoukoudian.

Mountainous areas north and west of Zhoukoudian had mixed forests of pine, cedar, elm, hackberry, and Chinese redbud. Bison, saber-toothed tigers, tigers, leopards, cheetahs, horses, woolly rhinoceroses, striped hyenas, sika deer, and elephants inhabited this region. A big river and possibly a lake located to the east contained various water species; along the shorelines grew reeds and plants. Buffalo, deer, otters, beavers and other animals inhabited the area.

Some of these species may have formed the diet for *H. erectus*. *H. erectus* at Zhoukoudian hunted big game; however, the daily meal consisted as much of small animals such as hedgehogs, frogs, and hare. Ash layers in the cave abounded with the bones of these animals and others such as hamsters, mice, black rats, and harvest mice. There are fragments of ostrich eggs. The inhabitants of Zhoukoudian may have exploited vegetable foods; however, these fossilize poorly. Charred hackberry seeds are the only evidence to date of possible plant food.

Skull bones and reconstructed face and Skull of *Homo erectus* from Zhoukoudian, China. A, First *H. erectus* skull found in Zhoukoudian in 1929 at Locality E; B, Anterior (mostly the frontal) and posterior (mostly the occipital) parts of Skull H found in 1966; C, Right-side view of the reconstructed skull H; D, Left-side view of skull H; E, Superior view of skull H; F, Bust of *H. erectus* from Zhoukoudian; G, Reconstructed face and skull.

Numerous quartz stone tools were uncovered from Locality 1. The characteristic tool was a pebble trimmed to make an edged chopping tool. Zhoukoudian yielded flake scrapers and points made with a crude beak. Many tools are large, massive, and crudely worked; few display an even, uniform retouch. Possibly some of the tools were used to work animal skins that could have been worn or used as enclosure walls to keep out the cold.

The tool manufacturers often selected oval-shaped pebbles to make single edged or doubled-edged axes. These were found in most layers in the cave deposit. Most tools were made from sandstone and varied in shape from round disks to irregular triangles and long slabs. These tools were blunted, and there is bountiful evidence of blunt and resharpened artifacts in the cave.

Binford and Ho (1985) and Binford and Stone (1986) questioned many of the notions associated with the finds at Zhoukoudian. Binford and Ho rejected arguments that *H. erectus* at Zhoukoudian was a cannibal, an argument posited at least in part because the skulls lack faces, supposedly removed to gain access to the brain. They argued that hominid remains at the site are instead the work of other scavenging carnivores. Once inside the cave, these remains were dispersed throughout the cave by secondary scavenging, probably by wolves and hyenas. Although many have stated that *H. erectus* used fire and perhaps cooked food, Binford and Ho and Binford and Stone suggested that the traces of fire at Zhoukoudian do not reflect purposeful burning and do not indicate the use of hearths.

Binford and Ho and Binford and Stone questioned whether bone deposits indicate that *H. erectus* at Zhoukoudian was a big-game hunter. They stated that the deposits, including the human bones, were the remains of carnivore activity. They also doubt that the hackberry seeds represent the remains of *H. erectus's* dietary practices. They have shown rather conclusively that earlier reports of bone tools at Zhoukoudian are incorrect, and that the so-called bone tools are instead bones that were modified by carnivore chewing and not human handiwork.

Jia (1989) reacted strongly against many of the propositions stated by Binford and Ho and Binford and Stone. Jia suggested that the broken bone at Zhoukoudian may be attributed to an attempt to extract marrow. He also stated quite adamantly that the human inhabitants at Zhoukoudian not only used but also controlled fire. Given the cold climates at the time when the cave was inhabited, fire use could have been an important means for keeping warm.

Another important Chinese *H. erectus* site is at Chenjiawo in Lantian District, Shaanxi Province. In 1963, Chinese paleontologists recovered a mandible from the site; in 1964 at Gongwangling Hill, about 20 kilometers away, they recovered facial bones, a tooth, and a fairly complete skullcap all from the same individual. Both sites predate Zhoukoudian. The cranial material may date to 800,000 to 750,000 years ago. The mandible may be dated to 590,000 to 500,000 years ago. The Lantian jaw is robust, and in many respects the skullcap is similar to material from the Javan Djetis faunal beds. The skullcap is low and excessively thick; the massive brow ridges are arched near the middle. The cranial capacity is estimated to be approximately 780 cc, close to the estimate for the skull from the Djetis faunal beds.

The mandible is well preserved except for small parts of the vertical ramus. All the teeth are present except the right first premolar, which was lost before death. There is no mandibular third molar on either side. This was a congenital condition. Indications in the bone and teeth suggest that Lantian suffered from periodontal disease and manifested a condition known as periodontoclasia.

According to Woo (1964), the Lantian material is allied with Zhoukoudian. However, Woo maintains that differences exist, and he first classified the Lantian material as "Sinanthropus lantianensis." Most paleonologists prefer the designation *H. erectus lantianensis*.

A heavily fossilized *Homo erectus* skull belonging to a young male was unearthed in southern China in 1980 from Longtang cave in Hexian Country, Anhui Province. Excavations in 1980 and 1981 exposed parts of the frontal and parietal bones from the skull, the left side of a fairly robust mandible (which may belong to a female), and nine teeth. As many as four individuals may be represented by the fossil sample. The skull has similarities and differences from the skulls found at Zhoukoudian. The fission track date is 190,000 to 150,000 years old and the amino acid racemization date is about 200,000 years old.

Mammalian fauna associated with the human remains consist largely of animals that probably inhabited forests and woodlands. There are indications that the climate at this time in the middle-late Middle Pleistocene was cool.

Fossil and archaeological material from Longgupo Cave (called the Wushan Hominid Site in Wu and Poirier, 1995), situated near the eastern border of Sichuan Province, lends support for the notion that hominids were established in Asia by 1.9 mya. (This is also suggested by the most recent dates for some *H. erectus* sites in Java). The Longgupo materials may support an early hominid lineage in Asia, and the morphology of the left side of mandibular piece from an adult suggests to Wangpo et al. (1995) that the materials may not be those of *H. erectus* but instead are from *H. ergaster*. The root structure of an isolated upper incisor from the site suggests likewise.

The dating of Longgupo is a crucial matter for understanding hominid evolution in Asia. Paleomagnetic and ESR dating and faunal associations are consistent with an old date.

The bifed root (a trait of *H. ergaster*) of the mandibular premolar and primitive features of the mandible lead some to exclude the materials from *H. erectus*. The in situ molar has five cusps like *H. habilis* and *H. ergaster* and is unlike the six cusps found in Asian *H. erectus*. If this is a species more primitive than H. erectus, it implies that the first heminid out of Africa was not *H. erectus* and that *H. erectus* may have evolved in Asia and spread back into Europe and Africa. Two items from the site that may be early stone tools were made of a type of rock not found elsewhere at the site.

Having said this, it must also be said that some doubt that the dental remains are even those of a hominid. Also recovered from the site were *Gigantopithecus* remains. Perhaps the Longgupo specimen was not hominid but was related to the Chinese hominoid *Lufengpithecus*. Longgupo may have no special affinity to early African hominids.

At the 700,000-to-1-million-year-old site of Xiaochangliang many artifacts and fossil animal bones, some with cut marks, were found. Taken together, these are the earmarks of a workshop or possible campsite. The ratio of waste flakes to artifacts that were used as tools is consistent with the ratios expected at sites where tools were actually made. An open-air campsite is a rare find in China. Until the Xiaochangliang finds, the only clearly occupied early sites in Asia were caves, notably Zhoukoudian, located 65 miles southeast.

The artifacts seem to be home-grown and show no indication of having been introduced from outside populations, such as a migration from Africa. Artifacts from Xiaochangliang show that relatively sophisticated technologies arose in localized parts of the Far East as early as, or earlier than, in Europe.

In 1989 a human skull was found in Yunxian County, Hubei Province, central China. Another skull was found in 1990. Both skulls are without mandibles, are big and robust, and have distinct postorbital constriction. There is no median sagittal ridge, the parietal bone is short and flat, and there is no angular torus.

These skulls have features usually seen in *H. erectus*, such as very distinct postorbital constriction and the low position of the maximum breadth of the braincase. However, there are other features concordant with archaic Chinese *H. sapiens*. The advanced features favour a *H. sapiens* attribution, but final taxonomic ascription must await further analysis.

Based on the associated fauna, the site dates to the middle part of the Middle Pleistocene. It has been suggested that the site may be at least 1 million years old. Some stone artifacts were surface collected and some appeared in situ.

Placing the Javan and Chinese H. erectus fossils in the same species affirms their close evolutionary relationship. The major features differentiating the Zhoukoudian and Javan specimens are in the skull and teeth; the cranial capacity of the Zhoukoudian fossils is larger, and the teeth are significantly smaller compared to earlier fossils. The Zhoukoudian skulls vary greatly in length, and cranial capacities range from 915 cc to 1,225 cc. Asian *H. erectus* samples do not differ in their postcranial anatomies and approximate *H. sapiens* in this respect.

Asian *H. erectus* specimens have equally thick skull bones, and all have large brow ridges, but there are major differences in cranial capacities. The mean cranial capacity of the Javanese material is estimated at 975 cc, or lower; the Lantian material at 780 cc; and the material from Locality 1, Zhoukoudian at 1,075 cc. The increased brain volume differentiating the Javanese and Chinese Locality 1 samples is matched by other modernizations of their skull morphology.

When compared with older Javanese material, the dental arch of the Zhoukoudian sample is shorter and more rounded in front, and there is no sign of a diastema in the upper tooth row. The Zhoukoudian mandible is also shorter and more compact. In the Zhoukoudian sample the first upper molar is the largest (it is the second largest in the Javan sample), but the lower third molar is shorter than the second.

Fossil evidence for the presence of Middle Pleistocene humans has been reported from Vietnam. There are human remains from the sites of Tham Khuyen and Tham Hai and tools from the sites of Nui Do and Quan Yen. The Vietnam fossils have a preliminary date of about 400,000 years ago.

Thailand. A rock shelter dating to approximately 700,000 years old has been found in northern Thailand. The Kao Pah Nam rock shelter, located in a heavily eroded limestone landscape, yielded animal bones, stone tools, and the remnants of a hearth. Faunal remains include forest-dwelling animals such as the hippopotamus, and giant ox, deer, bamboo rat, and porcupine—all or some of which may have been included in the diet.

Europe

The first recovered human skeletal material dating from Middle Pleistocene Europe is the Heidelberg or Mauer mandible. It was recovered in 1907 from a sandpit in the village of Mauer, Germany, located 6 miles southeast of Heidelberg. It lay 78 feet below the surface in soils containing bones of the spotted hyena, *Crocuta crocuta*, and is dated to between 300,000 and 250,000 years ago, although the date is insecure.

The mandible is well preserved; most of the teeth are moderately worn, except for four left molars and premolars, broken at the time of discovery and only partially restored. It is chinless and massive—one of the largest human mandibles yet recovered. Compared with other human mandibles, Mauer's most striking feature is the great width of the ascending ramus. This vertical strut of bone underlies the back of the cheek and carries at its posterior corner the condyle that articulates the jaw with the skull. The functional implications of the mandibular breadth are related to the fact that the temporalis muscle (the largest of the muscles that close the jaw) attaches to the coronoid process of the mandible and fans out on the side of the skull. The masseter muscle attaches to the outer surface of the lower portion of the ramus. Although the width of the ascending ramus is undoubtedly related to mandibular length, it also suggests powerful chewing muscles and wide-flaring, strong cheek bones. Despite the large size of the mandible, the teeth are not proportionately large.

Beyond guessing that it had a wide, not strongly projecting face, we cannot reconstruct the Heidelberg (Mauer) skull. Despite its large size, the jaw and teeth place the specimen in the genus *Homo*. Some scientists refer Mauer to a *H. erectus* subspecies, for example, *H. e. heidelbergensis*. Others include it with early *H. sapiens*. Increasingly. however, Mauer is referred to as *Homo heidelbergensis* and is touted as ancestral to Neandertals. Some newly recovered English) and Spanish (Atapuerca) fossils and other materials are also oeing referred to as *Homo heidelbergensis*.

Middle Pleistocene human remains have been recovered from the site of Vértesszöllös, situated west of Budapest, Hungary. Vértesszöllös appears to have been a campsite with a small saucerlike depression, no bigger than an average room. The area was apparently formed by rising waters of a hot spring, which may have made it an attractive gathering place during cooler weather. In and around the site were a number of pebble tools—the first undisputed early tools found in Europe. The only human skeltal remains were an occipital bone and quite a large amount of burned bone, indicating the use of fire. The site was first dated at approximately 400,000 years ago, but this date may be too old, and there exist estimates of 350,000 years to as recently as approximately 185,000 years ago.

The occipital is rather thick and has a well-marked angle and ridge for the attachment of the neck muscles. Despite its thickness, some argue tha the skull bone is from a population advanced over African and contemporaneous Asian *H. erectus*. The reconstructed angle formed by the roof and floor of the skull appears to have been wider than that in *H. erectus*, indicating that the cranial capacity was considerably larger. Thoma (1966) estimated that the cranial capacity was about 1,400 cc, close to the average of modern *H. sapiens* and well above the previously described *H. erectus* sample.

Because of the morphological differentiation from the Middle Pleistocene Asian sample, and because of its estimated large brain size, Thoma (1967, 1969) suggested that the Vértesszöllös occipital might be designated *H. sapiens paleohungaricus*. Thoma's suggestion would place a population of more progressive *H. sapiens* contemporaneous with the *H. erectus* population. It has been suggested that an evolutionary line ran from Mauer through Vértessöllös to later *H. sapiens* populations. According to Wolpoff (1977), the potential importance of Vértessöllös lies in the information it provides concerning evolutionary changes at the end of the *H. erectus* lineage or the very beginning of *H. sapiens*.

The occipital bone probably belonged to a large individual. It exceeds *H. erectus* occipitals in total height; however, it shows general affinities to *H. erectus*. Although its cranial capacity has been used to assess its phylogenetic position, direct determination of taxonomic status is impossible from the Vértessöllös occipital alone.

The site of Pozletice, Czech Republic, is probably older than either Mauer or Vértessöllös. Pozletice is located near Prague and was first recognized in 1938. The site is rich in early Pleistocene mammalian fauna. Stone tools were recovered, but no undoubted human remains appeared. Most of the bone assemblage was broken, and it has been claimed that one end of a long bone fragment from a deer had been worked to form a keel-like point. Simply on the basis of its time placement, Pozletice might possibly represent an early *H. erectus* site in Europe; but without human skeletal remains, one cannot be sure.

A significant *Homo erectus* find was made recently in the country of Georgia, at a site called Dmanisi. The find is a large and heavy mandible, with all its teeth in place. The mandible was found in conjunction with stone tools, the skulls of two saber-toothed tigers, and the ribs of an elephant.

The dating of this material is 1.6 mya or 900,000 years ago, based on paleomagnetism. Gabunia and Vekua (1995) date the site to between 1.8 and 1.6 mya. This is an important find because of the paucity of early *H. erectus* remains outside of Africa. Although there are *H. erectus* materials of 1 mya in Asia, most European remains are in the range of 500,000 years ago. The materials from Georgia will help focus the issue of when *H. erectus* migrated from Africa. The Dmanisi mandible seems more similar to African than Asian *H. erectus*.

A Middle Pleistocene site, Fontana Ranuccio in Italy, yielded limestone tools dating to 700,000 years ago and a bone industry dating to 450,000 years ago. The two human teeth (one an incisor) date to the latter level. This is the earliest well-dated association of Acheulian stone tools with human remains to be found in Italy. It is not clear if the material belongs to a member of *H. erectus*, although the 700,000-year-old date for the choppers is in the *H. erectus* time span.

Another site in Italy, Castel di Guido, is estimated to date about 300,000 years ago. The site yielded animal remains, Acheulian stone and bone tools, and small tools. Fragments of human remains, including a temporal bone, were recovered. The temporal bone shows traits of *H. erectus* and has a number of features associated with *H. sapiens*.

In 1994 a 900,000-to-800,000-year-old *Homo* skullcap called Ceprano Man was recovered in Italy. Unlike classic *H. erectus*, this find has no crest dissecting the midline of the skull, and brain is significantly larger than classic *H. erectus*. Italian researchers find more resemblance between this material and that from Algeria than with the Mauer find.

The Montmaurin mandible from France may date to about 300,000 years ago. It resembles the Mauer mandible in its robustness but is about the same size as the early *H. sapiens* mandible from Steinheim, Germany. Montmaurin differs from Mauer in having less broad rami. The mandible has no chin, and in other feature resembles H. erectus.

Some of the newest *H. erectus* materials are from Europe, and fossils from Spain are among the most exciting. Gran Dolina, one cave site from Atapuerca, yielded hundreds of hominid remains and tools paleomagnetically dated to at least 780,000 years ago. There are four to six individuals (perhaps victims of cannibalism), including a child and an adolescent. Some facial traits suggest the remains are more aligned with African fossils that some call *H. ergaster*. Others call the materials *H. heidelbergensis*.

Some Spanish researchers refer to the young boy's remains as *H. antecessor*, a species they consider ancestral to modern humans and Neandertals. The name *H. antecessor* comes from the Latin word meaning explorer or one who goes first. Unfortunately, this new species is based on a juvenile specimen. If it is accepted, it will bump *H.*

erectus and *H. heidelbergensis* off the main line of descent to modern humans, placing them among the proliferating side branches of the human family tree.

The young boy from Atapuerca had a face with modern features, sunken cheekbones, and a projecting nose and midface. There is a prominent brow ridge and multiple roots for the premolars. The multiple roots of the premolars are not found in any other European fossil, all having a single root. Only African *H. erectus* (=*ergaster*) has multiple roots. It is an unusual mosaic that de Castro et al. (1997) think cannot be accommodated within *H. heidelbergensis*.

De Castro and his colleagues suggest that *H. antecessor* is a close kin of *H. ergaster* in Africa and is ancestral to *H. heidelbergensis*, which many agree led to the Neandertals. De Castro et al. suggest that *H. ergaster* gave rise to *H. antecessor*, probably in Africa, although it is still only seen at Atapuerca. They think *H. antecessor* dispersed to Europe about 1 mya, leaving remains at Atapuerca about 780,000 years ago. It further evolved into *H. heidelbergensis* and to the Neandertals, but not to modern humans. The southern branch of *H. antecessor*, probably still in Africa, gave rise to modern humans through another, yet unidentified species. This form may be revealed in the Kabwe ad Bodo specimens. Bone fragments and stone tools cone from three sites at Orce in Spain. The sites are situated around and ancient lake basin. Disputed hominid bones date to 1.6 mya, making Orce possibly among the oldest European sites with hominid remains.

Dates at Orce are based on Paleomagnetism and micro- and macromamalian fauna. Fuentenueva 3 contains more than 100 flint artifacts from three distinct layers and nonhuman bones. Barranco Leon contains more than sixty artifacts similar to the Developed Oldowan of Africa. A cranial fragment of highly contentious affinity and two small pieces of humerus were found. The Venta Micena site yielded more than 15,000 bones comprising more than thirty species. The nearby Cueva Victoria site yielded four fragmentary hominid fossils.

While some claim the bones from Venta Micena are human, others claim they are from horses. The cranial fragment is the most contentious item, while the small pieces of humerus are more readily accepted as hominid. Immunological tests of proteins confirm both human and horse presence. Recent studies by Borja et al. (1997) that seem to confirm human albumin from the skull fragment have not stilled the controversy. While the bone remains are vigorously debated, the archaeological remains dated to 1.4 mya are generally accepted.

Although the migration route to Europe is usually thought to be a land route running north from the Middle East and then west across land bordering the Mediterranean or regions further inland, humans could have traveled across the Strait of Gibraltar from northern Africa or across the Bosporous from Turkey. Declines in sea level between 2.4 and 1.6 mya reduced the distance across the Bosporous to about 3 miles, with a small island approximately halfway. African fauna associated with tools and butchery support the notion that humans entered Europe prior to 1 mya.

European Middle Pleistocene human populations were possibly pursuing a big-game hunting way of life as is suggested at the sites of Torralba and Ambrona, located in the rolling countryside of north-central Spain and at the sites of Aridos 1 and 2 also located in Spain. The Ambrona Valley was part of a major migration route for herds of deer, horses, and elephants. Concentrated in a relatively small area at Torralba are the remains of approximately thirty elephants, twenty-five horses, twenty-five deer, ten wild oxen, and six rhinoceroses. Detailed stratigraphic and pollen analysis at Torralba and Ambrona yields a data of 500,000 years ago.

Binford (1987b) raised serious doubts concerning the interpretations of the Torralba and Ambrona sites. It is generally conceded that the Torralba data were overinterpreted. It no longer seems feasible to view Torralba as an elephant-hunting site. Aridos, a Middle Pleistocene locale containing elephant bones and stone tools, is located 18 kilometers southeast of Madrid and provides evidence for organized exploitation of elephant remains by hominids. The evidence appears quite different from the marginal scavenging and minimal involvement by hominids with the Torralba elephant remains.

Aridos 1 and 2 are 200 meters apart and are dated to the Middle Pleistocene based on microfauna and tool associations. Both are individual death sites containing elephant remains. There is no evidence of projectile points, spears, or natural traps to indicate hunting behavior. However, both sites clearly indicate meat acquisition through early access to the carcass. There is extensive evidence of knapping and tool resharpening. The Aridos finds conform to patterns document at butchery sites. There is no convincing evidence to support a hypothesis of elephant drives at Aridos, and the newer analysis at Torralba no longer supports the possibility of elephant drives there.

Potentially exciting and quite unexpected finds were recovered from the Siberian site of Diring Yuriakh by Y. Mochanov and S.

Fodoseeva (Morell, 1994). Diring Yuriakh, located above the flood plain of the Lena River, is at 62 degrees north latitude and had annual temperature fluctuations of -70 degrees F to +70 degrees F. Occupation of Diring may represent an opportunistic movement of people into the area during a warm climatic interval. Even then the climate would have required clothing, fire, and shelter to keep warm. It is not clear how long Diring was occupied or whether occupation in the Lena Basin was uninterrupted for a long period of time. Other sites in the region date from 10,000 to 30,000 years old.

Russian researchers thought the purported tools at Diring to be similar to East archaeological record has large rocks that look as if they were used as anvils. Flake debitage surrounding these "anvils" is supplemented by a few unifacially flaked choppers or scrappers. Some purported tools had only a few flakes struck from the core; others appear to be extensively worked. Richard Potts suggests that Diring was a quarrying site. While Waters et al. (1997) report the finding of 4,000 artifacts, other archaeologists remain skeptical and suggest the materials are geofacts.

Dating at the site is disputed (Gibbons, 1997b) with Russian researchers giving dates of 1.6 mya and M. Waters giving a dates of about 500,000 years ago. A TL date of 260,000 is also suggested. All these date are unexpectedly early for an area previously assumed uninhabited by early hominids. We are not considering the 1.6 mya date as possible. Water's date is within the *H. erectus* time span, previously thought to have migrated no further north than Zhoukoudian, 2,500 kilometers south of Diring.

The 14-inch-long tibial fragment from Boxgrove, England, is dated to 515,000–485,000 years ago. Based on the length and thickness of this lower leg fragment, it is estimated to come from a very muscular person about 6 feet tall and weighing more than 168 pounds. Since Boxgrove predates the Neandertals, some suggest it may be the common ancestor of modern humans and Neandertals.

Two human lower incisors found in the fine silt layer below the tibia may belong to another individual living after the owner of the tibia. The teeth look like teeth in the Heidelberg mandible, and some refer Boxgrove to *H. heidelbergensis*. Root damage suggests periodontal disease.

The human remains are associated with handaxes and remains of elephants, rhinoceroses, hyenas, deer, bear, and wolves. Cut marks on rhinoceros and horse bones show that the carcasses were skinned, disarticulated, and defleshed, prior to the smashing of the bones for the marrow. There is no evidence of secondary scavenging by other animals.

Although other European sites date to about the same time as Boxgrove, none shares its detailed archaeology. More than 125 flint handaxes are scattered about the dismembered animal remains.

Acheulian Tradition

The Acheulian tool tradition associated mostly with African *H. erectus* sites first appeared about 1.5 mya. It seems to be an advance over the preceding Oldowan tool tradition associated with *H. habilis*. Acheulian tools were made from a wider range of raw materials, which may indicate a greater awareness by the toolmakers of their environment. In contrast to the all-purpose Oldowan tools, different Acheulian tools were apparently tailored for different purposes. They appear to be more standardized and specialized than Oldowan tools.

The basic Acheulian tool kit includes a versatile rang of cutting, scraping, piercing, chopping, and pounding tools used to prepare animal and plant materials. Bones were broken and trimmed to use as tools, and wood was worked into spears.

The precision evident in the manufacture of Acheulian handaxes appearently resulted from a new flaking method, the *soft-hammer technique*, in which a stone flake is detached from the core with a wood, bone, or antler hammer. The use of these softer materials as a hammer allowed more control of the final product in terms of flake length, width, and thickness. The core surface was flaked all over, resulting in a bifacial flat and per-shaped tool with more regular, longer, and thinner edges. With this technique a new type tool—the cleaver, shaped like a "U" with a straight transverse cutting edge—was developed.

The evolution of Acheulian industries in *H. erectus* populations involved two factors: a marked improvement in manufacturing skills and an ability to translate the idea of the tool into a better-made product. Wolpoff (1980) argued that the second factor led to an increasing number of different tool types, the appearance of fire, and eventually fire-hardened spears.

Three waterlogged spears dated between 400,000 and 380,000 years ago were recovered in Germany. The spears are shaped with the thickest end toward the front and with a long tapering end, like modern javelins. They were apparently meant for throwing. Grooves on the spear suggest that flint tools may have been joined to make them a composite tool.

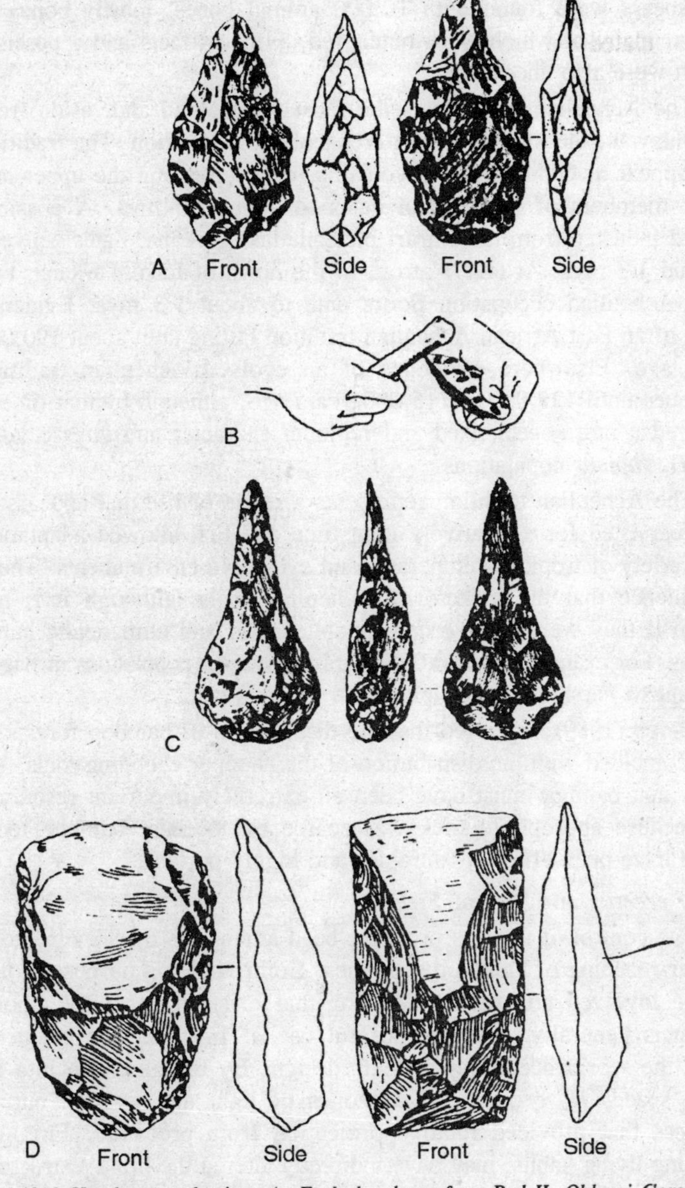

Fig. 12.6. Handaxe production. A—Early handaxes from Bed II, Olduvai Gorge; B—Use of an animal bone to make a handaxe; C—Acheulian handaxe from Swanscombe, England, one-third actual size; D—Acheulian cleavers, one-half actual size.

The spears were found with 10,000 animal bones, mostly bones of horses, many of which were butchered. Flint artifacts and a possible hearth were also uncovered.

The Acheulian tradition is the most widespread and, aside from the Oldowan, the longest-lasting archaeological tradition. The tradition first appear at Olduvai above volcanic tuffs separating the upper and lower members of Bed II and dates to about 1.5 mya. A possibly related industry from the Karari platean, lake turkana, dates between 1.5 and 1.3 mya. At lake Natron, to the north of Lake Turkana, two other Acheulian occupation floors date to about 1.3 mya. Evidence exists of an East African Acheulian tradition lasting until about 190,000 years ago. Elsewhere, evidence of an evolved Acheulian tradition continues until 125,000 to 115,000 years ago, although by that time it acquired a more specialized and regional character and is associated with *H. sapiens* populations.

The Acheulian tradition represents a mode of life that appears to have persisted for a relatively long time and that allowed adaptation to a variety of tropical, temperate, and even cold environments. There is evidence that the makers of Acheulian tools, although it is not known if they were *H. erectus*, inhabited arid and climatically harsh regions. For example, by 250,000 years ago human populations managed to adapt to harsh semiarid uplands in Tadzhikistan.

Pope (1989) suggested that the distribution of bamboo forests in Asia coincided with the distribution of the chopper-chopping tools. He thinks that bamboo must have been an extremely important resource, not because appropriate rock was scarce but because bamboo tools would have been efficient, durable, and highly portable.

Homo erectus and Use of Fire

The control of fire has probably been among the most momentous and far-reaching of human discoveries. Controlled and purposeful use of fire involved conquering the fear that exists toward fire among mammals generally. With purposeful use of fire, humans began to shape the world according to their design. By bringing fire into its living space, *H. erectus carved* zones of light and warmth out of darkness that provided relative protection from predators. Fire, by changing living habits, may have indirectly altered the brain's structure and enhanced the ability to learn and communicate.

The earliest evidence of fire with possible human connections comes from sites at Lakes Baringo and Turkana in Kenya. Burnt clay at these sites dates between 1.5 and 1.4 mya. Although some researchers suggest an association between this evidence and human occupation, most are unconvinced of such a purposeful association. It is not difficult to find traces of fire early in Africa. The problem is distinguishing between controlled fire used by humans and naturally occurring fires.

The Ethiopian sites of Gadeb and Melka Konture would have extreme daily temperature fluctuations, with freezing or near freezing temperatures at night. It is significant that some of the earliest evidence of hominid control of fire occurs at Gadeb at 1.4 mya.

If fire at Lakes Baringo and Turkana is not associated with humans, then the earliest possible evidence of fire use associated with humans may be the discoveries from the South African cave site of Swartkrans. The evidence from Swartkrans Member 3 dated between 1.4 and 1 mya, has been reported by Brain and Sillen (1988). The Swartkrans site contains skeletal remains of both *A. robustus* and *H. erectus*, and it is not certain who was using the fire. Evidence of fire use from other parts of the world is only found in association with *H. erectus*.

The evidence for controlled use of fire at Swartkrans is in the form of burned bone, which is found in several distinct layers of limestone within the cave. The 270 charred bone fragments were heated to a range of temperatures that occur in campfires. If controlled use of fire in South Africa is verified, this is among the oldest incidence of humans using fire in Africa.

Fire was probably originally obtained from such ready-made sources as volcanic eruptions, brush fires, or gas and oil seepages. Hunters may have camped near fire, which was a natural resource as were game, water, and shelter. From the beginning fire may have been used to keep predators away. Perhaps humans became regular cave dwellers only after they learned to use fire to drive predators away from the cave.

Sillen and Brain (1990) suggested that one possible use of fire at Swartkrans was to keep away predators. There is evidence from Swartkrans that leopards were human predators, and modern leopards at least are wary of fire. Sillen and Brain dismissed the use of fire for cooking food at Swartkrans because the burned bone does not indicate temperatures found in cooking of food. Conceivably, the bones were used as the fodder for the fire—they may have been burned for warmth.

Perhaps each social group took smouldering embers with them when they moved. Each group may have had its own fire-bearer, responsible for carrying and keeping the embers alive. Control of this potent material may have conferred an aura of respect and invincibility on its keeper. Could such an individual have become an important leader? Could the use and control of fire confer status on those who possess thc knowledge of where to find and how to use fire? Perhaps this control and knowledge were prized possessions guardedly passe from generation to generation.

Fire may have been used in hunting to stampede prey. Fire may also have been used to produce more effective spears. Fire hardens the core and makes the outer part crumbly and easier to sharpen. The earliest possible fire-hardened spear, however, dates to only approximately 80,000 years ago and was found in Germany.

The presence of fire-hardened wooden spears is important because they indicate an ability for rather complex problem solving. A set of primary stone tools was needed to chop, cut, carve, and scrape a secondary tool made of wood. That wood was chosen from among the best available varieties. The secondary product, the wooden spear, was then fire-hardened, representing still another technology and a tertiary process that altered the nature of the wood (Marschak, 1989).

Psychological changes may have accompanied the use of fire; for example, cooking may have produced behavioural restraint, that is, control of a tendency to do things on the spur of the moment. Cooking of food suggests that less food is eaten on the spot and more is carried back to the camp. If the assumption that food sharing requires elaborate social rules is correct, it follows that the more there is to share, and the more individuals there are to share with, the more rules are needed. A continuing elaboration of the communication system would be in order.

Introducing fire into the living spaces creates an artificial day independent of the sun's movements; evening hours could be illuminated and one's attention turned to other pursuits. Extra time could be used to make tools, and to think about food-getting and migration routes. Fire may be a potent stimulant that served early ceremonial functions.

We can raise a number of unanswerable questions at this time. For example, what did early cave dwellers think of shadows dancing on dark cave walls as they sat around their fires? Did willowy shapes on the walls lead to thoughts that may have had magical or ritual overtones? Where any of the shadows construed as friends or foes, as animals of the hunt? Did the place that fire holds in many of the world's religions have such beginning as these?

HOMO NEANDERTHELENSIS

Neandertal Age

When did the Neanderthal people live? Fossils of the "typical" representatives are dated from 70,000 to 35,000 years ago. This was the time when ice advanced from the north, down across much of Europe, during the pneultimate great Ice Age. One of the latest Neanderthal specimens from Saint Cezaire, France, could be as little as 31,000 years old. Several features of the Neanderthal skeleton indicate a body adapted to cold-climate survival.

Discoveries pertinent to their distant past have been made in France and Wales. In 1976, parts of a skull and jaw were recovered from a site at Biache, and estimated to be more than 100,000 years old. The lower rear of the skull and jaw were recovered from a site at Biache, and estimated to be more than 100,000 years old. The lower rear of the skull has the projecting Neanderthal-type shape. (This may be connected with anchorage of the neck muscles). Other French finds at La Chaise and Lazaret support this evidence, and suggest that Neanderthal people were living in Europe more than 130,000 years ago, another time of ice and cold.

From 1978, bits of teeth and jaws were found at Pontnewydd, a Welsh site of roughly the same age as Swanscombe. The teeth show similarities with those of typical Neanderthalers. Taken with the Swanscombe and Steinheim evidence, they could indicate people with Neanderthal features living in Europe more than 200,000 years ago.

A Neanderthaler Described

Reconstructions from the various fossilized bones and teeth of Neanderthal people give a good idea of their size and build. A typical adult was about 64 inches tall and weighed 155 pounds - shorter than today's Europeans, but stockily built. He or she had strong, thickset bones, large and firm joints, powerful muscles, a long trunk, and relatively short legs, especially in the calves. The posture was upright, as in ourselves, and not stooped and shuffling as in the earlier reconstructions.

Estimates of brain size for the Neanderthal people are mostly above 1,300 milliliters, and up to 1,700 milliliters in some large males. This is in the same broad range as for humans of today, and up to 10 percent bigger when taking into account the ratio of size of brain to body. In the head region the forehead receded, the prominent superciliary brow ridges arched over the eyes, and the face jutted forwards. The overall shape of the skull was low and long from front to back, with a bulge already mentioned, the "occipital bun," at the lower rear of the skull, above the neck. In the facial area, the size of the nasal aperture and cavity indicate a large, protuberant nose, while grooves in the bone for blood vessels suggest that the face received a good supply of blood. There was a midfacial prominence and lack of a chin.

Tools, Clothes, and other Artifacts

The tools associated with typical Neanderthal fossils are of a type known as Mousterian, named after the Le Moustier cave in the Dordogne region of France, where they were first recognized and described. They include knives, spearheads, cleavers, chisels, borers, and scrapers, mostly made from flakes chipped off a large "core stone." The core was selected for its hardness and splitting qualities, and the flakes struck from it were then finely chipped ("knapped") to the final cutting edge. Such workmanship indicates a high degree of "intelligence" and manual dexterity.

A general picture emerges of Neanderthal people living in small groups, hunting for meat and also gathering food plants. Evidence of fire, in the form of charred bones and hearth remains, occurs at many sites. In the bleak tundra of ice-age Europe, especially during the intense cold period of 60,000–40,000 years ago, fire would have helped survival by keeping low temperatures and marauding predators at bay, as well as being useful for thawing food and cooking. The Neanderthal people were probably able to hunt and roast large animals such as deer, wild ox, cave bears, mammoths, and woolly rhinoceros. They may have used animal skins in a rough-and-ready way for clothing, and possibly to make tentlike shelters. And they seemed to have cared for their old and sick. Some fossils are remains of individuals who were so diseased or aged that they would have been unable to fend for themselves.

Caves figure largely as sites of Neanderthal fossils. This may be partly because the Neanderthalers lived in them, sheltering from the bitter cold, ice, and thick snow. It is also because, as the evidence suggests, they buried their dead there. And this seems to have carried out with some form of ceremony or ritual. This is the first known evidence of humans indulging in ceremony.

Neanderthal Funerals

Possible burial sites occur across the Neanderthal range, from La Ferraise, Le Moustier, and La Chapelle-aux-Saints in France, to Kiik-

Koba near the Black Sea, South to Tabun and Amud in Israel and Shanidar in present-day Iraq, and east to Teshik-Tash near Tashkent.

At this last site, a shallow pit had been dug for the purpose. Ibex horns seem to have been arranged around the head of the dead person, who was probably a boy, and there is a charred area which could have been a graveside fire. At Le Moustier, the arrangement of fossil bones and artifacts indicates that an adolescent boy had been carefully placed on his side. His head rested on a "pillow" of flints, and a stone axe and charred bones were adjacent. This burial dates from about 50,000 years ago.

At Shanidar, pollen analysis suggests that flowers or other plants were placed over the body, as "grave gifts," 60,000 years ago. Skilled analysis of one of the skeletons shows that the owner was probably blind in one eye, and had a malformed arm. In an Italian find, a circle of stones surrounds the fossil skull, suggesting a ritual.

These various decorations and gifts, along with traces of what may have been food, could mean that the Neanderthal people held some type of superstitious or religious beliefs, and even the concept of an afterlife.

What Happened to the Neanderthaler?

After about 30,000 years ago, Neanderthal remains are conspicuously absent. Presumably these people died out. Why? In Europe, their disappearance occurred soon after a new type of human appeared. These were the Cro-Magnon people, known from their fossils to be anatomically fully modern *Homo sapiens sapiens*, and physically almost indistinguishable from people of today.

There are various proposals concerning the disappearance of the Neanderthal people. One is that they did not in fact die out-they evolved with great rapidity into the Cro-Magnons. This proposal has come in and out of favour over the years. It is supported by some Neanderthal fossils that supposedly show modern features, and a few Cro-Magnon-type remains that display certain Neanderthal characteristics. But the picture is not so simple. In the locality of Saint Cezaire, the later Neanderthal inhabitants were contemporaries of the earlier Cro-Magnons. This overlap means the later Neanderthal people, at least, could not be ancestral to the early Cro-Magnons.

Recent research from Israel adds more data for consideration. Fossils of modern humans from Jebel Qafzeh, near Nazareth, have been studied since 1935. They are admittedly somewhat "primitive," with slight brow ridges and other features. They have been dated by new techniques, such as electron spin resonance (ESR) and thermoluminescence at 100,000 years old. Neanderthal remains from the Kebara cave, on Mount Carmel, are much younger –about 60,000 years of age.

There are two other well-known cave sites on Mount Carmel. At Tabun, Neanderthal-type fossils were previously believed to be 45,000 years old. Animal teeth from the same level of excavation have also been dated recently, by ESR, to 120,000 years old. A few hundred yards away is the cave of Skhul, with modern-type fossil humans formerly estimated at 40,000 years of age. The new dating techniques on associated teeth give an age of around 100;000 years. So Neanderthal and modern people lived in the same area at almost the same period, with little sign of one evolving into the other, or with enough time for it to happen. Nevertheless, future discoveries could radically alter the picture, once again.

Evidence from tools is also inconclusive. The typical Neanderthal tool culture is Mousterian, while the main culture of the early Cro-Magnons in Europe is the considerably more sophisticated Aurignacian. However, at Saint Cezaire, some and more mousterian Neanderthalers seem to combine and more Mousterian advanced features. Conversely, at Jebel Qafzeh and Skhul, the fossilized human skulls of the modern type are associated with basically Mousterian tools.

Extinction by Epidemic

Another proposal centers on extinction from disease. Assuming the Cro-Magnon people came to Europe from elsewhere, they may have brought with them diseases to which the Neanderthalers had no resistance. Similar events have occurred during recorded history. When Europeans traveled to the Americas in the 16th century, they introduced diseases such as smallpox, measles, diphtheria, influenza and whooping cough to the largely nonresistant inhabitants. Historians estimate that tens of millions of native American peoples died as a result.

A further possibility is that the Neanderthal people suffered "passive displacement" by the more highly-organized and better-equipped Cro-Magnons. The newcomers were superior at hunting, gathering, finding shelter, and generally surviving. They out-competed the Neanderthal people, who succumbed through lack of food and shelter.

Yet another version is that the displacement was not passive but active. The Cro-Magnon killed off the Neanderthal people. It is fascinating to speculate, but firm evidence for one or more of these proposals is still awaited.

Subspecies or Species?

What of the Neanderthal people's status and position on the "tree of human evolution"? For some decades, many authorities have regarded them as a distinct subspecies, *Homo sapiens neanderthalensis*, of our own species. More recently, some experts have argued that they are sufficiently different not only from modern humans (*Homo sapiens sapiens*), but also from our overall species *Homo sapiens*, and that they deserve their own species status: *Homo neanderthalensis*.

For example, at the Israeli sites mentioned above, Neanderthal and modern people lived in the same localities, around 100,000 years ago. Yet the fossils from nearby Kebara, 40,000 years later, show no sign of combined or merged characteristics - they still retain their distinctive Neanderthal features. And modern human fossils from the same area, 30,000 years old, show no Neanderthal influence. If the two groups could not (rather than would not) interbreed, this could fulfil one of the criteria for calling them separate species. It would also mean that the Neanderthal people were not subsumed by breeding into larger groups of modern humans.

As to their position, it seems likely that Neanderthal people were not on the main evolutionary line to fully modern humans. Both these groups may have developed from a common ancestor, more than 200,000 years ago. A cold-adapted people, the Neanderthalers evolved further in ice-age Europe, surviving successfully until their relatively sudden end, without large-scale merger into *Homo sapiens sapiens*.

The Cave in the Dordogne

The 1860s were an exciting decade. In 1868, as the world debated Darwin's *On the Origin of Species* and the nature of the Neanderthal fossils, French workers were building a road near the village of Les Eyzies, in the Dordogne region. They chanced upon a cave-like rock shelter containing bones virtually identical to those of people today, yet dating from 30,000 years ago. Along with the bones and teeth were finely-worked stone tools and other objects, and signs of campfires. The name of the site was Cro-Magnon, and the discovery opened another chapter of human evolution.

Subsequent finds of similar bones, teeth, and artifacts have been made at many sites across Europe, particularly at Chancelade and Le Solutre, also in France; Grimaldi in Italy; Predmost and Dolni Vestonice in Czechoslovakia; and in the Middle East and North Africa. The name Cro-Magnon is now usually applied to all such peoples, who appeared fairly rapidly in the region about 40,000-30,000 years ago. They are included within our own modern subspecies, *Homo sapiens*. They continued to thrive through the last great Ice Age to the end of the Pleistocene epoch, 10,000 years ago, when the climate warmed. The many discoveries have given us extensive knowledge of what the Cro-Magnon people looked like, how they lived, and how they differed from the Neanderthal people before them.

A typical Cro-Magnon male was more than 70 inches tall, a woman more than 65 inches. They stood erect, with straight limbs. Muscle anchorage sites and other features of the bones show that they had a powerful physique. Their skulls lack brow ridges but have a chin, a flat face (though slightly broad by today's standards), a vertical forehead and a domed braincase. To paraphrase an anthropological saying: given a good wash, modern clothes, and modern behaviour, "Cro-Magnon Man" would pass unnoticed on any city street.

Tools of Increasing Variety

The tools of the Cro-Magnons showed great advances over those of Neanderthal people. They were beautifully made in stone, bone, and wood. Many of the stone implements were based on long, narrow pieces called blades, struck from a larger prepared "nucleus". These were shaped and sharpened either by direct striking with a hammer of stone, bone, or antler, or by indirect percussion, using the hammer to hit an intermediate "punch" antler or bone, which was applied to the workpiece. Cro-Magnon tools included spears, leaf-shaped spearheads, and spearthrowers; points with double barbs; double-edged knives, "backed knives" with on blunt edge for gripping; borers and scrapers, and fishing hooks and barbe harpoons.

One important type of tool was the chiselike burin. This came in various designs and was used to shape bones, antlers, ivory, and wood-not only to make other tools, but also for carved and sculpted objects of art. The earliest evidence of prehistoric art dates from Cro-Magnon times. The carvings, engravings, molded clays, sculptures, and paintings were remarkable new cultural developments.

The earlier types of Cro-Magnon tools, common around 30,000 years ago, are termed apurignacian. Their later tool "industries" include the Solutrean, around 20,000-17,000 years ago, and the Magdelenian, at about 15,000 years. These various industries belong to the Upper Paleolithic phase, or Upper Old Stone Age. (The Mousterian tools of the Neanderthal people are from the Middle Paleolithic stage.)

Hunting Techniques

Cro-Magnon people were proficient hunters. Around the Dolni Vestonice camp are deep piles of old animal bones, mostly mammoths. More mammoth remains, representing perhaps 600 of these extinct elephant relatives, occur at Predmost, along with remnants of tents. The remains of some 10,000 horses were uncovered at Le Solutre, in France. Other prey were reindeer and woolly rhinoceros. The great numbers of prey animals could indicate that each Cro-Magnon group had many mouths to feed. Perhaps more likely, the hunters trapped large herds of prey in ravines or made them stampede over cliffs. Then they cut up the meat, skins, and bones at their leisure.

The picture of Cro-Magnon life is well-established. These people are prominent in our history because they were the first of their kind to be scientifically discovered and investigated. Yet evidence of modern humans from similar times, from 50,000 years ago, has now been unearthed in many parts of the world. Where did these people come from?

Neanderthals had large brains. The neanderthal braincase is much larger than that of archaic *H. sapiens*, ranging from 1245 to 1740 cc, with an average size of about 1520 cc. In fact, Neanderthals had larger brains than modern humans, whose brains average about 1400 cc. It is unclear why the brains of Neanderthals were so large. Some anthropologists point to the fact that their bodies as a whole were much more robust and heavily muscled than those of modern humans and suggest that the large brains were simply a result of the fact that larger animals usually have bigger brains than smaller ones do.

Neanderthals had more rounded crania than H. erectus. The Neanderthal skull is long and low, much like the skulls of archaic *H. sapiens*, but relatively thin walled. The back of the skull has a characteristic rounded bulge or bun, and does not come to a point at the back like a *H. erectus* skull. There are also detailed differences in the back of the cranium.

Neanderthals had big faces. Like *H. erectus* and archaic *H. sapiens*, the skulls of Neanderthals have large browridges, but they are larger, rounder, and stuck out less to the sides. Moreover, the browridges of *H. erectus* are mainly solid bone, while those of Neanderthals are lightened with many air spaces. The function of these massive browridges is not clear. The face, and particularly the nose, is enormous—every Neanderthal was Cyrano, or perhaps a Jimmy Durante.

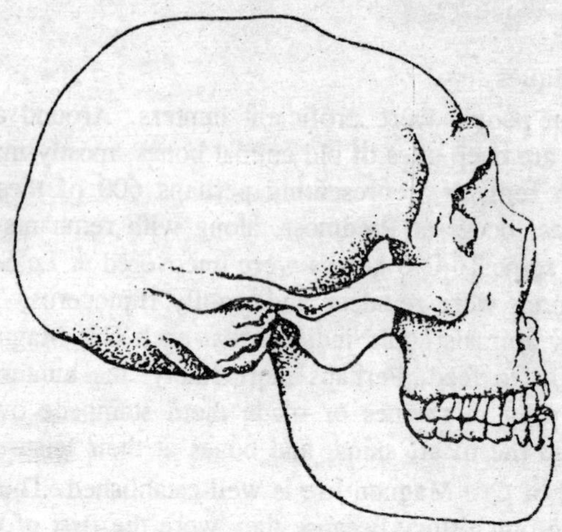

Fig. 12.7. The skull of Neanderthals, like this one from Shaindar Cave, Iraq, is large and long, with large browridges and a massive face.

Neanderthals had small back teeth and large, heavily worn front teeth. Neanderthal molars are smaller than those of *H. erectus*. Neanderthal molars had distinctive *taurodont roots* in which the pulp cavity expanded so that the roots merged, partially or completely, to form a single broad root. Neanderthal incisors are relatively large and show very heavy wear. Careful study of the wear patterns indicates that Neanderthals may have pulled meat or hides through their clenched front teeth. Also there are often microscopic, unidirectional scratches on the front of the incisors as if Neanderthals had held meat in their teeth while cutting it with a stone tool. Interestingly, the direction of the scratches suggets that most Neanderthals were right-handed, just as the Oldowan toolmakers were.

Neanderthals had robust, heavily muscled bodies. Like *H. erectus* and archaic *H. sapiens*, Neanderthals were extremely robust, heavily muscled people. Their leg bones are much thicker than ours; the load-bearing joints (the knees and hips) were larger; the *scapulae* (shoulder blades)show more extensive muscular attachments; and the rib cage is larger and more barrel-shaped. All these skeletal features indicate that Neanderthals were very sturdy and strong. They were a few inches shorter, on average, than modern Europeans and had relatively larger torsos and shorter arms and legs. This distinctive Neanderthal body shape may have been an adaptation to conserve heat in their glacial environemnt. In cold climates animals tend to be larger and have shorter and thicker limbs than do individuals of the same species in

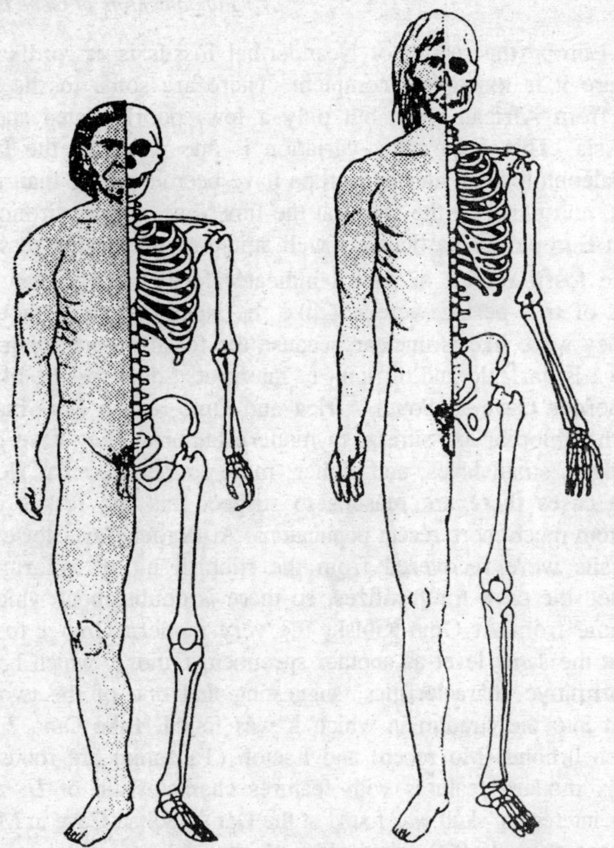

Fig. 12.8. The Neanderthals were very robust people. Contrast the bones of this skeleton (left) with the bones of a modern human (right).

warmer environments. This is because the rate of heat loss for any body is proportional to its surface area, and so any change that reduces the amount of surface area for a given volume will conserve heat. Animals can reduce their surface area in two ways; by increasing their overall body size or by reducing the size of their limbs. In contemporary human populations there is a consistent relationship between climate and body proportions. One way to compare body proportions is by calculating the *crural index*, which gives the ratio of the length of the shin bone (tibia) to the length of the thigh bone (femur). As figure shows, people in warm climates tend to have relatively long limbs in proportion to their height. Note that Neanderthals resembled modern peoples who live above the Arctic Circle.

In Europe the record of Neanderthal fossils is extrordinary, but elsewhere it is much less complete. There are some fossils for this period from African sites, but only a few, poorly dated specimens from Asia. This geographic variation is due partly to the fact that more paleontological sites in Europe have been explored than in other regions, and pardy to the fact that the limestone cave environments in western Europe are particularly well suited to preserving fossils

The fossil record in Africa indicates fairly clearly that African humans of trns period were not like the Neanderthals, although just what they were like is unclear because the fossil record for trns time in Africa is variable and difficult to interpret. Some fossils, like those from Border Cave in South Africa and Omo-Kibish For- mation in southern Ethiopia, are similar to modern humans They have reduced browridges, small faces, and higher, more rounded crania. However, in both cases there are reasons to suspect that the fossils actually come from much more recent populations. At Border Cave, for example, the fossils were recovered from the rubbish pit of a farmer who excavated the cave for fertilizer, so there is doubt about which level they came from At Omo-Kibish, the very modern Omo 1 fossil was found at the same level as another specimen, Omo 2, which has much more prirnjtive characteristics, suggesting that one of the two fossils intruded into the stratum in which it was found. Like Omo 2, fossils from Jebellrhoud (Mo rocco) and Laetoli (Tanzania) are robust skulls that rnjx modern features with features characteristic of *H. erectus*. Another interesting skull was found at the Dar es Soltan Cave in Morocco. It is more than 40,000 years old and shows none of the distinctive Neanderthal features, except for prornjnent browridges.

Despite the difficulties in interpreting the African record, three things are clear.

1. Like the Neanderthals, African hominids of the period show large cranial volumes, ranging from 1370 to 1510 cc.
2. None of the African fossils show the complex of features that are diagnostic of the European Neanderthals
3. While the fossils are variable and some are quite robust, they are all more like modern humans than the Neanderthals are.

Mousterian and Middle Stone Age Tools

Archaeological sites that date to the Neanderthal period are dominated by flake tools. This means that Neanderthals and their contemporaries struck flakes from cores and then used the flakes as their tools. In contrast, as we saw earlier in the chapter, Acheulean sites are dominated by large hand axes and choppers made from cores. Hand axes are almost completely absent from most Neanderthal sites. During this period the stone-tool industry found in Europe is called the Mous- terian (or sometimes the Middle Paleolithic), and the very similar industry found in Africa and in southern and eastern Asia is called the *Middle Stone Age* (MSA). You may recall from that some archaeologists believe Oldowan hominids used mainly flake tools as well. However, unlike the small, irregular Oldowan flakes, the Neanderthals and their contemporaries produced quite symmetric, regular flakes using sophisticated techniques. One method, called the Levallois technique (after the Parisian suburb where such tools were first identified) , involves three steps. First the knapper prepares a core having one pre- cisely shaped convex surface. Then the knapper makes a striking platform at one end of the core. Finally, the knapper hits the striking platform, knocking off a flake whose shape is determined by the original shape of the core. A skilled knapper could produce a variety of tools by modifying the shape of the original core.

Mousterian and MSA tools are more variable than Acheulean tools. One of the most influential students of Mousterian tools, the French archaeologist Francois Bordes, classified tools into a large number of distinct types based on their shape and inferred functions: points, side scrapers (flake tools bearing a smooth edge on one or both sides), denticulates (tooth-shaped tools with a serrated edge), and so on. Instead of finding a random mix of tool types at each site, Bordes found that all sites fell into one of four categories that he called Mousterian variants, and that each variant had a different mix of tool types. Bordes thought that different sites from the same period showed different Mousterian variants, and that different levels of the same site were also characterized by different variants. He concluded that these sites were the remains of four wandering groups of Neanderthals, each preserving a distinct tool tradition over time and structured much like modem ethnic groups.

Recent studies give reason to doubt Bordes's interpretation. The use of better dating techniques seems to indicate that the four Mousterian variants tended to follow one another chronologically;" that is, they differed over time but not by geographic area. Moreover, tool assemblages from some sites excavated after Bordes's analysis was completed do not match any of the recognized variants.

Many archaeologists believe that the variation among sites results from differences in the kinds of activity performed at each locale. For example, Lewis Binford (of Southern Methodist University) and Sally

Binford argued that differences in tool types depend on the nature of the site and the nature of the work performed.

Some sites may have been base camps where people lived, while others may have been work camps at which people performed subsistence tasks Different tools may have been used at different sites for woodworking, rude preparation, or butchering prey To test this idea, the Binfords assigned functions to the tools *hifore* doing their analysis, and then showed that different sites are characterized by functionally related clusters of tools. Thus, several tools that seemed to be well suited for butchery were found at the same site More recently, however, microscopic studies of wear patterns on Mousterian tools suggest that the majoriry of the tools were used only for woodworking As a result, there seems to be no association between a tool rype (such as a point or side scraper) and the task for which it was used.

A related explanation for the variabiliry in Mousterian tools focuses on the availabiliry of raw materials for toolmaking When suitable raw materials are close by and easily available, people don't spend much time repairing a particular tool—they simply make a new one When raw materials are difficult to acquire, perhaps because they come from a distant source, people mend and retouch tools over and over again. University of Pennsylvania archaeologist Harold Dibble has argued that many of the differences in tool types recognized by Bordes and others are really just different stages in the life histories of individual tools. For example, a toolmaker might start out by miaking a scraper with a single sharp edge. When the sharp edge was dulled, she might sharpen the other side, and when both sides were worn, she might retouch both sides again. Archaeologists like Bordes might classify each stage as a different tool type. If so, they would inadvertently label tools from those sites where tools had been extensively retouched as different from those at sites where tools were used briefly and discarded.

Stone-tipped spears and stone axes are made by *hafting* (attaching) a wooden handle to a stone tool. Hafting is an extremely important innovation because it greatly increases the efficiency with which humans can apply force to stone tools. (Imagine using a hammer without a handle.) Microscopic analyses of the wear patterns on Mousterian and MSA tools suggest that some tools were hafted, probably to make spears. The discovery of a 3.4-m-long (10ft) wooden spear in the rib cage of an elephant indicates that Neanderthals also used these weapons. The Neanderthals made virtually no use of bone, and they did not decorate their tools in any way that has survived in the archaeological record.

Neanderthals usually made tools from stone acquired locally. When archaeologists excavate a Mousterian site, they typically find that the stone used to make the tools can be found within a few kilometers of the site. As we will see in chapter, this pattern is altered considerably when modern humans arrive on the scene.

Neanderthal Lifeways

Neanderthal sites contain the bones of many animals alongside Mousterian tools. European sites are rich in the bones of red deer (called elk in North America), fallow deer, bison, aurochs, wild sheep, wild goat, and horse, while eland (a large antelope that looks like a Brahman cow with corkscrew horns), wildebeest, zebra, and bushpig are found at African MSA sites. Archaeologists find few bones of very large animals such as hippopotamus, rhinoceros, and elephant, even though they were plentiful in Africa and Europe.

By itself this evidence does not mean that the Neanderthals killed these animals, and this has provoked as much debate as similar ambiguity has for earlier hominids. Some archaeologists, like Lewis Binford, believe that Neanderthals and their contemporaries in Africa never hunted anything larger than small antelope, and even these prey were acquired opportunistically, not as a result of planned hunts. Any bones of larger animals were acquired by scavenging. As evidence in support of this view, Binford points to the fact that skulls and foot bones of eland predominate at Klasies River Mouth, a MSA site in South Africa, and these body parts are commonly available to scavengers. Binford believes that the hominids of this period did not have the cognitive skills necessary to plan and organize the cooperative hunts necessary to bring down large prey.

Others, such as Stanford archaeologist Richard Klein, believe that Neanderthals were proficient hunters who regularly killed large animals. He points out that animal remains at sites that date from this period are often made up almost exclusively of the bones of one or two prey species. For example, at the Klasies River Mouth site, the vast majority of bones are from eland. Similarly at Mauran, a site in the French Pyrenees, over 90% of the assemblage is from bison and aurochs. The same pattern occurs at sites across Europe. The prey animals are large creatures, and they are heavily overrepresented at these sites compared with their occurrence in the local ecosystem. It is hard to see how an opportunistic scavenger would acquire such a nonrandom sample of the local fauna. Moreover, the bones of prime-aged adults are well represented at such sites, a pattern not associated with contemporary African scavengers like hyenas, who prey on old or young animals more often than prime-aged adults. Instead, a large quantity of remains are from mature, healthy individuals—the pattern associated with catastrophic events in which whole herds of animals are killed. Several sites consist of huge jumbles of bones and tools at the bottoms of cliffs, which suggests that Neanderthals and their contemporaries hunted large game by driving them over cliffs. It also turns out that the pattern at Klasies River Mouth is far from universal. There are several sites, like Combe Grenal in France, where the bones from the meatiest parts of prey animals are overrepresented. Moreover, the cut marks on these bones suggest that neanderthals stripped off the fresh flesh.

There are many Neanderthal sites, some very well preserved, where archaeologists have found concentrations of tools, abundant evidence of toolmaking, many animal remains, and concentrations of ash. Most of these Mousterian sites occur in caves or in *rock shelters*, places protected by overhanging cliffs. This does not necessarily mean that neanderthals were all cavemen or cavewomen, however. Cave sites are less likely to have been destroyed by erosion, and they are also relatively easy to locate because the openings to many caves from this period are still visible. Most archaeologists believe that cave sites represent home bases—semipermanent encampments from which Neanderthals sallied out to hunt and to forage.

Most sites at which Mousterian tools or Neanderthal fossils are found lack postholes or other features that archaeologists interpret as the remains of hearths or shelters. The few exceptions to this rule occur near the end of the Mousterian period. For example, evidence of hearths occurs at Vilas Ruivas, a site in Portugal period. For example, evidence of hearths occurs at Vilas Ruivas, a site in Portugal dated to about 60,000 years ago.

The wealth of complete Neanderthal skeletons suggests that, unlike their hominid predecessors, Neanderthals frequently buried their dead. Burial protects the cropse from dismemberment by scavengers, and thus preserves the skeleton intact. Detailed study of the geological context at sites like La Chapelle-aux-Saints, Le Moustier, and La Ferrassie in southern France also supports the conclusion that Neanderthal burials were common.

It is not clear whether these buriasl had a religious nature, or if Neanderthals buried their dead just to dispose of the bodies. Anthropologists used to interepret some sites as ceremonial burials in which Neanderthals were interred along with symbolic materials. In recent years, skeptics have cast serious doubt on such interpretations. For example, anthropologists used to think that the presence of fossilized pollen in a Neanderthal grave in Shanidar Cave, Iraq, was evidence that the individual had been buried with a garland of flowers. However, more recent analyses revealed that the grave had been disturbed by burrowing rodents, and it is quite plausible that they brought the pollen into the grave.

Detailed study of Neanderthal skeletons indicates that Neanderthals didn't live very long. The human skeleton changes throughout the life cycle in characteristic ways that can be used to estimate the age at which fossil hominids died. For example, human skulls are made up of separate bones that fit together in a three dimensional jigsaw puzzle. When children are first born, these bones are still separate, but later they fuse together forming tight, wavy joints called *sutures*. As people age, these sutures are slowly obliterated by bone growth. By assessing the degree to which the sutures of fossil hominids have been obliterated, anthropologists can estimate how old the individual was at death. Several other skeletal features can be used in the same way. All of these features tell the same story—that Neanderthals died young. None lived beyond the age of 40 to 45 years.

Many of the older Neanderthals sufferend disabling disease or injury. For example, the skeleton of a Neanderthal man from La Chapelle-aux-Saints shows symptoms of severe arthritis that probably affected his jaw, back, and hip. By the time this fellow died at about age 45, he had also last most of this teeth to gum disease. Another individual, Shanidar 1 (from the Shanidar site in Iraq), had suffered a blow to his left temple that crushed the orbit. Anthropologists believe that his head injury had led to partial paralysis of the right side of his body, which in turn had caused his right arm to wither and his right ankle to become arthritic. Other Neanderthal specimens display bone fractures, stab wounds, gum disease, withered limbs, lesions, and deformities.

In some cases, Neanderthals survived for extended periods after injury or sickness. For example, Shanidar 1 lived long enough for the bone surrounding his injury to heal. Some anthropologists have proposed that these Neanderthals would have been unable to survive their physical impairments—to provide themselves with food or to keep up with the group—had they not received care from others. Some researchers argue further that these fossils are evidence of the origins of caretaking and compassion in our lineage.

Katherine Dettwyler, an anthropologist at Texas A & M University, is skeptical of this claim. The existence of individuals who recovered from severe injuries and lived with serious physical impairment does not necessarily imply that they had received care from others, or that other members of their communities showed compassion for them. In some contemporary societies, disabled individuals are sometimes able to support themselves, and they do not necessarily receive compassionate treatment from others. In addition, nonhuman primates sometimes survive despite permanent disabilities. At Gombe Stream National Park, a male chimpanzee named Faben contracted polio, which left him completely paralyzed in one arm. He managed to feed himself, traverse the steep slopes of the community's home range, keep up with his companions, and even climb trees. We might also question the premise that caretaking originates with the Neanderthals. There are several examples of solicitous treatment of blind and physically impaired infant monkeys and adoption of orphaned monkeys and apes. If these primates are capable of caring for one another, it seems likely that early hominids were as well.

INDEX